Glass Machines

Construction and Operation of Machines for the Forming of Hot Glass

Prepared by numerous specialists

edited by

W. Giegerich · W. Trier

Springer-Verlag Berlin Heidelberg GmbH 1969

Title of the original edition:
"Glasmaschinen"

The book contains 300 figures

ISBN 978-3-662-27193-3 ISBN 978-3-662-28676-0 (eBook)
DOI 10.1007/978-3-662-28676-0

Originally published by Springer-Verlag Berlin Heidelberg in 1969
Softcover reprint of the hardcover 1st edition 1969

Library of Congress Catalog Card Number 68-56941

Title No. 1499

Translated by

Professor Dr. NORBERT J. KREIDL
University of Missouri
Rolla, Missouri, USA

Editors

Director Ing. W. GIEGERICH, Heilbronn
Dr.-Ing. W. TRIER, Frankfurt/M.

With the support of the
Deutsche Glastechnische Gesellschaft, Frankfurt/M.

Contributors

Dipl.-Ing. H. ALBRECHT, Düsseldorf
Director Dr.-Ing. K. BECKER, Obernkirchen
Director Dipl.-Ing. K. FLESSA, Weiden
Dipl.-Ing. H. GANSS, Essen
Director Ing. W. GIEGERICH, Heilbronn
Ing. J. GOEDICKE, Berlin
Dipl.-Ing. K. GROOTHOFF, Mainz
Obering. W. GRUNDMANN, Mainz
Dipl.-Ing. F. HOLL, Zwiesel
Dr. G. KIPER, München
Director Ing. L. MUSCHALEK, Bad Godesberg
Dr. F. OBERLIES, Würzburg
Director Dipl.-Ing. W. v. REIS, Gelsenkirchen
Dipl.-Ing. H. RIEDE, Düsseldorf
Director Dipl.-Ing. E. SCHULZE, Essen
Dr.-Ing. H. SPIEKERMANN, Düsseldorf
Dr.-Ing. W. TRIER, Frankfurt/M.
Ing. H. J. WALTHER, Siebenstern
Professor Dr.-Ing. R. WILLE, Berlin

Preface

The processes involving the working of viscous hot glass—the only topic of this book—are basically different from those customary in other technologies. The strong temperature dependence of the working properties of glass, particulary of its viscosity, require the exact coordination of cooling and forming. For this reason the temperature conditions in the forming parts of the machine and the timing of the process are of prime importance. This particularity of glass as a material as well as the diversity in type and quality of glass products explain the specificity of glass machines in the field of machine design.

The literature on glass machines is sparse and scattered. Because of the increasing specialization in the field of glass machines, it is not only the beginner but also the expert who finds it hard to obtain insight into the entire field of glass forming. In response to this situation Committe IV ("Glasmaschinentechnik") of the "Deutsche Glastechnische Gesellschaft", the German glass society, decided to suggest and sponsor the publication of a book on glass machines aiming at the comprehensive presentation of the state of the art as well as the historical development.

Because of the great specialization in the field the publishers had to secure the help of several colleagues whom they wish to thank for a decisive role in this book. Thanks are also due to the glass and machine factories providing material.

Regarding the English edition the publishers and authors are particularly indebted to Professor Norbert J. Kreidl, Rolla, who has undertaken the translation of the German manuscript.

For the table "The Relationship of Viscosity to Processing of Glass" on the inside of the back cover we are obliged to Owens-Illinois, Toledo, Ohio, for their permission to use it and, once more, to Dr. H. H. Holscher who had compiled it some time ago. This table permits to read out with great facility the viscosity temperature relations characteristic of various process phases.

This translation would have been impossible without the expert and thorough editing work of Professor C. J. Phillips, Rutgers, The State University, and his simultaneous command of style and glass terminology. The translator is also greatly obliged to the following experts for aiding in the establishment of terminology and careful reading of large sections pertaining to their fields: Mr. Helmut Glaser, Owens Corning Fiberglass Co., Granville, O.; Mr. H. P. Hanneken, PPG, Pittsburgh, Pa.; Dr. H. H. Holscher, Owens-Illinois; Dr. Aaron K. Lyle, Emhart Corp., Hartford, Conn. Dr. Holscher also gave freely of his expert advise on many facets of the English edition.

The book contains so vast an amount of machine terms and functions that the translator, in spite of the generous aid of these men, assumes full responsibility for remaining errors in the translation.

The following companies gave financial assistance to the task of terminological review and finalization of the manuscript: Brockway Glass, Brockway, Pa., Ball Bros., Muncie, Indiana, Emhart Corp., Hartford, Conn., Owens-Illinois, Toledo, Ohio. Dr. C. J. Phillips and Dr. N. J. Kreidl are indebted to them for this generous aid. Barbara Fleckenstein Mickewich, Karen Packan and Georgette Henderson produced the difficult manuscript of the translated text and figures.

The book addresses itself to all persons interested in the technology of glass. It should facilitate the initiation of the beginner as well as the expert's survey of the entire field. The single sections were carefully coordinated. However, overlaps and differences of opinion could not be avoided and might positively contribute to a lively, objective presentation.

The book deals with the forming principles of hot glass; the elements of glass machines such as molds, gears, cooling organs etc.; the glass machines proper in the form of their most important types; and their production characteristics. The history of glass machines is treated in a special, hopefully instructive, section. Numerous references to the literature were provided to facilitate the study of special areas.

Unfortunately the editors were unable to secure a collaborator for a sufficiently thorough inclusion of plate glass. Since the literature in this field seems to indicate only a state of art around 1930 no bibliography of this specialty was attempted.

The editors hope that the book will fulfill its purpose and become a valuable aid for everyone who seeks advice in the field of glass machines.

Heilbronn and Frankfurt (Main)
December 1968

W. Giegerich W. Trier

Contents

Tables in pocket at end of the book:

Conversion Table

1 atmosphere (atm)[1]	760 millimeters of mercury (mm Hg)
1 atmosphere (atm)[1]	14.7 pounds-force per square inch (lbf/in^2)
1 British thermal unit	.252 kilocalories (kcal)
1 British thermal unit	778 foot pounds-force (ft lbf)
1 British thermal unit	.292 Watt hours (Wh)
1 British thermal unit per cubic foot (Btu/ft^3)	8.90 kilocalories per cubic meter ($kcal/m^3$)
1 centimeter (cm)	.394 inches (in)
1 cubic centimeter (cm^3)	.0610 cubic inches (in^3)
1 cubic foot (ft^3)	.0283 cubic meters (m^3)
1 cubic foot (ft^3)	7.48 gallons US (gal US)
1 cubic inch (in^3)	16.4 cubic centimeters (cm^3)
1 cubic meter (m^3)[2]	35.3 cubic feet (ft^3)
1 cubic meter (m^3)	1.307 cubic yards (yd^3)
1 cubic yard (yd^3)	.765 cubic meters (m^3)
1 centigrade degree (grd)[3]	1.8 Fahrenheit degree (deg F)
1 Fahrenheit degree (deg F)[4]	.556 centigrade degree (grd)
1 foot (ft)	.3048 meters (m)
1 foot pound-force	1.36 Joules (J)
1 foot pound-force	.00129 British thermal units (Btu)
1 gallon US	3.785 liters (l)
1 gallon US	.134 cubic feet (ft^3)
1 grain (gr)	.0684 grams (g)
1 gram (g)	15.44 grains (gr)
1 gram (g)	.002205 pounds (lb)
1 horsepower (hp)	.707 British thermal units per second (Btu)/sec)
1 horsepower (hp)	.178 kilocalories per second (kcal/sec)
1 horsepower (hp)	.746 kilowatts (kW)
1 inch (in)	2.54 centimeters (cm)
1 inch of mercury (in Hg)	13.6 inch of water (in H_2O)
1 inch of mercury (in Hg)	.490 pounds-force per square inch (lbf/in^2)
1 inch of water (in H_2O)	.0735 inch of mercury (in Hg)
1 inch of water (in H_2O)	.0361 pounds-force per square inch (lbf/in^2)
1 Joule (J)	.738 foot pounds-force (ft lbf)

[1] x atü $= (x - 1)$ atm.
[2] $Nm^3 = m^3$ related to 760 mm Hg and 0 °C.
[3] 0 °C = +32 °F.
[4] 0 °F = −17.78 °C.

1 Joule (J)	.000239 kilocalories (kcal)
1 kilocalorie (kcal)	3.97 British thermal units (Btu)
1 kilocalorie (kcal)	4187 Joules (J)
1 kilogram (kg)	2.205 pounds (lb)
1 kilopond (kp)	2.205 pounds-force (lbf)
1 kilopond per square centimeter (kp/cm^2)	14.22 pounds-force per square inch (lbf/in^2)
1 kilopond per square meter (kp/m^2)	.205 pounds-force per square foot (lbf/ft^2)
1 kilometer (km)	.6214 miles (mi)
1 kilowatt (kW)	.949 British thermal units per second (Btu/sec)
1 kilowatt (kW)	.239 kilocalories per second (kcal/sec)
1 kilowatt (kW)	738 foot pounds-force per second (ft lbf/sec)
1 kilowatt (kW)	1.341 horsepower (hp)
1 kilowatt (kW)	1.323 Pferdestärke (PS)
1 liter (l)	.0353 cubic feet (ft^3)
1 liter (l)	.264 gallons US (gal US)
1 meter (m)	3.281 feet (ft)
1 meter (m)	39.37 inches (in)
1 meter (m)	1.094 yards (yd)
1 mile (mi)	1.609 kilometers (km)
1 mile per hour (mi/h)	1.467 feet per second (ft/sec)
1 millimeter (mm)	.03937 inches (in)
1 ounce/avoir (oz)	28.35 grams (g)
1 ounce/troy (oz)	31.09 grams (g)
1 Pferdestärke (PS)	.986 horsepower (hp)
1 pound-force per square foot (lbf/ft^2)	4.88 kilopond per square meter (kp/m^2)
1 quart (qt)	.946 liters (l)
1 radian (rad)	57.3 angle degrees (°)
1 revolution per minute (rpm)	1 Umdrehung pro Minute (Upm)[1]
1 square centimeter (cm^2)	.155 square inches (in^2)
1 square inch (in^2)	6.45 square centimeter (cm^2)
1 square foot (ft^2)	.093 square meters (m^2)
1 square meter (m^2)	10.76 square feet (ft^2)
1 square meter (m^2)	1.196 square yards (yd^2)
1 short ton (shtn)	907 kilograms (kg)
1 yard (yd)	.9144 meters (m)

[1] Sometimes abbreviated *n*.

A. Principles of Forming

1 Properties of Glass

By W. Giegerich, Heilbronn

Glass forming aims at shaping the desired object from the hot, viscous glass with simultaneous withdrawal of heat so that the glass retains a stable shape whether formed by hand or by machine. Viscosity and heat transfer are the most important properties involved. Surface tension also plays a significant part in free forming. To a limited extent, expansivity is of concern. This is particularly so in the fabrication of hollow ware where sharp local cooling may induce stresses from thermal gradients thus leading to surface checks. Since heat transport in glass is treated in the section "Heat Transmission and Heat Transfer in Forming" (A 2) this section is limited to data on viscosity, surface tension, and thermal expansion.

1.1 Viscosity

Viscosity [1, 2, 3] is by far the most important physical property of glass in high temperature forming. It depends strongly on temperature and on composition. In Fig. A 1 the relationship between the logarithm

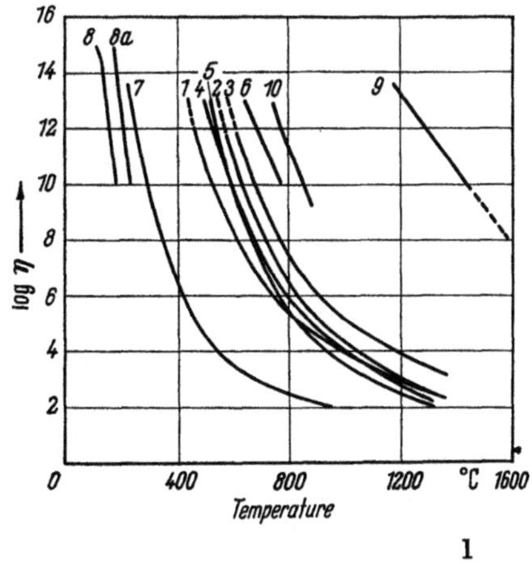

Fig. A 1. Viscosity of different glasses according to W. Oldekop [1].

1: Lead glass [14]; *2*: Soda lime [14]; *3*: Borosilicate [14]; *4*: 79.8 SiO_2, 18.3 Na_2O [15]; *5*: 69.7 SiO_2, 21.0 Na_2O, 9.1 CaO [15]; *6*: 69.8 SiO_2, 4.5 B_2O_3, 8.3 CaO, 5.5 BaO, 5.3 Al_2O_3, Rest Na_2O, K_2O, MgO, Fe_2O_3; *7*: B_2O_3 [17]; *8*: 63.7 BeF_2, 8.2 AlF_3, 28.1 KF; *8a*: 44.4 BeF_2, 31.8 AlF_3, 28.8 NaF (*8* and *8a* measurements by G. Hach); *9*: SiO_2 [18]; *10*: Osram-Hartglas [19].

of viscosity (η) and the temperature (Θ) is plotted for various glasses. It is obvious that the curves are not very different, except for extreme glasses.

The strong dependence of the working properties of glass upon its viscosity has led to a specific nomenclature for temperatures corresponding to particular viscosities, such as working point ($\log\eta = 4$), flow point ($\log\eta = 5$), and softening point ($\log\eta = 7.6$).

No completely successful theoretical relationship between η and Θ has been found. The best to date is the VOGEL-FULCHER-TAMMAN equation

$$\log\eta = A + B/(\Theta - \Theta_0),$$

where A, B, and Θ_0 are constants best determined from three fixed points on the viscosity-temperature curve for a given glass. For a clear container glass (73.1 wt. % SiO_2, 15.07 $Na_2O + K_2O$, 9.14 CaO, 1.45 MgO, 0.68 Al_2O_3, and 0.09 Fe_2O_3) the following was obtained:

$$\log\eta = -0.351 + 3164.6/(\Theta - 301).$$

For an iron-bearing chromium green glass (65.67 wt. % SiO_2, 14.24 $Na_2O + K_2O$, 7.89 CaO, 4.68 Al_2O_3, 2.86 MgO, 2.3 MnO, 1.54 Fe_2O_3, 0.08 Cr_2O_3, and 0.23 TiO_2) the equation is

$$\log\eta = -0.606 + 3370.2/(\Theta - 279).$$

For a bulb glass [3] the relationship is

$$\log\eta = -1.746 + 4530/(\Theta - 194.8).$$

Depending on the temperature-viscosity behavior one can distinguish in practice between "sweet" (slow setting) and "short" (quick setting) glasses. "Sweet" glasses have large temperature intervals between softening and working points, "short" glasses have small ones. A plot of viscosity versus temperature for an assumed working time is shown schematically in Fig. A 2 for both types of glass. Assuming equal cooling rates one obtains for glass I a viscosity η_2 as early as in time t_K, while glass II requires the longer time t_L since the viscosity η_2 is attained only at the lower temperature Θ_{2L}. In practice these relation ships are not always so clearly recognized. Starting temperatures do not always coincide precisely and cooling rates may differ. On fast cooling a "sweet" glass may set faster than the "short" glass.

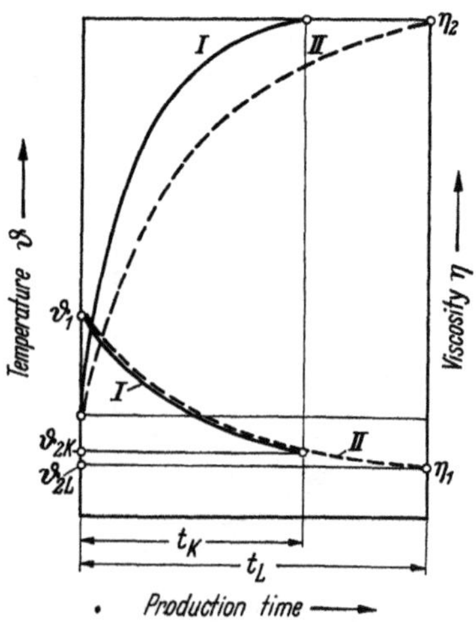

Fig. A 2. Comparison of working of (I) short and (II) sweet glass.

Other than temperature, chemical composition is the principal influence on the viscosity of glass. Many years ago GEHLHOFF and THOMAS [5] systematically investigated the influence of the most important glass components. The temperature changes required to hold the original viscosity, when SiO_2 is replaced by other components, are given in Table A 1, line 3, for glass of a particular composition. This table shows that the influence of the various components differs widely. Na_2O and Al_2O_3 exert particularly large effects: Na_2O lowers the viscosity whereas Al_2O_3 raises it.

In spite of the importance of viscosity values in the different phases of forming, the literature on this is sparse [6][2]. Table A 2 gives an approxi-

Table A 1. *The effect of variation in composition on viscosity in the working range*

Viscosity (poises)	10^3	10^4	10^5	10^6
Temperature (°C)	1175	995	880	795
Viscosity diff. per 5 °C	229	3530	57500	552000
Temperature change[1] for				
1% Al_2O_3 (0—5%)	+12.8	+10	+ 8	+ 6.6
1% CaO (5—10%)	— 8	— 7	— 4.3	— 1.8
1% MgO (0—5%)	—10.4	—11.2	— 8.2	— 4.3
1% Na_2O (10—15%)	—14.8	—13	—11.5	—10.2
Viscosity change per 1% oxide change in %				
Al_2O_3	29.2	34.8	44.5	36.8
CaO	18.3	24.4	23.8	10.1
MgO	23.8	39	45.5	24.0
Na_2O	33.8	45	63.5	56.5

Table A 2. *Glass viscosity for different processes and manufacturing stages*

Manufacturing Process	Stage	Viscosity (log η)
Bottle-suction process	After suction	3.2
	Transfer	3.9
	Delivery	5.8
Bottle-feeder process	Gob	3.3
	Transfer	3.8
	Delivery	5.3
Press blown	Gob	3.8
	Transfer	5.4
	Delivery	8.1
Press	Gob	3.6—3.7
	Transfer	5.3—7.3
Hand gather and blow	Gathered	2.7—3.7
	Working	4.0—5.6

[1] GEHLHOFF and THOMAS [5].

[2] Table "The Relationship of Viscosity to Processing of Glass" in pocket at end of the book.

mate survey of these values for the fabrication of hollow-ware. There are large differences in viscosity between the skin and the interior of the glass gob [7].

1.2 Surface Tension

The surface tension of molten glass is three to four times greater than that of water at room temperature and equals in order of magnitude that of molten metals. It plays a part in free forming in which no tools are used, as for instance in some flat glass processes. Under the influence of surface tension the glass tries to assume a shape with a minimum surface for a given volume. In practice this results in a tendency to round off surface unevenness. In the flat glass processes the glass tries to neck in at the drawing zone which must be prevented by specific arrangements for edge cooling.

The surface tension of glass depends upon composition and temperature. The temperature dependence is not nearly as strong as in the case of viscosity. However, the effect of surface tension is much stronger at high temperatures where the viscosities are low. For two container glasses KEPPELER and ALBRECHT [8] give the values listed in Table A 3.

Table A 3. *Surface tension of glasses* (Dyn/cm)

Temperature (°C)	Household glass	Green bottle
1150	308	344
1250	304	326
1370	301	322

The influence of composition was thoroughly investigated by DIETZEL [9]. He determined factors f_i (in dynes/cm) which permit the additive calculation, at 900 °C, of the surface tension by multiplying them by the percentages q_i of each oxide in the composition. For temperatures above 900 °C he found the surface tension decreased so that at Θ (> 900 °C) the surface tension γ was given by

$$\gamma = \sum^{n} q_i f_i - (\Theta - 900)\,(0.04) \text{ dynes/cm}.$$

The f_i factors may be taken from Table A 4.

Table A 4. *Surface tension. Factors for* 1% *by weight of different glass components*

Component	Li_2O	Na_2O	K_2O	MgO	CaO	BaO	PbO	ZnO	B_2O_3
f_i	4.6	1.5	6.6	6.6	4.8	3.7	1.2	4.7	0.8
Component	Al_2O_3	SiO_2	TiO_2	V_2O_5	CaF_2	Fe_2O_3	CoO	NiO	MnO
f_i	6.2	3.4	3.0	−6.1	4.5	4.5	4.5	4.5	4.5

Later investigation [6] demonstrated that SO_2 and SO_3 decrease the surface tension very strongly, but numerical data were not given.

1.3 Thermal Expansion

Thermal expansion has some significance in glass forming. It is suspected that glasses with high expansivity are prone to surface checks, particularly in the fabrication of containers. It is thus important, particularly for the final stage of forming, to know the thermal expansion of the glass to be fabricated.

The coefficients of thermal expansion can be determined additively from the composition as in the case of surface tension. The values determined by English and Turner [10] are considered most exact. From their constants approximate coefficients of expansion can be calculated from the composition. If, however, a numerical expression for the effect of compositional changes is desired the factors of Gehring and Knight [11], shown in Table A 5, are recommended.

Table A 5. *Factors for coefficient of linear thermal expansion of bottle glass* ($\times 10^{-7}$)/*weight* %) (Courtesy Owens Illinois)

Component	Na_2O	K_2O	CaO	CaO · MgO	MgO	LiO_2	BaO	SiO_2	Al_2O_3	B_2O_3
Factor	3.86	3.20	1.36	1.10	0.73	6.56	1.08	0.28	0.24	−0.60

If, for example, 1 % SiO_2 is to be replaced by 1 % Na_2O the coefficient of expansion increases as follows:

Na_2O: $+1.0 \times 3.86 \times 10^{-7} = +3.86 \times 10^{-7}$
SiO_2: $-1.0 \times 0.28 \times 10^{-7} = -0.28 \times 10^{-7}$
Total: $+3.58 \times 10^{-7}$

This simple procedure often permits a good estimate of the practical effect of a batch change on the coefficient of expansion.

From ambient temperature to the transformation point the density of glass decreases almost linearly and so the coefficient of expansion remains nearly constant. In the transformation region a sudden decrease of density is observed. Above this region most glasses again exhibit a linear decrease of density, but a much larger decrease than is shown below the transformation region [12]. An exception is the Al_2O_3—B_2O_3—SiO_2 system where special conditions prevail due to the Al_2O_3—B_2O_3 anomaly. (See A 2.1 — Heat Conduction.)

References

[1] Oldekop, W.: Theoretische Betrachtungen über die Zähigkeit von Gläsern. Glastechn. Ber. 30 (1957) 8—14.

[2] Boow, J., and W. E. S. Turner: The viscosity and working characteristics of glass. J. Soc. Glass Technol. I. 26 (1942) 215—237. II and III. 27 (1943) 94—112, 207—237. IV and V. 29 (1945) 199—232, 233—249.
[3] Müllensiefen, W., and K. Endell: Zähigkeitsmessungen an technischen Gläsern und ihre Bedeutung für die Wirtschaftlichkeit der maschinellen Formgebung. Glastechn. Ber. 11 (1933) 161.
[4] Dietzel, A., and R. Brückner: Ein Fixpunkt der Zähigkeit im Verarbeitungsbereich der Gläser. Schnellbestimmung des Viskositäts-Temperatur-Verlaufes. Glastechn. Ber. 30 (1957) 73—79.
[5] Gehlhoff, G., and M. Thomas: Die physikalischen Eigenschaften von Gläsern in Abhängigkeit von der Zusammensetzung. III. Die Viskosität der Gläser. Z. techn. Phys. 7 (1926) 260—278.
[6] Jebsen-Marwedel, H.: Glastechnische Fabrikationsfehler. Eine pathologische Technologie des Werkstoffes Glas, 2nd Ed., Berlin/Göttingen/Heidelberg: Springer 1959.
[7] Trier, W.: Zähigkeitsverteilung und Fließen des Glases in Vorformen. Glastechn. Ber. 32K (1959) II/17—II/25.
[8] Keppeler, G., and A. Albrecht: Zur Kenntnis der Oberflächenspannung von Gläsern. Glastechn. Ber. 18 (1940) 201—213, 236—245, 275—280.
[9] Dietzel, A.: Praktische Bedeutung und Berechnung der Oberflächenspannung von Gläsern, Glasuren und Emails. Sprechsaal 75 (1942) 82—85.
[10] English, S., and W. E. S. Turner: Relationship between chemical composition and the thermal expansion of glass. J. Amer. ceram. Soc. 10 (1927) 551—560.
[11] Gehring, L., and M. P. Knight: Properties and diagnosis of cords in pressed and blown glassware. J. Amer. ceram. Soc. 27 (1944) 260—266.
[12] Hänlein, W.: Über das spezifische Gewicht von Glas in Abhängigkeit von der Temperatur. Glastechn. Ber. 10 (1932) 126—130.
[13] Giegerich, W.: Ergebnisse und Probleme der maschinellen Flaschenfertigung. Glastechn. Ber. 30 (1957) 299—308.
[14] Lillie, H. R.: The viscosity of glass. Proc. Internat. Comm. Glass 2 (1955) 11—22.
[15] Eitel, W., M. Pirani, and K. Scheel: Glastechnische Tabellen, Berlin: Springer 1932, pp. 96—100. See contribution "Ein Tabellenwerk über die physikalischen und chemischen Konstanten der Gläser." Glastechn. Ber. 9 (1931) 459—463.
[16] Morey, G. W.: The properties of glass, New York: Reinhold Publ. Corp., p. 157. (Compare also 2nd Ed., 1954, pp. 162/63.)
[17] Jenckel, E.: Das Transformationsintervall. Kolloid-Z. 120 (1951) 160—174.
[18] Eitel, W.: Physikalische Chemie der Silikate, Leipzig: Barth 1941, p. 91.
[19] Osram-Gläser. Catalog of Osram GmbH KG, Berlin 1939, pp. 38/39.

2 Heat Transmission and Heat Transfer in Forming

By W. Trier, Frankfurt/Main

It is typical of all glass fabricating processes that the glass is cooled at the same time that it is formed. Only proper cooling and thus proper control of the rapidly rising viscosity will permit successful production of glass objects of a desired shape at maximum efficiency. This tight coupling of forming and cooling clearly implies the great importance of heat transmission and heat transfer in glass fabrication.

2.1 Heat Conduction

The thermal conduction of a material is characterized by its thermal conductivity λ. The conductivity is a fundamental material property which is independent of the direction of heat flow in isotropic materials such as glass. It does, however, often depend on temperature. If an infinitely extended wall of thickness d with boundary temperatures Θ_{W1} and Θ_{W2} is considered, and if a quantity of heat Q flows through an area F in time t, then the thermal conductivity (over the temperature interval Θ_{W1} to Θ_{W2}) is given by

$$\lambda = \frac{Q\,d}{F\,(\Theta_{W1} - \Theta_{W2})\,t}. \tag{1}$$

See Fig. A 3. In the technical system of units the thermal conductivity has the dimensions kcal/(m h degree).

Reliable data on the thermal conductivity of glasses is available only for temperatures below 200 °C. At higher temperature measure-

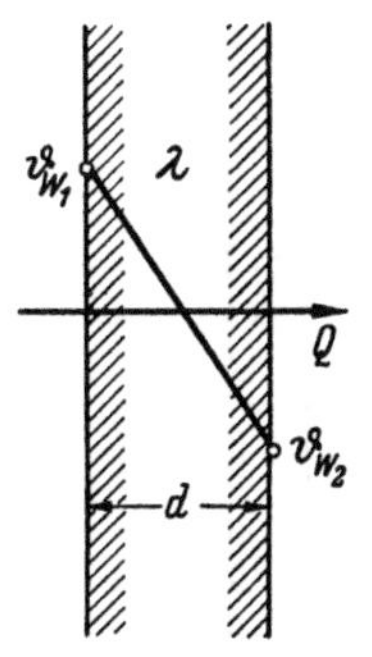

Fig. A 3. Stationary thermal conduction through a wall.

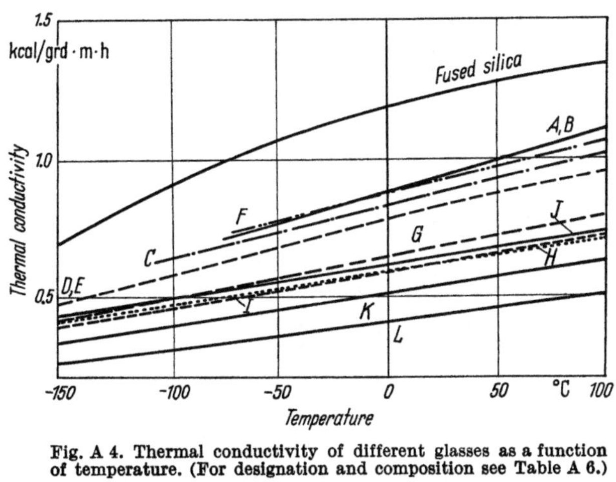

Fig. A 4. Thermal conductivity of different glasses as a function of temperature. (For designation and composition see Table A 6.)

ments are difficult because of the influence of radiation. So far there is no completely satisfactory method for the reliable determination of λ at elevated temperatures.

The thermal conductivity of glass increases with temperature and is also influenced by composition. Recent data obtained by the plate method have been published by Ratcliffe [1]. Values measured between -150 °C and $+100$ °C can be taken from Fig. A 4. The composi-

Table A 6. *Chemical composition of the examined glasses in weight* (see Fig. A 4)

Glass	Density (kg m^{-3})	SiO_2	K_2O	NaO_2	PbO	Sb_2O_3	As_2O_3	B_2O_3	Al_2O_3	ZnO	CaO	BaO	F_2	MgO
A	2.22	80.8	—	4.2	—	—	—	12.8	2.2	—	—	—	—	—
B	2.27	79.5	—	5.3	—	—	—	13.1	2.1	—	—	—	—	—
C	2.45	72.7	14.5	4.0	—	0.4	—	7.7	0.4	0.4	—	—	—	—
D	2.52	71.2	8.9	7.6	—	0.8	—	2.9	0.1	3.0	5.5	—	—	—
E	2.42	67.7	1.8	14.6	—	—	<1.0	4.0	1.8	—	5.4	3.3	—	1.3
F	2.55	57.9	2.3	9.4	—	—	—	1.9	11.1	9.7	4.9	—	2.7	—
G	3.19	54.2	7.1	2.4	34.2	2.0	0.2	—	—	—	—	—	—	—
H	3.18	49.8	8.5	2.9	5.9	0.7	0.2	—	—	7.8	—	24.2	—	—
I	3.55	46.0	9.2	—	44.8	—	0.1	—	—	—	—	—	—	—
J	3.56	40.5	—	—	—	0.3	0.2	6.5	1.8	7.7	—	42.9	—	—
K	4.29	35.6	4.4	—	59.7	—	0.2	—	—	—	—	—	—	—
L	6.10	20.0	—	—	80.0	—	—	—	—	—	—	—	—	—

Table A 7. *Coefficients* $\lambda_{0,i}$ *of important glass components*

Components (a_i)	SiO_2	Na_2O	K_2O	MgO	CaO	BaO	ZnO	B_2O_3	Al_2O_3	PbO	Fe_2O_3
λ_{0i}	0.974	0.282	−0.105	0.790	0.406	0.276	0.423	0.894	0.644	0.459	0.600

tions of these glasses are given in Table A 6. Fused silica has the highest λ while the extreme (80%) lead glass L has the lowest conductivity of the glasses investigated. The influence of composition is considerable, that of temperature moderate. Near room temperature λ increases about 0.15% per degree.

Russ [1 a] has measured the thermal conductivity of numerous glasses between 0—100 °C and derives the following

$$\lambda_t = \sum_0^n a_i \lambda_{0i} (1 + .00095\, \Theta) \tag{2}$$

where λ_t is the conductivity at temperature Θ (°C), a_i is the weight per cent of oxide i, and λ_{0i} is the coefficient for oxide i and is taken from Table A 7.

There are no exact measurements of thermal conductivity at elevated temperatures. McGraw [2] gives the following relationship for a bottle glass (feeder glass with 0.05% Fe_2O_3), for temperatures above the transformation range:

$$\lambda_\Theta = 0.9 + 0.00128\, \Theta \text{ [kcal/m h degree]}. \tag{3}$$

If the heat flow and the temperature profile vary with time one deals with a variable heat flux. In the simple case of a plane with a one dimensional temperature field vertical to the wall surface the following equation holds

$$\frac{\partial \Theta}{\partial t} = \frac{\lambda}{\varrho c} x \frac{\partial^2 \Theta}{\partial x^2}. \tag{4}$$

Here x is the coordinate of length perpendicular to the wall surface, ϱ is the density and c the specific heat of the wall material. The factor $\lambda/\varrho\, c$ is called the "temperature conductivity" or the "thermal diffusivity, a". It measures the rate of equalization of temperature differences in a body. The dimensions of a are m^2/h. The calculation of the "temperature conductivity" from the heat conductivity presupposes a knowledge of density and specific heat in the proper temperature range.

For cylindrical heat flow the following relation holds

$$\frac{\partial \Theta}{\partial t} = a \left(\frac{\partial^2 \Theta}{\partial r^2} + \frac{1}{r} \frac{\partial \Theta}{\partial r} \right). \tag{5}$$

The partial differential Eqs. [4] and [5] can be solved analytically or by an approximation method given by Schmidt [3]. For certain simple cases (wall, cylinder, sphere) there are ready solutions in diagram form [4].

The density of glasses usually decreases linearly with rising temperature. In the transformation range a sudden decrease is observed which is related to structural changes. At temperatures above the transformation range the density decreases more than below. For four glasses, whose compositions are listed in Table A 8, density changes with temperature have been measured by HÄNLEIN, DIETZEL, and BABCOCK [5, 6]. The results are shown on Fig. A 5.

Table A 8. *Chemical composition of* HAENLEIN'S, DIETZEL'S *and* BABCOCK'S *glasses in weight* %

Glass	SiO_2	R_2O_3	CaO	MgO	R_2O	BaO	PbO	B_2O_3
I	71.0	0.5	5.5	3.5	17.5	2.0	—	—
II	69.9	2.2	0.2	—	5.1	—	6.1	13.3
III	74.0	—	10.0	—	16.0	—	—	—
IV	71.24	1.47	10.8	2.92	12.87	0.69	—	—

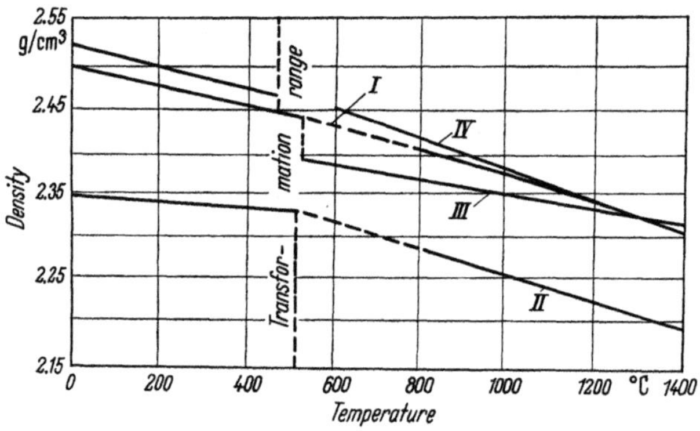

Fig. A 5. Decrease of density with increase in temperature. (Compare Table 8.)

The mean specific heat of various glasses is shown in Table A 9. It increases with temperature and shows an anomaly in the transformation (softening) range. The molecular configuration of glass changes in this range. Further discontinuities may appear in the range where devitrification may occur. The transition to the ordered crystalline state causes a decrease in energy content which is expressed in a decrease of c.

Table A 9. *Mean specific heat of technical glasses* (kcal/kg grd)

Temperature range (°C)	Sheet	Green bottle	Household	Lead crystal	Colorless bottle
20—100	0.198	0.212	0.202	0.176	0.208
20—300	0.233	0.236	0.232	0.192	0.224
20—500	0.253	0.253	0.251	0.207	0.240
20—700	0.267	0.264	0.265	0.219	0.256
20—1000	0.278	0.280	0.280	0.234	0.273
20—1300	0.295	0.291	0.294	0.258	0.288

The thermal diffusivity of conventional glasses lies in the range from 0.01 to 0.0015 m^2/h. If one wishes to calculate a from λ, ϱ, and c one must take care to use values valid for the selected temperature, not mean values.

2.2 Heat Radiation

2.2.1 Heat Radiation, General

Glass is more or less transmittant to heat radiation, depending upon composition, particularly iron content. This property causes considerable

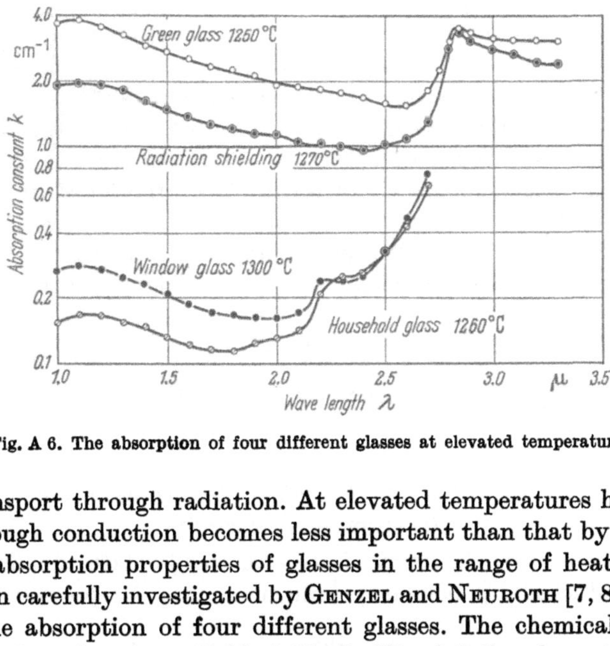

Fig. A 6. The absorption of four different glasses at elevated temperatures.

heat transport through radiation. At elevated temperatures heat transport through conduction becomes less important than that by radiation.

The absorption properties of glasses in the range of heat radiation have been carefully investigated by Genzel and Neuroth [7, 8]. Fig. A 6 shows the absorption of four different glasses. The chemical compositions can be taken from Table A 10. In Fig. A 6 the absorption coeffi-

Table A 10. *Chemical composition of glasses in Fig.* A 6

Glass type	SiO_2	Na_2O	CaO	Al_2O_3	K_2O	Fe_2O_3	MgO	SO_3	TiO_2	PbO	BaO	Mn_3O_4
Household glass	74.9	14,2	8.34	1.59	0.61	0.046	—	—	—	—	—	—
Window glass	71,78	14.82	9.56	0,47	—	0.13	2.53	0.461	0.08	—	—	—
Radiation shielding	34.0	2.0	—	—	—	—	—	—	—	61.3	3.0	—
Green bottle	68.38	15.18	10.20	2.66	—	2.48	0.14	—	—	—	—	0.94

cient $K(\lambda)$ is given as a function of the wave length, according to the law of absorption:

$$J = J_0 \exp(-K(\lambda)\, d).$$

J_0 is the intensity of the incident monochromatic radiation of wavelength λ; J is the intensity after passing a layer of thicknes d, which is given in cm. The steep rise of absorption for wavelengths above 2.75 μ should be noted. For transmittance of heat radiation the content of the glass in iron (Fe_2O_3), chromium (Cr_2O_3) and nickel oxide (NiO) is significant. Between 800 and 1200 °C the change of the absorption coefficient with temperature can usually be neglected. At lower temperatures one has to expect large changes.

A peculiarity of heat transport by radiation in glass as compared with heat conduction is that the flux at a particular place does not depend upon the local temperature gradient and the thermal conductivity, but on the difference of the radiation flow through the glass layers before and behind this place. This peculiarity makes the thickness of the glass and the boundary characteristics very important.

If the thickness is large and the surfaces "far" apart, it has been found useful to work with the term "radiative conductivity" $\varkappa_s$ which, in analogy to the thermal conductivity is given by the relation

$$\varkappa_s = Q_s \Big/ \left(F\left(\frac{d\Theta}{d\varkappa}\right) t\right), \tag{7}$$

where $\varkappa_s$ has the same dimensions as λ. F is the mean free path of the radiation quanta.

2.2.2 *Heat Flux by Radiation, One-Dimensional, Disregarding Boundaries*

The radiation conductivity of glass for steady-state one-dimensional flow, undisturbed by boundary influences, may be presented by the relation

$$\varkappa_s = \frac{16}{3}\sigma\, n^2\, T^3\, F \tag{8}$$

as was shown by Czerny, Genzel and Heilmann [9]. In Eq. 8 σ is the radiation constant (1.39×10^{-12} cal/cm^2 sec degr^4), n the refractive index (for conventional flat and hollow glass about 1.5) and T the glass temperature in °K. F, the mean free path of the radiation quanta, depends upon glass type and temperature. Values for F, depending upon iron and chromium (Fe_2O_3, Cr_2O_3) content are summarized in Fig. A 11. See references [10, 10a].

If the glass is thicker than the mean free path, the radiation conductivity $\varkappa_s$ can be calculated by means of Eq. (8) and the corresponding value of F from Table A 11. From this, via Eq. (7), the heat flux from radiation can be determined. If, however, the thickness of the glass is less than the mean free path F, $\varkappa_s$ must not be calculated from Eq. (8).

Table A 11. *Mean free path F for different glasses* (cm)

Glass	% Fe_2O_3	% Cr_2O_3	$F_{1000\,°C}$	$F_{1200\,°C}$	$F_{1000\,°C}$	$1/k$[1]	Melting condition
28	0.02	—	5.12	6.40	7.26	10.1	Oxidizing
30[2]	0.05	—	3.70	4.80	5.30	6.90	Oxidizing
30[3]	0.05	—	3.60	4.30	5.05	6.90	Oxidizing
22	0.10	—	1.82	2.22	2.65	3.30	Reducing
34	0.20	—	1.26	1.56	1.82	1.99	Oxidizing
24	0.20	—	1.10	1.38	1.72	1.83	Reducing
36	0.50	—	0.73	0.95	1.03	1.07	Oxidizing
26	0.50	—	0.62	0.77	0.86	0.86	Reducing
12	—	0.01	5.60	4.82	4.48	—	Oxidizing
43	—	0.1	1.90	1.52	0.97	—	Slightly reducing
71	—	0.1	2.80	1.70	0,75	—	Strongly reducing

The influence of the boundaries may now become so great that totally different relations prevail. A complete mathematical representation of this case has not yet been achieved. There are some experimental results, which, however, can be used only with great caution as will be shown in Sec. 2.3.

The case of non-stationary heat flux radiation has not been solved satisfactorily. First theoretical attempts were presented by GEFFCKEN [11] and GARDON [11a], and there is some experimental data from VAN ZEE and BABCOCK [12]. LÜCK [13] has investigated the pecularities arising when glasses of various shapes (plates, cylinders) are irradiated from external sources.

2.3 Superimposition of Heat Conductivity and Heat Radiation

While below 200 °C heat is transported through glass predominantly by conduction, heat transport by radiation increases at higher temperatures. It must, however, be noted that the portion due to radiation does not depend alone upon temperature and absorptivity but also upon thickness and shape effects. The strong influence of thickness and geometry upon heat transport in glasses at high temperatures is a pecularity of glass which makes the calculation of heat fluxes very difficult. In spite of this it has been found useful, in analogy to Eqs. [1] and [7], to connect the flux from conduction and radiation with the local tempera-

[1] k absorption coefficient (cm^{-1}).
[2] 25,4 mm path length.
[3] 6,3 mm path length.

ture gradient via an "effective heat conductivity coefficient" $\varkappa$ by means of the equation:

$$Q_{L+S} = \varkappa\, F(d\Theta/dx)\, t. \tag{9}$$

This equation, however, should be used only after careful examination of prevailing conditions and is in no way as generally valid as Eq. (1).

NEUROTH [14] in several publications has shown how large an influence is exercised on the effective heat conductivity of glasses by

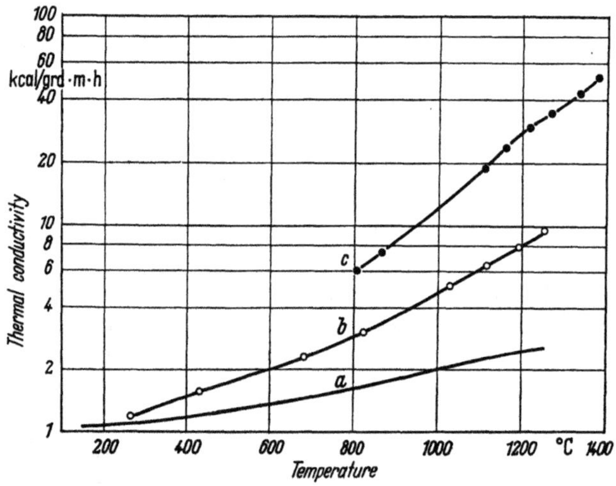

Fig. A 7. Thermal conductivity of window glass. Measurement at different temperatures. *a* ECKHARDT (9 mm thickness); *b* ELIGEHAUSEN (17.5 mm); *c* VANZEE and BABCOCK (165 mm diam.). (Instead of conventional midthickness values maximum values are plotted which are about 10–15% higher.)

the thickness. Fig. A 7 shows the dependence of heat conductivity $\varkappa$ upon temperature and thickness. An increase in thickness from 9 mm (*a*) to 165 mm (*c*) at 1000 °C causes an increase of the effective heat conductivity $\varkappa$ from 2 to 12 kcal/m h degr. These numbers show how dangerous it is to use measurements of heat transport on glasses at high temperature if the geometry is not the same. To this one must add that it is not the absolute difference in dimensions that matters but the mean free path of the radiation quanta which depends upon the absorptive behavior and the geometrical dimensions of the particular glass.

The influence of the absorptive behavior of glass on its conductivity, according to NEUROTH [14], is represented in Fig. A 8. In addition to the steep rise of the curves with increasing temperature, the strong difference between different glass types is striking. Nevertheless, comparing Figs. A 7 and A 8, one recognizes the overwhelming influence of thickness.

For non-uniform heat flow, time represents one additional parameter. For instance, during temperature equalization the temporary appearance

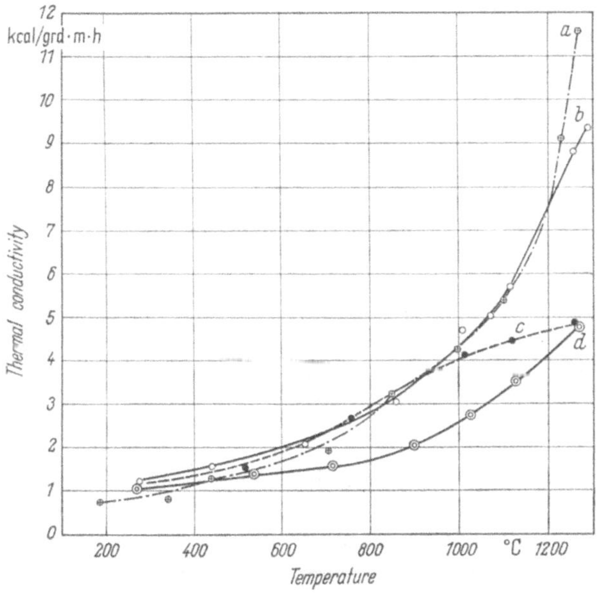

Fig. A 8. Thermal conductivity of different glasses after measurements by ELIGEHAUSEN [15] (glass cylinder 17.5 mm thick).
a Lead crystal; *b* Colorless bottle; *c* Black glass; *d* Green glass. (Instead of the conventional midthickness values maximum values are plotted which are about 10—15% higher.)

of negative effective thermal conductivities $\varkappa$ is definitely conceivable. If the temperature distribution at time t is the one illustrated in Fig. A 9, there appears at location A an integrated heat flux from left to right in spite of the negative temperature gradient. This is because the radiative heat flux from the hot glass layers on the left to the glass layer on the right may well surpass the conduction flux from right to left.

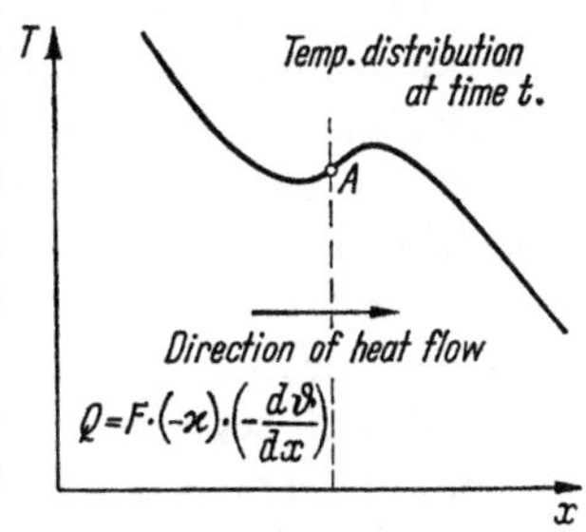

Fig. A 9. Appearance of negative effective thermal conductivity coefficients during unsteady heat flow in glass.

The less the temperature distribution deviates from the steady-state and the slower the externally impressed temperatures changes are, the more the effective heat conductivity will approach values valid for the steady-state case. Since temperature changes are relatively easy to measure, the non-stationary heat flux permits the determination of the thermal diffusivity. (VAN ZEE and

BABCOCK [12].) Thermal diffusivities of soda lime glasses determined on a cylindrical glass layer of 165 mm thickness are reproduced in Fig. A 10.

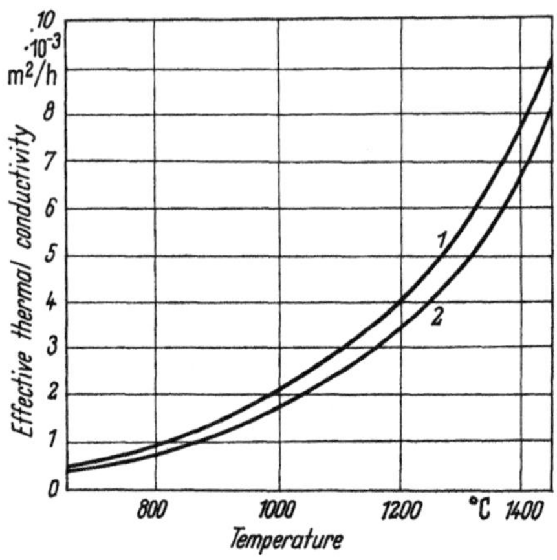

Fig. A 10. Influence of temperature on thermal conductivity of two soda lime glasses.

The chemical compositions of these glasses are given in Table A 12.

Table A 12. *Chemical composition of* BABCOCK's *glasses* [12]

Glass	SiO_2	$Al_2O_3 + TiO_2$	Fe_2O_3	CaO	CuO	Na_2O	SO_3
1	67.98	4.01	0.038	10.82	0.0009	16.63	0.12
2	72.30	0.35	0.033	14.40	0.0005	12.50	0.14

2.4 Heat Loss from Glass During Forming

The cooling of glass during fabrication has been investigated in regard to how it depends upon glass color and weight and upon the kind of forming machine. Calorimeters adapted for small objects were described by BABCOCK [6], a larger one suited for entire bottles, by UNGER [16].

Table A 13 summarizes the heat loss of bottle glasses of various kinds during fabrication in suction and feeder machines. The data listed come from UNGER [16] and GIEGERICH [17]. UNGER's calorimeter was used. The mean cooling rate of the glass during working was quite high — about 10—20 degr/sec. Several runs of the cooling process are illustrated in Fig. A 11. The cooling rate of glass in feeder machines is on the average greater than in suction machines. The reason for this phenomenon is possibly associated with the larger volume and the commensurately larger surface of the parison molds in feeder machines. There is also a stronger counter-blow of the parison in the feeder machines.

As will be shown in more detail in Sec. 2.5 certain conditions lead, depending on external circumstances, to a more or less large temperature

Table A 13. *Heat loss of bottle glasses in forming (after calorimetric measurements)*

	Wt. (g)	Total time t_{tot} (sec)	Gob temp. (°C)	Cooling (°C/sec) Blank mold	Blow mold	Total (°C/sec)
Vac. blow						
a Owens green	531	27.8	1145	15.2	14.4	14.9
b Owens green		28.0	1135	14.3	15.9	15.5
c Owens brown	616	29.2	1051	5.3	15.1	12.1
d Owens brown		27.5	1135	7.9	16.8	14.1
e Roirant A 6 brown	538	27.0	1110	11.1	10.9	11.0
f Roirant A 6 colorless	848	25.0	1135	16.5	8.3	12.0
g Roirant B green	388	27.3	1140	17.3	10.5	12.6
h Roirant B brown	533	32.0	1070	10.9	9.3	10.2
i Roirant B colorless	347	24.3	1090	8.1	15.0	12.6
Mean	543	27.6	1112	11.8	12.9	12.8
Feeder blow						
n Roirant R 7 green	346	13.7	1130	26.5	13.9	18.8
o Roirant R 7 green	650	20.6	1090	19.8	8.5	12.6
p Roirant R 7 green	379	14.8	1140	25.8	15.0	18.8
q Roirant R 7 green	617	19.2	1115	14.8	15.8	15.4
r Roirant R 7 green	868	27.0	1075	12.5	8.2	9.8
s Roirant R 7 green	690	19.0	1095	12.1	15.0	13.8
t Roirant R 7 brown	525	18.8	1063	10.5	17.7	14.6
k JS-machine brown	338	19.0	1108	19.8	11.4	19.4
l JS colorless		14.3	1160	20.8	20.6	20.6
m Semi-automat colorless	540	17.0	1075	10.0	15.0	13.2
Mean	550	18.3	1105	17.3	14.1	15.7

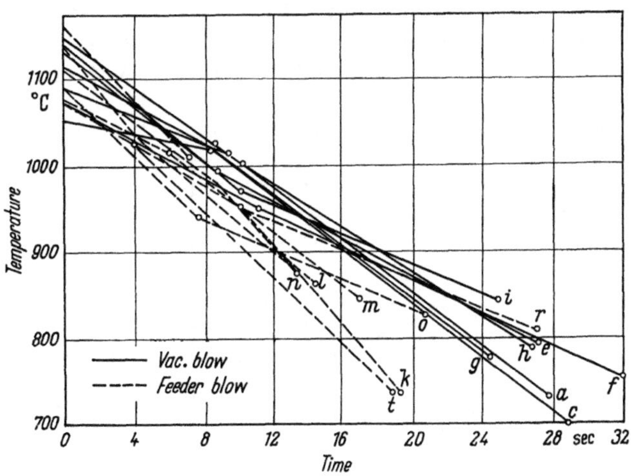

Fig. A 11. Average glass temperature during forming. (Calorimetric measurement, labels, see Table A 13.)

difference between glass and mold wall (100—200 degr.). One now believes that glass touches only protrusions of the mold but can not penetrate valleys in the inevitably uneven surface of the mold. In this way glass and mold wall are separated by a thin layer of air which strongly impedes heat transfer from glass to mold. New measurements by COENEN [17a] suggest that glass shrinks on cooling and so pulls away from the mold surface, strongly decreasing heat contact between glass and metal.

If the mold surface exceeds a certain temperature limit — for conventional glasses and mold materials about 610 °C — contact conditions between glass and mold seem to change suddenly. The glass begins to stick to the mold wall. The reason for this phenomenon remains largely unexplained. Possibly at this temperature the lowered viscosity permits the glass to penetrate fine mold cracks. The solution of a fine oxide skin at the mold surface might also contribute to the phenomenon.

If a heat transfer coefficient α between glass and mold is defined by

$$\alpha = Q_{G-F}/\big(F(\Theta_{OG} - \Theta_{OF})\big),$$

where Θ_{OG} and Θ_{OF} designate the surface temperatures of glass and mold, measurements show that α depends upon time and pressure.

When 4.76 mm thick glass lids from colorless glass were press-formed at 8.3 atm in 1.4 sec, a mean temperature difference of $\Theta_{OG} - \Theta_{OF} = 190$ °C was observed. The coefficient α calculated from the measured heat flux and the difference of temperature is plotted in Fig. A 12. The steep decrease with time of contact is noteworthy.

TRIER [18] determined mean heat fluxes of 280×10^3 kcal/m² h for colorless glass 45 mm thick. The temperature discontinuity has to be assumed as 200 °C. The contact time was 4 sec, pressing pressure 2 atm and the inner mold temperature 470 °C. The heat transfer thus was much smaller than the one indicated in Fig. A 12. It must be as-

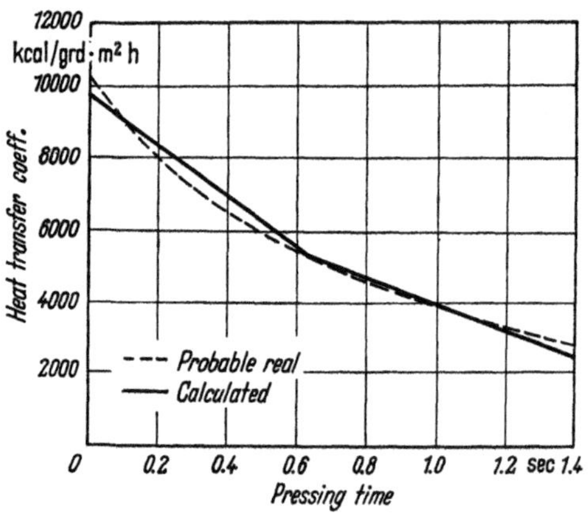

Fig. A 12
Heat transfer coefficient glass-mold in pressing. (According to MCGRAW [2].)

sumed that the longer contact (4 sec vs 1.4 sec) and the lower pressure (2 vs 8.6 atm) were responsible.

For green glass somewhat smaller values were obtained, but the difference was surprisingly small.

2.5 Temperature Distribution in the Glass During Forming

Although the temperature distribution in glass at forming is extremely important data are scarce. A satisfactory calculation was attained only for a large glass thickness (1 m) and the steady-state (7). The non-stationary case has been treated only once, which indicates the difficulties associated with it.

Experimental data have been published by BABCOCK [6] and TRIER [18]. BABCOCK used fine thermoelements to measure the temperature variation with time and thickness in thin glass layers during pressing. Fig. A 13 shows the data published by BABCOCK.

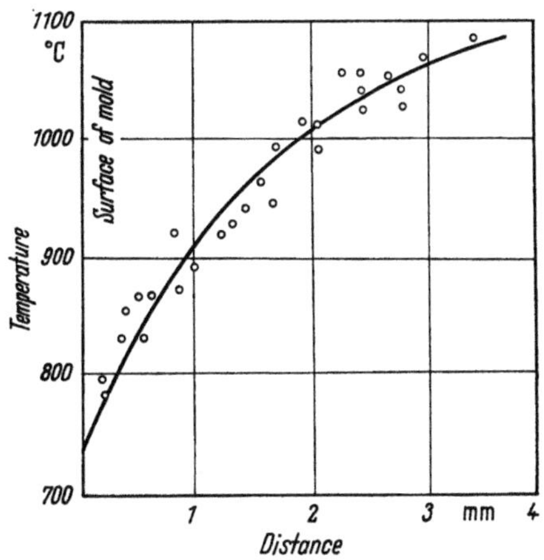

Fig. A 13. Temperature distribution in glass for thin ware (for a pressing time of 2.2 sec in colorless glass).

TRIER determined glass temperatures for colorless and green glass by means of a new puncture method via viscosity. In contrast to BABCOCK he used 44.6 mm thick cylindrical layers. The results of these measurements are summarized in Figs. A 14 and A 15. While in colorless glass remarkably strong and nearly uniform cooling of the inside of the glass was observed, cooling was slower in the interior of the green glass. At the surface temperature drops steeply. Here the differences between colorless and green glass are remarkably small. Unfortunately this method does not permit finding the temperature of the glass surface. It seemed, however, to be of the same order of magnitude as in BABCOCK's measurements.

McGraw derived theoretically that in pressing thin glass layers (4—5 mm) heat transport is almost exclusively by conduction, while transport by radiation is much less important [18a]. Consequently the

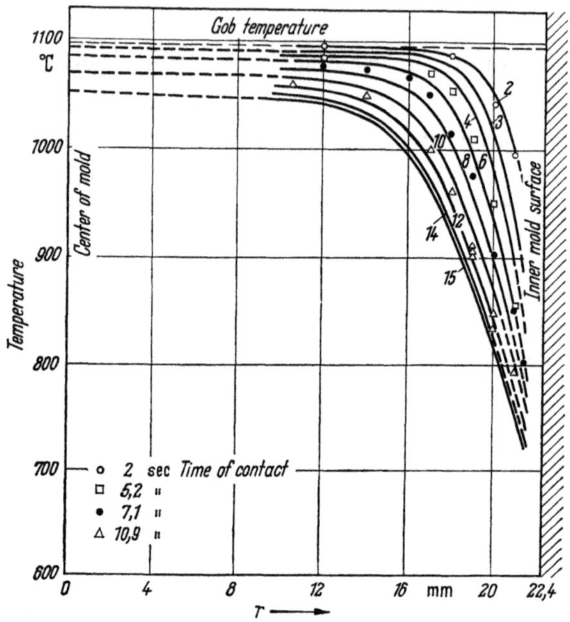

Fig. A 14. Temperature distribution in colorless glass (mold temperature 470 °C).

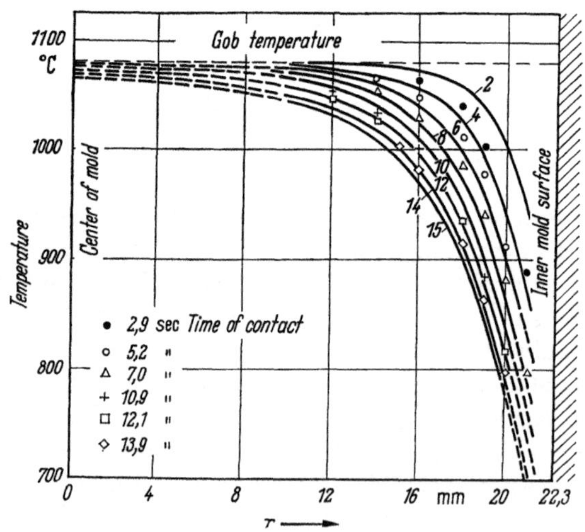

Fig. A 15. Temperature distribution in green glass (mold temperature 470 °C).

temperature distribution could be determined analytically from the basic equations (4) and (5) as long as layers were thin and heat transfer is large. For thicker layers, however, such procedure is not permissible because of the noticeable influence of heat radiation.

2.6 Heat Transmission in the Mold Wall

Because of the periodic working of the molds of container glass machines, the concomitant periodic heat supply on the inside, and the almost uniform heat loss by air cooling on the outside, a quasi-stationary temperature field becomes established inside the mold wall. Temperature fluctuations are particularly pronounced on the inside of the mold and decline toward the outside. According to measurements by TRIER [19]

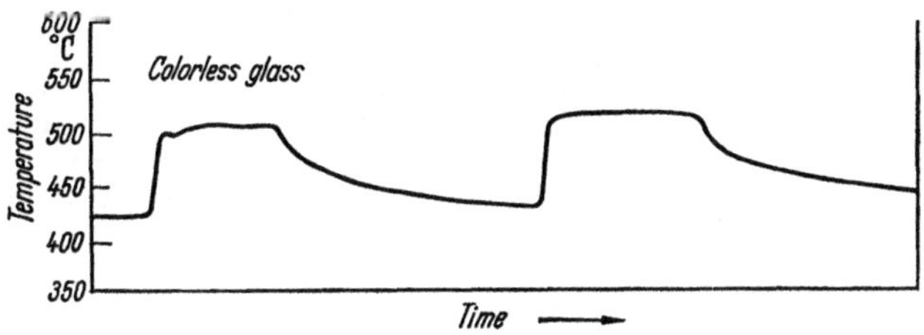

Fig. A 16. Inside temperature of a blank mold at half height in the forming of colorless glass.

and HOLSCHER [20] temperature fluctuations in bottle parison molds are 50—80 and in blow molds 30—50 degrees centigrade. The higher the temperature of the glass, and the lower the thermal conductivity of the mold material, the larger the amplitude of these fluctuations. Fig. A 16 shows a temperature behavior typical of the inside of a parison mold. For green glass practically the same curves result, with somewhat smaller amplitudes. The amplitudes can be calculated approximately as shown by TRIER [18] and COENEN [17a].

While the temperature distribution along the circumference of molds should be as uniform as possible, there are technological reasons (quick solidification at the finish) in favor of differences along the vertical axis. In practice it was shown that the inside temperature of molds should be highest at half elevation, with a desirable decrease above and below this location. The temperature distribution in parison and blow molds of a machine (according to TRIER) is given in Fig. A 17.

The correct mold temperature has great significance in the plant. If the mold is too cold, bottles exhibit a hammered surface deriving from uneven wall thickness; if the mold temperature is too high, the glass begins to stick at the mold wall. The correct working temperature is between 500—600 °C. In Fig. A 18 are plotted the inner temperatures of the parison mold of a semiautomatic unit for various working times. Short glass times produce a soft, unstable, parison while for long glass times the glass cools down too much in the mold and can not be worked

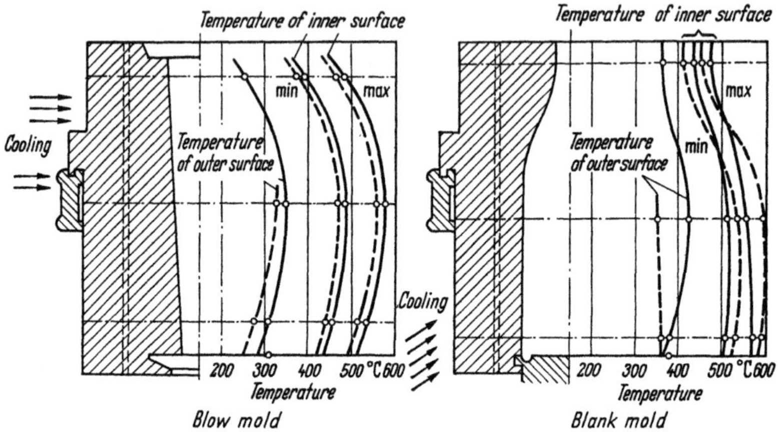

Fig. A 17. Temperature distribution in blank and blow mold on the IS machine in the middle range ———, near the parting line - - - -.

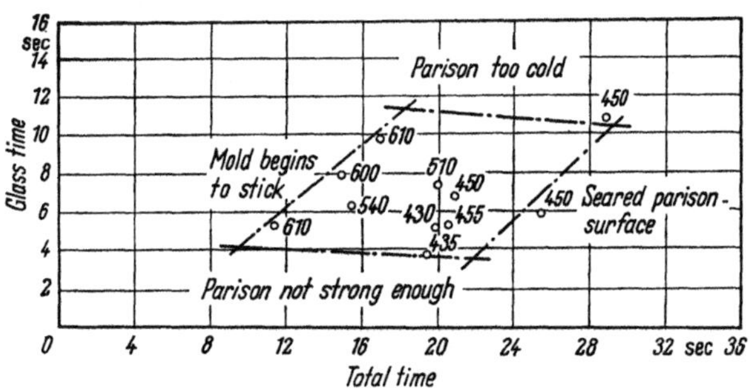

Fig. A 18. Inside temperature of a blank mold for different working times (colorless glass, outside cooling).

further. Short total time leads to overheating of molds and sticking, while long total times lead to cold molds.

From the theoretical viewpoint E. Deeg [21] dealt extensively with the temperature distribution in glass molds.

A high heat conductivity favors heat transport in the mold and decreases the temperature of the inner wall. A strong influence on the temperature of the inner wall is also exerted by the wall thickness of the mold. Thick walled molds generally run cooler than thin walled ones and have a more uniform temperature distribution. In thick molds strong local differences in the cooling of the outer mold surface give less disturbance of the inner surface temperature gradient [22].

A high heat capacity and a high heat conductivity of the mold material diminish the periodic fluctuation at the inner wall of the mold and so allow running at a higher mean interior temperature without sticking [17a, 24].

References

[1] RATCLIFFE, E. H.: Preliminary measurements to determine the effect of composition in the thermal conductivity of glass. Phys. and Chem. Glasses 1 (1960) 103/04.

[1a] RUSS, A.: Die Wärmeleitfähigkeit von Gläsern in Abhängigkeit von der chemischen Zusammensetzung. Sprechsaal 61 (1928) 887—891, 907—912.

[2] MCGRAW, D. A.: The transfer of heat in glass during forming. J. Amer. ceram. Soc. 44 (1961) 353—363.

[3] SCHMIDT, E.: Einführung in die Technische Thermodynamik, 8th Ed., Berlin/Göttingen/Heidelberg: Springer 1960.

[4] GROEBER/ERK/GRIGULL: Die Grundgesetze der Wärmeübertragung, 3rd revised Ed., by ULRICH GRIGULL, Berlin/Göttingen/Heidelberg: Springer 1955.

[5] HÄNLEIN, W.: Über das spezifische Gewicht von Glas in Abhängigkeit von der Temperatur. Glastechn. Ber. 10 (1932) 126—130.

[6] BABCOCK, C. L., and D. A. MCGRAW: Application of glass properties data to forming operations. Glass Ind. 38 (1957) 137—142, 144—146, 148—151, 161. (Travaux du IVᵉ Congrès International du Verre, Paris 1956, pp. 164—176.)

[7] GENZEL, L.: Messung der Ultrarot-Absorption von Glas zwischen 20 und 1360°C. Glastechn. Ber. 24 (1951) 55—63.

[8] NEUROTH, N.: Der Einfluß der Temperatur auf die spektrale Absorption von Gläsern im Ultraroten. Glastechn. Ber. 25 (1952) 242—249.

[9] CZERNY, M., L. GENZEL, and G. HEILMANN: Über den Strahlungsstrom im Inneren von Glasschmelzwannen. Glastechn. Ber. 28 (1955) 185—190.

[10] (a) CHARNOCK, H.: Experimental and theoretical comparison of radiation conductivity predicted by steady-state theory with that effective under periodic temperature conditions. J. Am. Cer. Soc. 44 (1961) 313.
(b) BEATTIE, J. R.: Acta Imeko 1961, Budapest, pp. 1—23.

[10a] HUHMANN, I.: Über den Energietransport in Glasschmelzen. Glastechn. Ber. 32 (1959) 189—197.

[11] GEFFCKEN, W.: Radiale Strahlungsleitung in langen Zylindern. Glastechn. Ber. 32K (1959) VII/28.

[11a] GARDON, R.: Calculation of temperature distributions in glass plates undergoing heat-treatment. J. Amer. ceram. Soc. 41 (1958) 200—209.

[12] VAN ZEE, A. F., and C. L. BABCOCK: A method for measurement of thermal diffusivity of molten glass. J. Amer. ceram. Soc. 34 (1951) 244—250.

[13] LÜCK, G.: Bestrahlung strahlungsdurchlässiger Körper. Chem.-Ing.-Techn. 32 (1960) 29—33.

[14] NEUROTH, N.: Zusammenstellung einiger Meßergebnisse der Wärmeleitfähigkeit von Glasschmelzen. Glastechn. Ber. 32 (1959) 197/98.

[15] ELIGEHAUSEN, H.: Über die Wärmeleitfähigkeit von Gläsern und einschmelzenden Gemengen. Diss. of Rhein-Westf. Techn. Hochschule Aachen 1957.

[16] UNGER, L.: Bestimmung des Wärmeinhaltes und der mittleren Temperatur von Flaschen während des Formgebungsprozesses. Glastechn. Ber. 32 (1959) 153—157.

[17] GIEGERICH, W.: Über technische Grundlagen der vollautomatischen Flaschenherstellung. Glastechn. Ber. 33 (1960) 441—449.

[17a] COENEN, M.: Messungen und Berechnungen über den zeitlichen Temperaturverlauf an der Grenzzone Metall-Glasschmelze. Advances in Glass Technology. Technical Papers of the VIth International Congress on Glass, Washington, New York: Plenum Press 1962.

[18] TRIER, W.: Temperaturverteilung und Wärmefluß im Glas in Vorformen von Hohlglasmaschinen. Glastechn. Ber. 33 (1960) 449—456.

[18a] NAUGHTON, T. J., and D. A. MCGRAW: Analysis of heat transfer in glass-pressing operations. Advances in Glass Technology. Technical Papers of the VIth International Congress on Glass, Washington, New York: Plenum Press 1962.

[19] TRIER, W.: Temperaturmessungen in Glasformen. Glastechn. Ber. 28 (1955) 336—351.

[20] HOLSCHER, H. H., J. G. COLEMAN, and C. C. COOKE: Messung von Glas- und Formtemperaturen beim betriebsmäßigen Formvorgang von Glasgefäßen. Glastech. Ber. 32K (1959) II/1—II/7. [Reprinted Glass Ind. 41 (1960) No. 2, 74—81, 113/14; No. 3, 142—147, 181/82.]

[21] DEEG, E.: Berechnung der Temperaturverteilung in Vorform und Külbel bei Hohlglasmaschinen. Glastechn. Ber. 32 (1959) 402—413.

[22] WILLE, R.: Methoden der Formenkühlung an Glasverarbeitungsmaschinen. Glastechn. Ber. 28 (1955) 351—359.

[23] SCHRECK, C.: Beitrag zur Ermittlung des Wärmeflusses in den Wänden von Glasformen. Glastechn. Ber. 28 (1955) 359—368.

3 Flow of Glass

By W. TRIER, Frankfurt/Main

3.1 Flow Behavior of Glass as a Material

In contrast to crystalline substances glass possesses no definite melting point but softens gradually over a broad temperature interval. For the technical glasses used in the manufacture of hollow and flat glass softening ($10^{7.6}$ poise) starts around 700 °C. With rising temperature the viscosity of glass decreases continuously. A viscosity range from $10^{7.6}$ to 10^2 corresponds to a temperature range from 700 to 1400 °C; that is the viscosity changes over several powers of 10. Consequently, flow behavior depends strongly upon temperature via viscosity. In particular the temperature distribution in the body of glass and the associated local viscosity differences play a big part in the flow process.

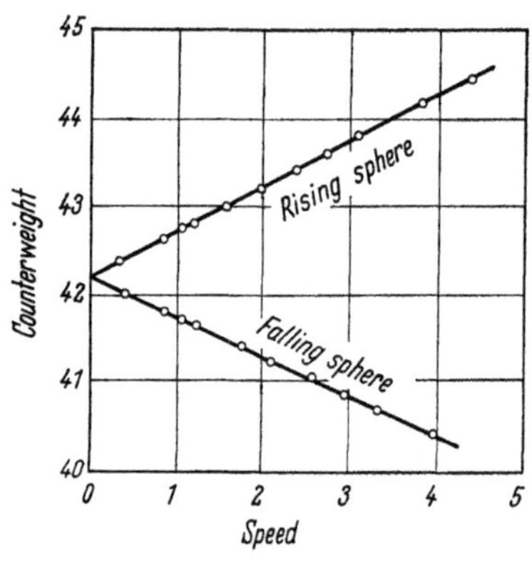

Fig. A 19. Speed of rising and falling sphere in a lithium borate glass melt as a function of counterweight (CAPPS [1]). (Unfortunately the reference does not indicate units.)

The flow behavior of glasses at constant temperature has been studied frequently to determine if glass acts like a Newtonian liquid or if, instead, there are indications which point to a structural viscosity. Investigations by CAPPS [1] on spheres rising and falling in glass melts revealed a linear connection between the velocity of the sphere and its resistance to shear (Fig. A 19). This means that in molten glass the shear stress is proportional to the shear gradient $\gamma = dv/dn$, a chief characteristic of Newtonian liquids. Thus, it appears that for glasses at temperatures above the softening range and to higher temperatures the simple formula for Newtonian flow

$$\tau = \eta \, dv/dn$$

is valid. It is, therefore, sufficient to characterize the flow behavior of glasses by the specification of a time and apparatus-independent, but temperature dependent viscosity η.

If, however, pronounced devitrification or immiscibility occurs in glasses, one has to expect non-Newtonian behavior, as can be seen from investigations by HARRISON, STEVENS and SHELTON [2] (Fig. A 20). They used a rotational viscometer and measured the deflection of the viscometer vessel as a function of rotatory velocity. They investigated various glass frits used for enameling metals. For some frits a linear

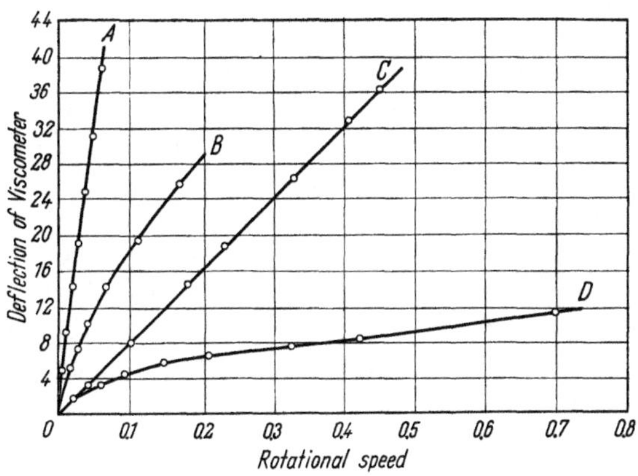

Fig. A 20. Deflection of a viscometer as a function of rotational speed (CAPPS [1]). (Unfortunately the reference does not indicate units.)

relation between velocity of rotation and deflection was found indicating the Newtonian behavior of these frits, while for other frits curved dependencies indicated non-Newtonian behavior. For non-Newtonian liquids a simplified relation

$$dv/dn = \varkappa \, \tau^m$$

can be used, where $\varkappa$ and m are empirical constants.

In summary one may say that homogeneous glasses above the softening range can be considered as Newtonian liquids.

For the transformation range of glasses in the viscosity range between 10^9 and 10^{14} poise OEL's [3] investigations have demonstrated a time dependence of viscosity. The magnitude of the time effects depends not only upon composition but also on pre-treatment.

3.2 Flow of Glass During Working in Hollow Glass Machines

The flow of glass in filling the parison molds (blank molds) of hollow glass machines, and during forming, is of great importance for the correct evolution of the forming process and the final properties of the hollow ware. Unfortunately the measurement of flow characteristics in the molds of hollow glass machines is difficult, so that there are only a few concrete foundations for this process.

3.2.1 Flow of Glass in Parison Molds of Suction Machines

In suction machines the parison mold dips into the viscous (10^3 poise) glass with its open lower end to a depth of several mm. By the application of a vacuum the glass is sucked upwards into the parison mold.

Unfortunately no one has yet succeeded in measuring this flow. From the striae structure of the gob, first investigated by EITEL [4, 5, 6], then by GIEGERICH [7], it may be assumed that this influx is a hot spot flow which causes an onion structure in the glass. The incoming glass forms on its surface a cap of rotational symmetry, the lower rim of which ends with the edge of the parison mold opening. It is suspected that between the glass and the parison mold surface little or no slippage occurs. The glass layers stretch like an elastic membrane; the domed surface contacts the parison mold surface at the base of the rising dome. The glass thus is strongly stretched perpendicular to the direction of flow. The chemical inhomogeneities present in the glass and appearing in the form of striae thus are strongly stratified. The stratification becomes visible as a fine line structure within the parison.

The probable flow of glass in the parison mold of a suction machine is indicated in Fig. A 21. A single glass particle follows an arc shaped flow path. Therefore, there is no Hagen-Poiseuille flow, in which single particles of a liquid exhibit a flow direction parallel to the axis of a tube. Another deviation from the Hagen-Poisemlle flow is caused by the local cooling of the glass near the mold surface and the associated strong rise in glass viscosity.

The formation of the hot spot flow pattern and the attachment of the dome-shaped upsurging glass at the mouth of the mold and around the plunger, may be seen in Fig. A 22 (according to GIEGERICH). Some particularly beautiful pictures are contained in EITEL's study [4].

The knife cutting the glass blank causes a strong deformation of the striae in the lower part of the parison. Irregular suction through thermal inhomogeneities or large temperature differences in the mold

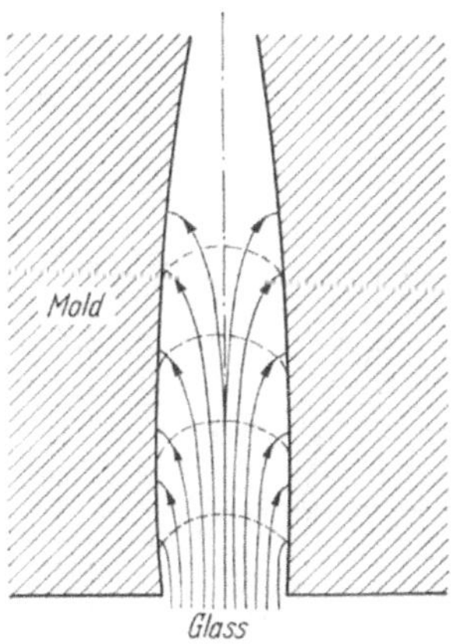

Fig. A 21. Probable flow of glass in a suction blank mold.

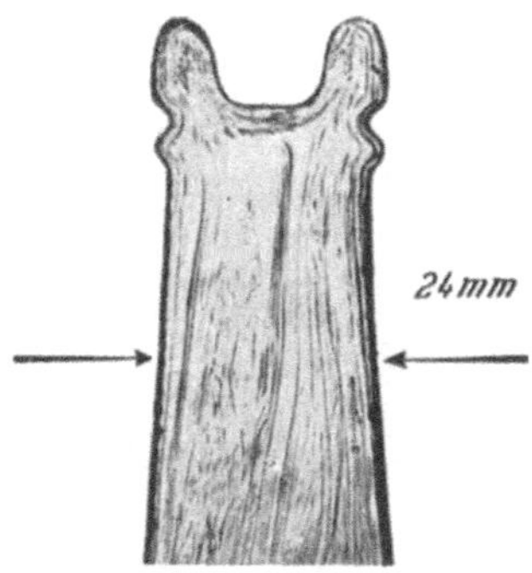

Fig. A 22. Striae comparison in the neck region of a parison (GIEGERICH [7]).

may disturb the rotational symmetry of the stream causing faults in the glass. Details can be found in the papers by EITEL [6] and GIEGERICH [7].

After complete filling, the blank mold lifts above the glass line in the trough, the knife cutting the glass also closes the mold, and the glass is blown down by compressed air while the plunger is withdrawn. At this point there is practically no glass flow. During the following phase, in which the bubble is blown, the knife frees the suction opening of the mold and the glass spills a little under the pressure of the compressed air which streams in from above. Within the mold, except for the mouth and suction parts, the glass flows as in a long tube. A nearly parabolic distribution of velocities is established, which, however, is strongly distorted near the surface of the mold where local cooling induces a strong increase in viscosity.

Flow patterns of glasses registered under similar conditions, that is, in the blank mold during puffing, have been studied intensively by GIEGERICH [7]. It becomes evident that the cooling of glass near the mold surface depends not only upon the mold and glass temperature, but also upon the time spent by the glass in the mold. The influence of contact time on flow pattern for hollow flint ware is represented in Fig. A 23, depicting measurements by Trier [8]. In all cases the same quantity of glass is flowing.

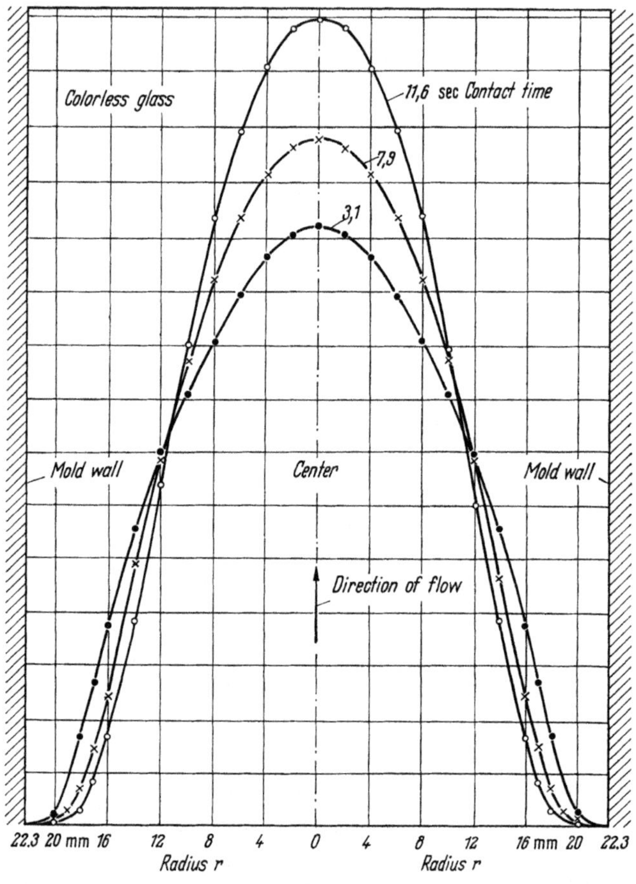

Fig. A 23. Flow figures in colorless glass. Effect of contact time (3—12 sec).

While for short contact times the viscous surface layer is relatively thin, and a broad parabola appears in the middle area, the viscous edge layer becomes much thicker at longer contact times. This causes the formation of a slender flow parabola with longer flow path.

Similar curves obtain for green glass. The flow pattern however, is more slender and deviates more from the parabola than in the case of flint glass [8].

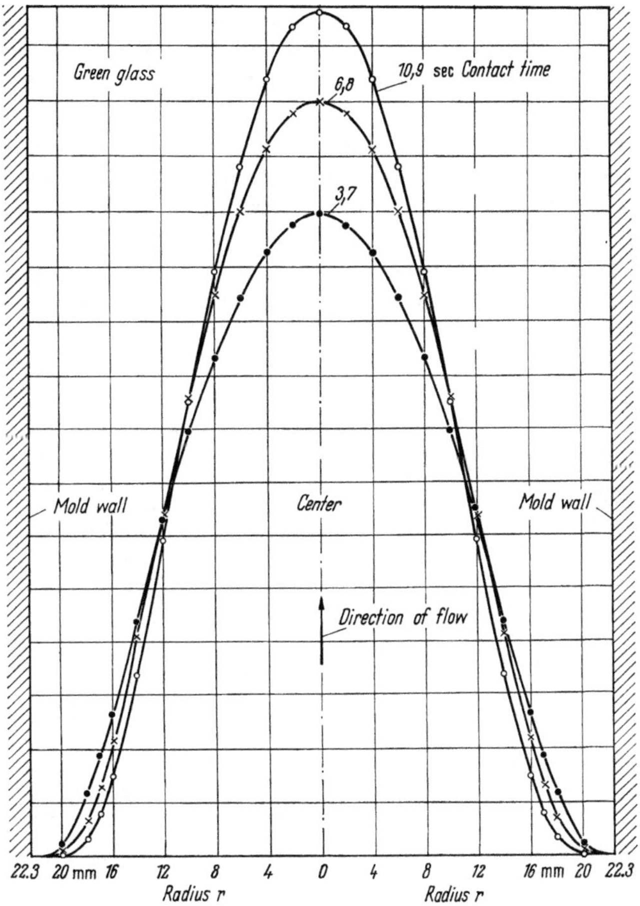

Fig. A 24. Flow figure in green glass. Influence of contact time (3—11 sec).

3.2.2 Flow of Glass in Parison Molds of Feeder Machines

In feeder machines parison molds are filled by a gob of glass falling from above. One tries to influence the shape of the gob by suitable adjustment of feeder and glass to facilitate the sliding of the gob into the mold. By giving the parison a funnel shape this sliding is facilitated. In contrast to the suction process the glass probably glides on the inner wall of the mold. Such gliding is encouraged by swabbing the molds with graphite during operation. In this way a thin, very slippery graphite film is formed between glass and mold. In contrast, suction parison molds are not lubricated. It must be assumed — measurements unfortunately are not yet available — that in the end phase of the filling of parison molds for feeder machines a similar hot spot pattern (see Fig. A 25) develops as in suction molds. The surfaces are fixed in relation to the middle parts due to their contact with the mold surface, and the gob, in addition, is deformed.

In the gobs of feeder machines distinctly stratified streaks also appear, and there are distortions in top and bottom areas caused by the shear cuts [6, 7]. It has to be assumed that alignment of striae has already been initiated in the feeder.

During counterblow of the gob in the parison mold one has to expect the formation of a hot spot flow pattern (Fig. A 26) similar to that in

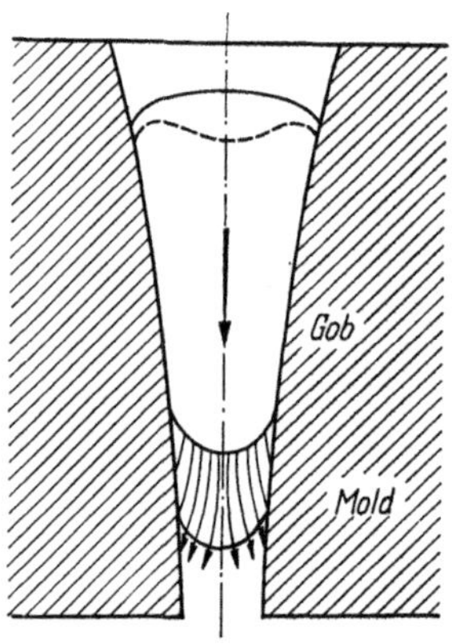

Fig. A 25. Gob fall in the blank mold of a feeder machine.

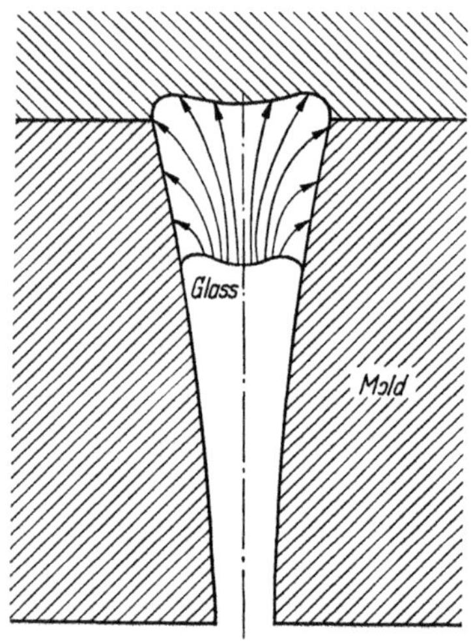

Fig. A 26. Probable flow diagram in a feeder blank mold.

the suction process. The upper, surging part of the gob is formed from the hot glass in the interior of the not yet counterblown gob. In the subsequent blowing in the blow mold the parison further expands and so leads to the well known "settle wave".

3.2.3 *Flow of Glass in Blow Molds*

While one may well speak of flow in parison molds particularly suction molds, the motion of glass in blow molds is more like the stretching and dilating of the preformed gob. There are no principal differences between suction and feeder machines, apart from the variation of gob shape.

By marking the parison wall the stretching process in the blow mold may be followed, as shown by Hahnel [9]. The results of such investigations show that in general the glass in the blow mold moves down and toward the outside. The flow paths determined by experimentation can be seen from Fig. A 27. In feeder machines the bottom part of the gob is already quite well preformed, while in suction machines

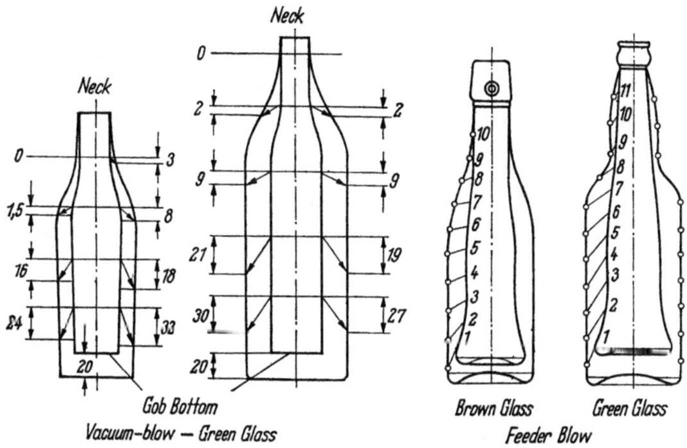

Fig. A 27. Flow paths obtained by means of marked sidewalls (HAHNEL [9]).

considerable stretching in this area is required with the result that there is a tendency toward the formation of thick and slanted bottle bottoms.

Unfortunately the gob does not always stretch symmetrically, but frequently more on one than on the other side, as can be seen easily from the shape of the parison seam on the finished bottle. Differential expansion of the gob easily causes irregular wall thickening with all its disadvantageous consequences.

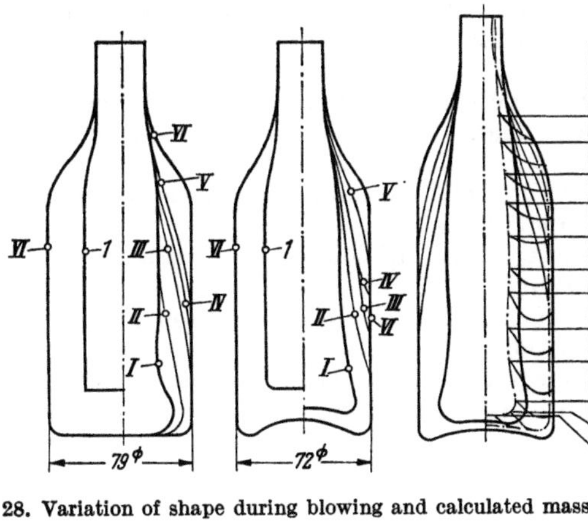

Flow Paths in mm			
I-II	II-III	III-IV	IV-V
2	–	2	2
3	1	3	4
4	2	5	4
5	3	5	2
6	5	5	1
8	10	–	–
13	6	–	–
16	–	–	–
14	–	–	–
12	–	–	–
14	–	–	–
6	–	–	–

Fig. A 28. Variation of shape during blowing and calculated mass motion (GIEGERICH [7]).

If one removes the gob in the various blow stages assumed in the blow mold, indications of the stretching process may be obtained. Curves so obtained in one case by GIEGERICH can be seen in Fig. A 28. One notices that at the beginning of the blow process the gob stretches downward until the bottom sits on the mold bottom. Next, the gob is blown up from below. The glass attaches to the mold wall proceeding from bottom up. The strongest stretches appear at the shoulder of the bottle, which explains the frequent occurrence of thin walls in this area.

3.3 Flow of Glass in the Manufacture of Flat Glass

Details about flow in the manufacture of flat glass are unfortunately not available at this time. In the rolling process (plate glass process) as well as in the diverse drawing processes, flow occurs which can not yet be described. It should ultimately be possible to develop experimental details about the flow processes for these methods of manufacture. KITA [10] recently treated model methods and results.

References

[1] CAPPS, W.: Viscosity of Glass. J. Colloid Sci. 7 (1952).

[2] HARRISON, W. N., R. E. STEPHENS, and S. M. SHELTON: Consistency of eight types of vitreous enamel frits at and near firing temperatures. J. Res. Nat. Bur. Stand. 20 (1938) 39.

[3] OEL, H. J.: Das Verhalten der Zähigkeit von technischen Gläsern zwischen 10^9 und 10^{14} Poise. Glastechn. Ber. 35 (1962) 57—60.

[4] EITEL, W.: Studien über die Strömungsvorgänge bei der vollautomatischen Glasverarbeitung im Owens-Prozeß. Glastechn. Ber. 10 (1932) 469—477.

[5] EITEL, W.: Studien über die Strömungsvorgänge bei der vollautomatischen Glasverarbeitung im Feeder- und Kipp-Prozeß. Glastechn. Ber. 11 (1933) 201—205.

[6] EITEL, W.: Studien über die Strömungsvorgänge bei der vollautomatischen Glasverarbeitung zu Flaschen im Feeder-Prozeß mit der Owens Bottle Co.-Maschine. Glastechn. Ber. 12 (1934) 222—227.

[7] GIEGERICH, W.: Ergebnisse und Probleme der maschinellen Flaschenfertigung. Glastechn. Ber. 30 (1957) 299—308.

[8] TRIER, W.: Zähigkeitsverteilung und Fließen des Glases in Vorformen. Glastechn. Ber. 32K (1959) II/17—II/25.

[9] HAHNEL, K.: Fließvorgänge bei der Herstellung von Flaschen (Abstract). Glastechn. Ber. 28 (1955) 82.

[10] KITA, H.: Some defects in rolled glass sheet. Reprints of the annual meeting of ICG, Sept. 1966, Tokyo, "Symposium on Defects in Glass".

4 Practical Problems in Forming

By W. Giegerich, Heilbronn

The basic processes in forming glass are:

1. Pressing
2. Blowing
3. Rolling

By means of displacement — plunger, air or rollers — pressure is exerted on the glass which causes a forced adjustment of the surface and body.

4. Drawing

Through an opening — for instance an outlet in the flat glass pull drawing method — and by means of tensile forces a continuous flow of glass is stretched to the desired thickness and width.

In practice, forming of glass depends not only on the application of these exterior pressing and stretching forces but also on the force of gravity. Practical forming is a complex process and is strongly influenced by shearing forces from the viscous edges. In the transition regions of the viscous, plastic and solid states, not only the elastic deformations are determinant, but particularly also the technical flow processes. It is always assumed that during a change in shape the heat of the glass is extracted by the molds and to the outside air in order to increase the glass viscosity and to obtain the desired shape after removal of the molds. A forming machine by using appropriate tools has thus the task of effecting the change in shape and directly or indirectly achieving heat abstraction from the glass to the outside air. In the past 10 years efforts have been made to approach the investigation from various sides of the complicated interplay that takes place in the glass making machine, but without any final clarification.

For a better understanding of what follows the known requirements with regard to the quality of the molded product are mentioned here:

(a) Uniform wall thickness and distribution,

(b) Smoothness of the surface,

(c) Proper size forming, especially precise control of the finish dimension.

4.1 Glass Forming Temperature

The glass temperature during loading of the machine — in the feeder process, the gob temperature; in the vacuum process and gathering by hand process the temperature at the takeout position — varies according to the type of object produced, its weight, and the manufacturing process. Using the same glass, it can vary, depending on the article, by more than 100 degrees. Thus it is the task of the feeder or the pot to adjust to these temperature differences and provide a homogeneous temperature. This is especially necessary when several machines having different jobs use the same tank. The feeder is able to do this in a relatively wide range. Difficulties arise nearly always because of limited cooling possibility ot the glass in the forehearth. The revolving pot, on the contrary, can carry this equalization out to a slight degree only, for it has only a small cooling and heating range. Besides it has the hard task of melting the cut-offs (getting rid of blisters and homogenizing) on the way from the filling point to the new incoming glass through the trough. Since the temperature of the working tank, even with separate heating, can only be changed to a small degree, and the feeder setup likewise does not make arbitrary temperature adjustments possible, limits on a tank making several types of ware simultaneously are set up by choice of the working temperatures. But different weights of the individual items also upset the tank operation because of uneven glass removal, especially melting, refining and homogenizing, if these cannot be compensated by other machines using the same tank. Due to the interplay of several machines, open take-out positions directly from the tank are endangered in an especially bad way such as when several suction machines having single arms are taking glass at the same time. Because of the unavoidable air blasts, temperature and atmosphere of the work tank are disturbed in a noticeable way. With widely different items, the installation of covers and their adjustment often causes unsolvable problems.

4.2 Mold Surface

A further general task, after the thermal conditioning of the incoming glass, is the preservation of the surface quality and temperature of the mold. Due to periodical heating by the glass, the variable intensity of contact, and the structural composition of the mold itself, individual oxide pockets are formed during operation that impede the flow of glass and can cause sticking and checks. These difficulties can be partially eliminated by using special mold dope materials but these also eventually leave residues behind and again disturb the forming operation. Common to all mold doping materials is the formation of a vapor layer. The individual processes taking place in the boundary layer have not been clarified as yet and thus the use of the swabbing material is based entirely on practical experience. The surface of the mold wall has been studied thoroughly and its effect on the glass flow has been examined by using cord patterns [2]. The metallic surface layer consists mainly of metallic iron, iron oxide, iron phosphate, iron nitrite [1], and graphite. The contact with glass takes place, especially in the case of the popular cast iron, with a boundary layer that deviates strongly from the basic material of the mold.

4.3 Individual Processes of Importance

The heat flow from the glass to the mold wall and to the outside air is influenced by the temperature of glass and mold, the contact time, and the pressure under which the contact takes place [1]. These factors are particularly effective in the blank mold. The blow mold on the other hand is essentially used to form the shape. Heat, of course, has to be removed here too in order to give the glass body sufficient rigidity. In the glass layers near the surface a strong cooling of the glass takes place on contact with the mold wall. Thus setting takes place faster the stronger the heat extraction in the blank mold. Ultimately the machine output is determined by it. Of great importance for forming and for production times is the shaping of the parison with respect to the shape of the finished glass container. For this every factory has its own empirically determined standards which are difficult to generalize.

The parison is thermally very inhomogeneous when leaving the mold, and a certain temperature equalization has to take place in the glass during the contact-free time during which the parison is transferred from the blank mold to the blow mold. The skin layers have to be heated from the inside by heat transport. The time necessary for this is called "reheating time". Dwell time in the blank mold and reheat time, in their mutual exchange, have a great influence on the formation of the bottom and the surface of the finished object. Short blank mold times generally also result in good wall thickness distribution, whereas a smooth glass surface is obtained neither with too short nor too long a reheat time.

Forming has been examined in past years from various sides:

(a) By measurements of the temperature distribution in glass and in the mold wall (Babcock and McGraw [3], Trier [4]),

(b) Measurements of the total heat content in the various manufacturing stages (Unger [5], Giegerich [6],

(c) Change in shape during forming (Hahnel [7], Krüssmann [8], Giegerich [9]),

(d) Time exposure in the manufacturing course of various machines (Grundmann [10], Giegerich [6]),

(e) Schlieren presentation of flow (Eitel [11], Giegerich [12]),

(f) Model experiments with other materials than glass (Heuer [13]),

(g) Theoretical and experimental observation of flow processes (Trier [4], Wille [14]).

These investigations have brought about much new knowledge, but have not as yet completely clarified the situation.

References

[1] TOOLEY, F. V.: Handbook of Glass Manufacture, New York 1953, Section: HOLSCHER, H. H.: Feeding and Forming, pp. 303—387.
[2] GIEGERICH, W.: Die Oberfläche der Formenwand und ihr Einfluß auf die Strömung des Glases. Glastechn. Ber. 26 (1953) 333—341.
[3] (a) BABCOCK, C. L., and D. A. MCGRAW: Application of Glass Properties Data to Forming Operations. Glass Ind. 38 (1957) 137—142, 144—146, 148 to 151. [Ref. Glastechn. Ber. 31 (1958) 67.]
(b) MCGRAW, D. A.: Transfer of Heat in Glass during Forming. J. Amer. ceram. Soc. 44 (1961) 353—363.
[4] (a) TRIER, W.: Temperaturmessungen an Glasformen. Glastechn. Ber. 28 (1955) 336—351.
(b) TRIER, W.: Zähigkeitsverteilung und Fließen des Glases in Vorformen. Glastechn. Ber. 32K (1959) II/17—II/25; 33 (1960) 449—456.
[5] UNGER, L.: Bestimmung des Wärmeinhalts und der mittleren Temperatur von Flaschen während des Formgebungsprozesses. Glastechn. Ber. 32 (1959) 153—157.
[6] GIEGERICH, W.: Über technische Grundlagen der vollautomatischen Flaschenherstellung. Glastechn. Ber. 33 (1960) 441—449.
[7] HAHNEL, K.: Fließvorgänge bei der Herstellung von Flaschen (Abstract). Glastechn. Ber. 29 (1956) 32.
[8] KRÜSSMANN, G.: Vorgänge bei der Formgebung von Flaschen (Abstract). Glastechn. Ber. 28 (1955) 82.
[9] GIEGERICH, W.: Ergebnisse und Probleme der maschinellen Flaschenfertigung. Glastechn. Ber. 30 (1957) 299—308.
[10] GRUNDMANN, W.: Arbeitsvorgänge an Glasmaschinen. Glastechn. Ber. 25 (1952) 426/27; 27 (1954) 32.
[11] EITEL, W.: Studien über die Strömungsvorgänge bei der vollautomatischen Glasverarbeitung. Glastechn. Ber. 10 (1933) 121—125, 469—471; 11 (1933) 201—205; 12 (1934) 222—227.
[12] GIEGERICH, W.: Fließvorgänge bei der Verformung ... mit Hilfe von Schlieren bildern. Travaux du IVe Congrès International du Verre, Paris 1956, pp. 177 to 183.
[13] HEUER, P.: Entwicklung eines Verfahrens zur Untersuchung von Fließvorgängen (Abstract).
(a) Glastechn. Ber. 33 (1960) 35; (b) Glastechn. Ber. 34 (1961) 593.
[14] WILLE, R.: Fließvorgänge bei der Herstellung von Flaschen (Abstract). Glastechn. Ber. 29 (1956) 32.

B. Machine Elements

1 Drives and Gears

By H. RIEDE, Düsseldorf

Drives and gears of hollow glass machines show some unique features when compared with those in conventional machine construction. These are connected with drive problems characteristic of the glass forming process [1—6]. The best known types of drives in hollow glass machines are presented schematically in Fig. B 1. Gears and drives used in glass machines are:

(a) Mechanical drives: Gear and worm drives [8—12],
Geneva drives [13—16],
Crank shafts [17],
Spatial curve gears,
Curve gears [18—20],

(b) Pneumatic and hydraulic drives,

(c) Electrical drives and

(d) Combinations of a, b and c.

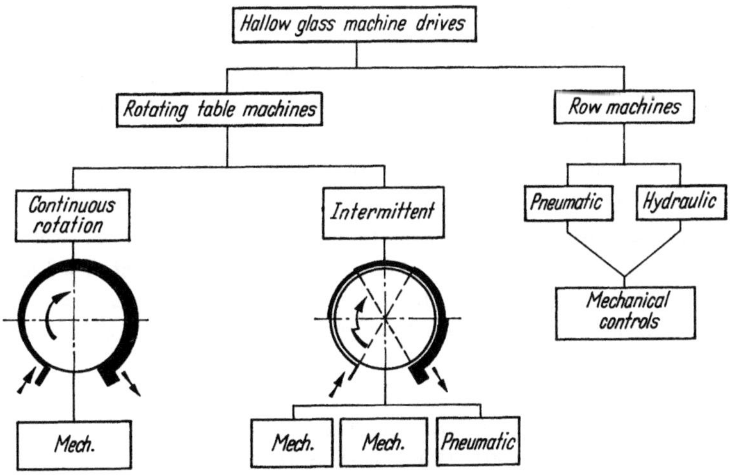

Fig. B 1. Diagram of drives for hollow glass machines. (Progress in forming indicated by increasing thickness of line. Numbers indicate stationary periods.)

1.1 Rotating Table Drives

Rotating machines are characterized by one or several turntables, or a rotating machine frame. On the turntables or frames are found the forming tools required for the forming of hollow glass as well as accessory activating tools. The type of turntable drive is very important in the construction of the machine and for production considerations.

1.1.1 Machine with Uniform Speed of Rotation

In continuous rotating machines with mechanical drives work motions are controlled by centrally arranged cams. These machines are driven with uniformly constant revolution, resulting in a relatively shock free run. For the drive cog wheels are always used.

1.1.2 Machines with Intermittent Rotation

The principal characteristic of intermittent machines is the periodic acceleration and deceleration of the turntable. The considerable variations in the drive momentum caused by mass action together with the reactive forces in bearings, etc. cause harmful shocks, which may be reduced to a tolerable level by special construction, narrow tolerances and good lubrication. Motion should be as vibration-free as possible, and extreme values of acceleration and deceleration must be minimized. The entire scheme of motion is important, with factors influencing the forming of the glass gob added to the postulates of machine construction. Forming processes are carried out while the turntable stands still. From this postulate and those pertaining to machine construction the following guidelines are derived:

(a) Stationary periods must be sufficiently long for forming,

(b) Uniformly quiet motion of the machine table, with favorable deceleration and acceleration relations,

(c) Maintenance of stand still periods,

(d) High productivity.

Since these requirements are, in part, contradictory to each other, a compromise must be found. This has to assure at least freedom from shock, since steadily increasing wear and steady acceleration have more favorable effects both on the forming of glass and on the machine than does intermittent wear.

1.2 Construction and Properties of Intermittent Turntable Drives

1.2.1 Mechanical Drives

Mechanical drives work in such a way that a uniformly rotating gear member, called the driver, is in only temporary connection with another member during which it transmits an impulse to this member, termed the driven wheel. The best known gear system of this sort, applied in most intermittent machines, is the Geneva drive. The method of operation is essentially characterized by the generally constant revolution of the driver which is in only temporary connection with one of the slits of the driven wheel. By the proper shaping of both wheels the driven wheel is automatically stopped for the standstill period ("dwell time").

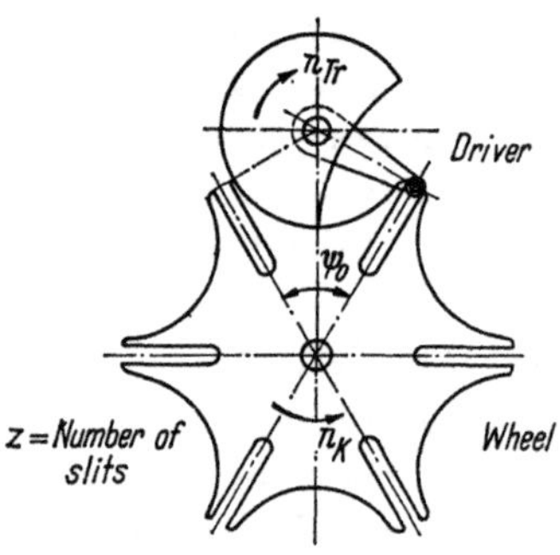

Fig. B 2. Six-fold Geneva drive.

In Fig. B 2 a six part Geneva gear for direct drive of an intermittently rotating six station turntable is represented. Regarding the kinetic relations such as the dependence of angular speed and acceleration upon the dwell time to coupling ratio, upon the number of stations etc., reference is made to the pertinent literature [8—10, 13—15]. Of special importance for the glass expert are, however, the dwell time to coupling ratios, especially since in most Geneva driven hollow glass machines the various production phases can only be altered during the dwell time. These phases include pneumatic control of press, parison, and blow mold processes. In Table B 1 the dwell time to coupling ratios are listed in

Table B 1. *Coupling/dwell time ratio as a function of slot number*

Slot number (z)	Coupling Angle ψ_0	$\frac{\text{Coupling}}{\text{total}}$ time	$\frac{\text{Coupling}}{\text{dwell}}$ time
3	120°	0.1667	0.8333
4	90°	0.2500	0.7500
6	60°	0.3330	0.6667
8	45°	0.3570	0.6250
9	40°	0.3899	0.6111
10	36°	0.4000	0.6000
12	30°	0.4167	0.5833

dependence upon the slit number z of the driven wheel. The coupling angle Ψ_0 here is $360°/z$.

The coupling to dwell time ratios of z-part Geneva gears are diagrammed in Fig. B 3. If in a Geneva gear the driven wheel is directly connected with the turn table, the number of stations determines the slit number

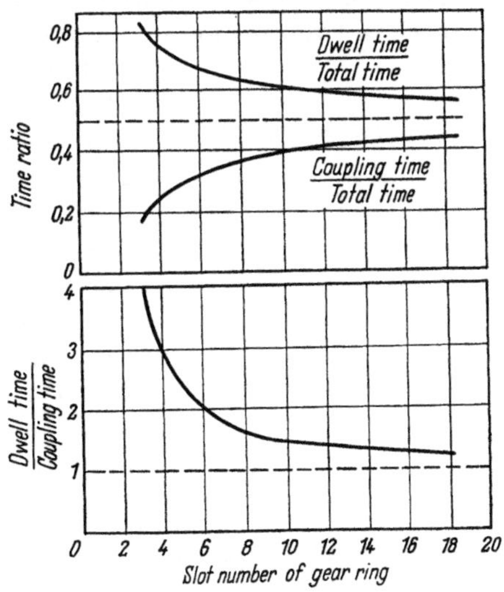

Fig. B 3. Coupling/dwell time ratios in a z-part Geneva drive.

of the driven wheel, the dwell time to coupling ratio, and all other properties of the gear. The gear of a two table machine with Geneva drive is shown in Fig. B 4. The motor works via a screw upon the driver of a 10 part

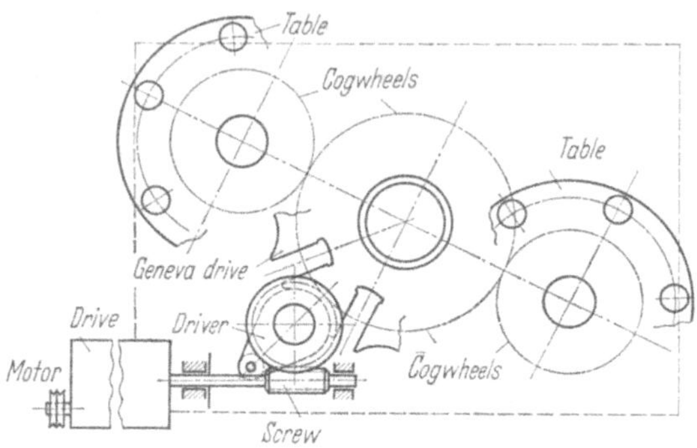

Fig. B 4. Drive of a two-table machine with Geneva drive.

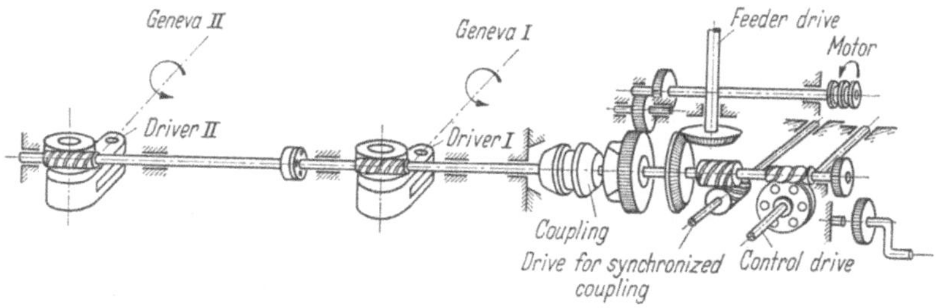

Fig. B 5. Two-table machine with two coupled Geneva drives.

Geneva gear. The stepwise rotation of the star wheel is transferred via 6 : 10 cog wheel transmission in equal phase and direction to the two 6 station turn tables. Another solution of the drive of a 2 table machine with Geneva drive is shown in Fig. B 5. In this machine two synchronized Geneva gears with a common axle are used. A gear system used in automatic bulb machines is also entirely mechanical and is represented in Fig. B 6. In this construction a special variant of the Geneva drive is used. It should be noted also that Geneva drives with more than two tables are not suited for the tasks considered here because they have less favorable dwell time to coupling ratios.

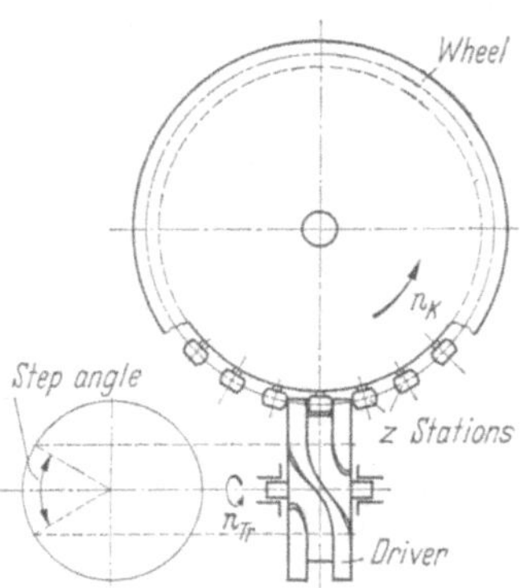

Fig. B 6. Space cam drive for intermittent round table.

1.2.2 Mechanical-Pneumatic Drives

In Fig. B 7 the essentials of a mechanical gear with pneumatic braking are illustrated. In this drive, which is used in a similar form to activate the turn table of heavy presses, a cam lever action like a crank is employed. Movement and braking of the turn table are effected by air activated stop bolts.

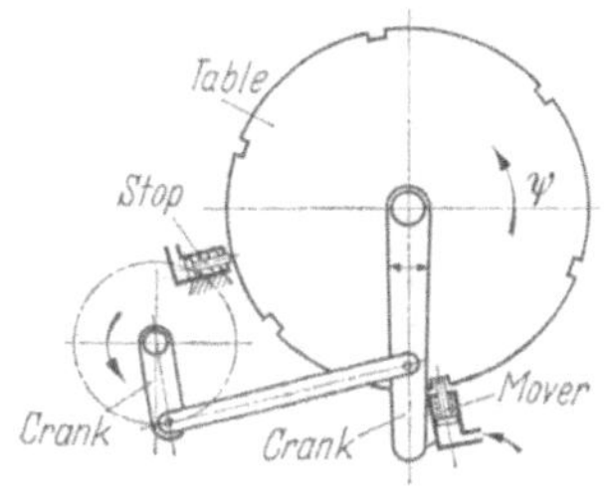

Fig. B 7. Crank lever change mechanism with pneumatic stops.

1.2.3 Pneumatic Drives

Pneumatic drives possess pistons activated by compressed air supplied periodically either through a roller valve belonging to the machine or via a control valve activated by the feeder mechanism. The transport of the turn table from station to station is done by transfer members such as roller valves or rack and pinion gears. For simple working pistons only one stroke is the working stroke; the reverse stroke towards the starting

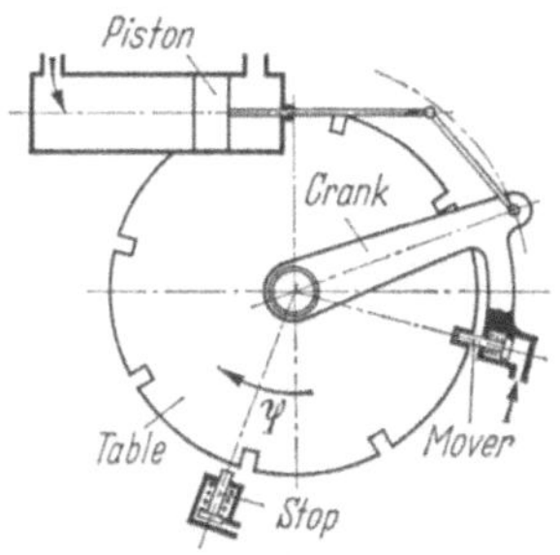

Fig. B 8. Pneumatic rotating table drive with pneumatic stops.

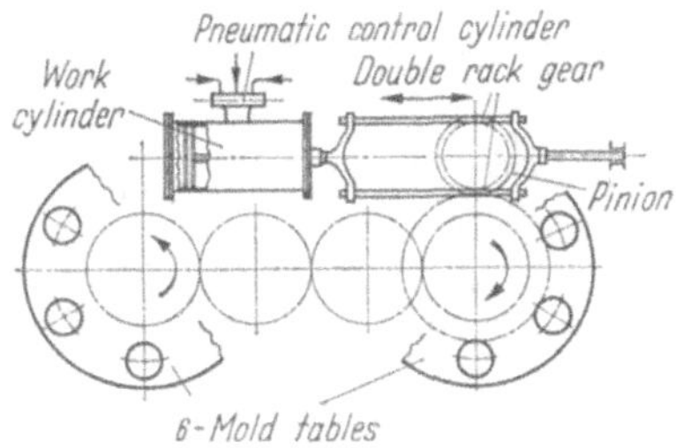

Fig. B 9. Pneumatic drive of a two-table machine with alternating rack gears (Lynch L 10).

point does not change the position of the table. Fig. B 8 shows a simple pneumatic table drive with controlled braking.

The drive scheme of a pneumatic two table machine can be seen from Fig. B 9. The motion of the piston is transferred to the table by two rack and pinion gears which alternately connect on their strokes with a lock so that the table is advanced stepwise for both motions of the piston. By means of a wheel gear the motion of one table is transferred to the second in equal phase but opposite sense. The necessary height adjustments of the lock pinion which transfers the motion of the rack to the turn table are accomplished by means of a piston activated by compressed air during the dwell time (Fig. B 10).

The drive scheme of a two table machine with two working cylinder acting upon a rack is represented in Fig. B 11. In this drive only one stroke is utilized to turn the tables; the reverse occurs during standstill.

The pinion transferring the motion to the turntable is lifted upwards by a piston activated by compressed air during standstill and thus frees the rack. The pneumatic lift of the pinion simultaneously blocks the tables. A disadvantage of pneumatic drives is the limited dimensional control over the motion. Thus the quality of the surfaces moving against each other in friction, the lubrication of gliding planes, small variations

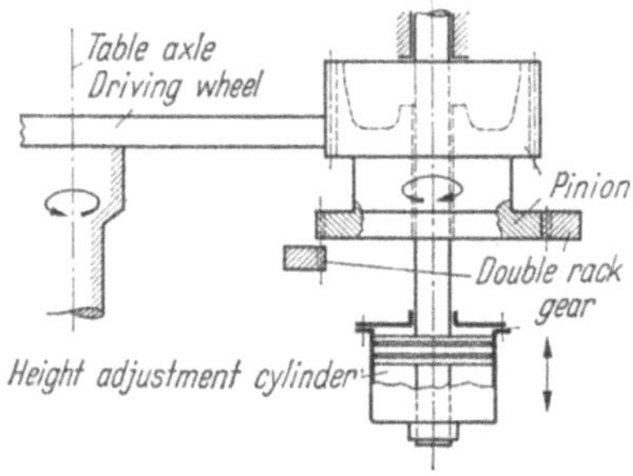

Fig. B 10. Pneumatic height adjustment of a drive pinion (Fig. B 9 unit).

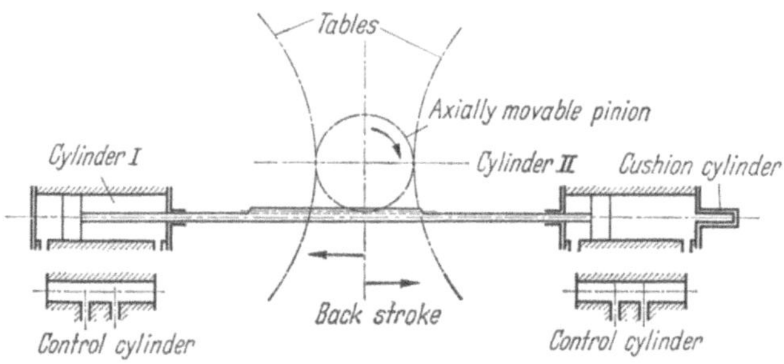

Fig. B 11. Pneumatic drive of a two-table machine with two cylinders.

in the pressure of the driving compressed air, the condition of gaskets etc. may cause irregularities in the motion, necessitating frequent adjustments of the pneumatic controls. To obtain a maximum of vibrationless motion in pneumatic drives the working piston is cushioned in the end positions. Cushioning is obtained either in the working cylinder proper or by means of an accessory cushion cylinder into which the prolonged piston rod goes. By influencing the entering or leaving air in the cylinder cushioning can be changed and regulated.

1.3 Drives in Row or Sectioned Machines

One can not speak of drives in the sense of gear technology in the case of sectional or similar machine entities which are fixed and independent. In the best known sectional machines for hollow ware, motion and activation required for production is activated pneumatically, with execution of the return stroke by spring power. There are also hydraulically operated glass machines. The extensive control mechanism required for these motions and activations is driven mechanically and it is essential to synchronize it with the feeder.

1.4 Special Drives

The feeder mechanism proper which contributes essentially to the formation of the feeder gob and thus carries the process name (feeder process) is just as interesting in regard to gear construction as the machine working the gob. The best known and most widely used installation of this kind, the Hartford Feeder, is represented as to construction and action of its gear mechanism in Fig. B 12. The mechanical drive of the

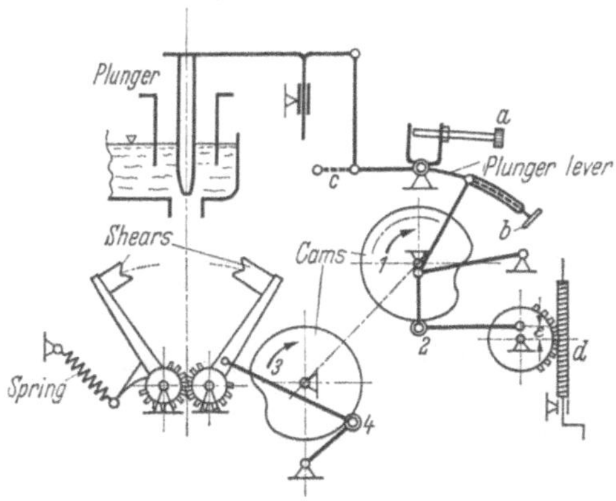

Fig. B 12. Schematic of Hartford feeder drive.

clay plunger and the shears is effected by two cams *1* and *3* which are synchronized with the machine. The plunger drops under its own weight. The feeder shears are activated by a spring or pneumatically. Rolls *2* and *4* will lift themselves in the case of disturbances (plunger break, tube break etc.) so that major damage to the gear mechanism is avoided. To change gob weight and shape the gear has the following adjustements:

1. Adjustment of the mean height of the plunger stroke (*a*),
2. Adjustment of plunger stroke, normal setting (*b*), large stroke (*c*),
3. Time adjustments of shear motion,
4. Cams *1* and *3* are exchanged depending upon production speed and gob weight. The shape of the shear cam is entirely a function of production speed.

The lever activating the plunger consists of rotatory arms supported centrally. By using the upper lock-screw "a" the angle of the two arms, in relation to each other, can be changed, and the turning point of the plunger can be raised or lowered. The lower lock-screw "b" changes the lever relationship of the two arms and consequently changes the length of the plunger stroke. Heavy articles demand a long stroke, light articles a short stroke. (Particulars about the feeder mechanism are given in Sec. D 3).

1.5 Survey

In Table B 2 the types of drive for the most important hollow glass machines are collected. It may be seen instantly that the mechanical and pneumatic drives are now dominant.

As to lubrication of glass working machines reference is made to the literature [21—23]. Information on bearings on glass machines is given in [24, 25]. Particular attention should be given the selection of the proper materials for single machine parts [26, 27].

References

[1] Breuer, L.: Zur Getriebetechnik der Flaschenblasmaschinen (Lecture given at Tagung der Getriebetechnik am 20. Okt. 1931 in Karlsruhe). [Glastechn. Ber. 10 (1932) 509.]

[2] Thierby, J.: Getriebetechnik bei Glasblasemaschinen (Abstract). Glastechn. Ber. 11 (1933) 99.

[2a] AWF-Getriebeblätter. Ed. by Ausschuß f. wirtschaftliche Fertigung. Berlin/Köln/Frankfurt: Beuth-Vertrieb.

[3] Alt, H.: Getriebetechnik bei Glasverarbeitungsmaschinen. Glastechn. Ber. 24 (1951) 200/01.

[4] Kiper, G.: Getriebefragen bei Verarbeitungsmaschinen in Karussellbauweise. Z. VDI 95 (1953) 1006—1013.

[5] Kammel, E.: Angewandte Getriebetechnik im Glasmaschinenbau. Getriebetechnik, VDI-Tagungsheft 1, Düsseldorf: Dtsch. Ingenieur-Verlag 1953.

[6] Kiper, G.: Drehtischantriebe für Glasverarbeitungsmaschinen. Fachausschußber. No. 54 der Dtsch. Glastechn. Ges. (1955) 467—477.

[7] Träkner, G.: Die Bedeutung der Getriebetechnik für die Konstruktion automatischer Maschinen. Maschinenbautechnik 2 (1953) 343.

[8] Alt, H.: Kurventafeln zur graphischen Ermittlung von Sternradgetrieben. Maschinenbau 8 (1929) 716.

[9] Alt, H.: Der Übertragungswinkel und seine Bedeutung für das Konstruieren periodischer Getriebe. Werkstattstechnik 26 (1932) 61.

[10] Bock, A.: Sternradgetriebe. Z. VDI 73 (1929) 397—401.

[11] Hoecken, K.: Sternradgetriebe. Z. VDI 74 (1930) 265—271.

[12] Lichtwitz, O.: Getriebe für aussetzende Bewegung, Berlin/Göttingen/Heidelberg: Springer 1953.

Table B 2. *Drives of the most important hollow glass machines*

Hollow glass machine		Rotating Table Machines						Row Machines		Number of stations
		Continuous	Intermittent Drive			1 Table	2 Tables	Pneumat.	Hydr.	
		Mech.	Mech.	Mech.-pn.	Pneumat.					
Hartford	IS							×		2—6
	"28"	×				×				12
Knox	WD	×				×				10
Lynch (Miller)	BB, L 10				×		×			6
	L 16		×				×			8
	IP, IPM etc.		×			×				6—12
	PBM, PGT etc.				×	×				6—12
	MDP etc.		×			×				6—12
Owens	AR, AQ	×				×				10; 15
Pöting	Presses and machines				×	×				
Roirant	BB[1]									1
	A 6		×			×				6
	R 7	×				×				7
Schwartzkopff OS, KS 6		×				×				6
Westlake		×				×				Double stations

[1] Machine for large (up to 60 l) glass ware, cam controlled, mech. drive, fixed frame.

[13] ALT, H.: Malteserkreuzgetriebe. Werkstattstechnik 10 (1916) 229.
[14] ALT, H.: Regelmäßige und unregelmäßige Malteserkreuzgetriebe. Maschinenbau 9 (1930) 202.
[15] BEYER, R.: Winkelbeschleunigung an Malteserkreuzgetrieben. Masch.-Bau Betrieb 5 (1937) 478.
[16] KLAUS, R.: Die Beeinflussung der Bewegungsverhältnisse bei Malteserkreuzgetrieben. Werkstattstechnik 25 (1931) 314.
[17] GRODZINSKI, P.: Schaltgetriebe mit Kurvensteuerung. Maschinenbau 6 (1927) 655.
[18] KIPER, G.: Kurvengetriebe für Glasverarbeitungsmaschinen. Glastechn. Ber. 29 (1956) 381–386.
[19] ALT, H.: Über ein regelbares Tauchhub-Getriebe einer Owens-Maschine (Abstract). Glastechn. Ber. 24 (1951) 142.
[20) RIEDE, H.: Kurvengetriebe bei Glasverarbeitungsmaschinen (Abstract). Glastechn. Ber. 29 (1956) 344.
[20a] BEYER, R.: Zur Synthese ebener Kurvenscheibengetriebe. Konstruktion 4 (1952) 208–210.
[20b) FLOCKE, K. A.: Zur Konstruktion von Kurvenscheiben bei Verarbeitungsmaschinen. VDI-Forsch.-Heft No. 345 (1931) 1–20.
[21] Anon.: Lubrication of glassmaking machinery. Lubrication 36 (1950) No. 8, pp. 85–96. Amer. ceram. Soc. (Ceram. Abstr.) 33 (1950) 225.]
[22] Anon.: The lubrication of glass-making machinery. Amer. Glass Rev. 76 (1956) No. 12, pp. 10–12.
[23] STANDER, B. B.: The value of correct lubrication in the glass industry. Glass Ind. 33 (1952) 126/27.
[24] HAHNEL, K., F. SUHRKAMP, O. KNOTT, and F. RITTMANN: Lagerung an Glasverarbeitungsmaschinen (Abstract). Glastechn. Ber. 24 (1951) 262.
[25] FLESSA, K.: Lager- und Antriebsfragen bei Fourcault-Ziehmaschinen (Abstract). Glastechn. Ber. 24 (1951) 142.
[26] WATERBURY, N. M.: Metal problems in the container field. Ceram. Ind. 54 (1950) No. 4, pp. 112, 172.
[27] WATERBURY, N. M.: Metal problems in the container field. Glass Ind. 31 (1950) 453/54.

2 Controls

By H. RIEDE, Düsseldorf

The control mechanism of hollow glass machines gives the proper sequence and course of the individual working steps. In Fig. B 13 the single production phases and the associated working procedures are listed for the three important processes: pressing, press-blow, and blow-blow.

The development of hollow glass machines is intimately connected with the corresponding development of suitable control systems. Requirements are flexibility of settings and massive construction so that there is positive control of each individual step of the operation even under difficult working conditions. Fig. B 14 gives a survey of control installations for various hollow glass machines.

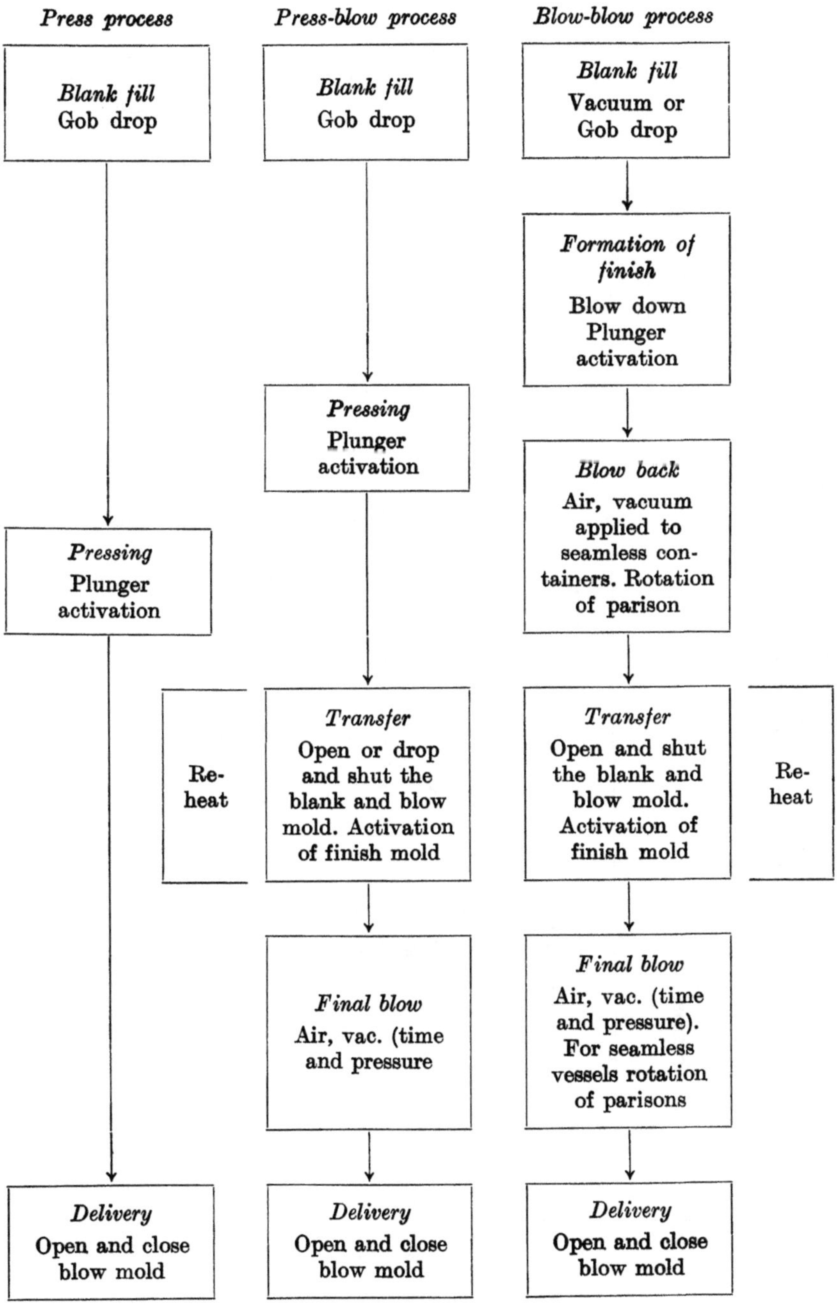

Fig. B 13. Process and control operations for various manufacturing processes for hollow and pressed glass.

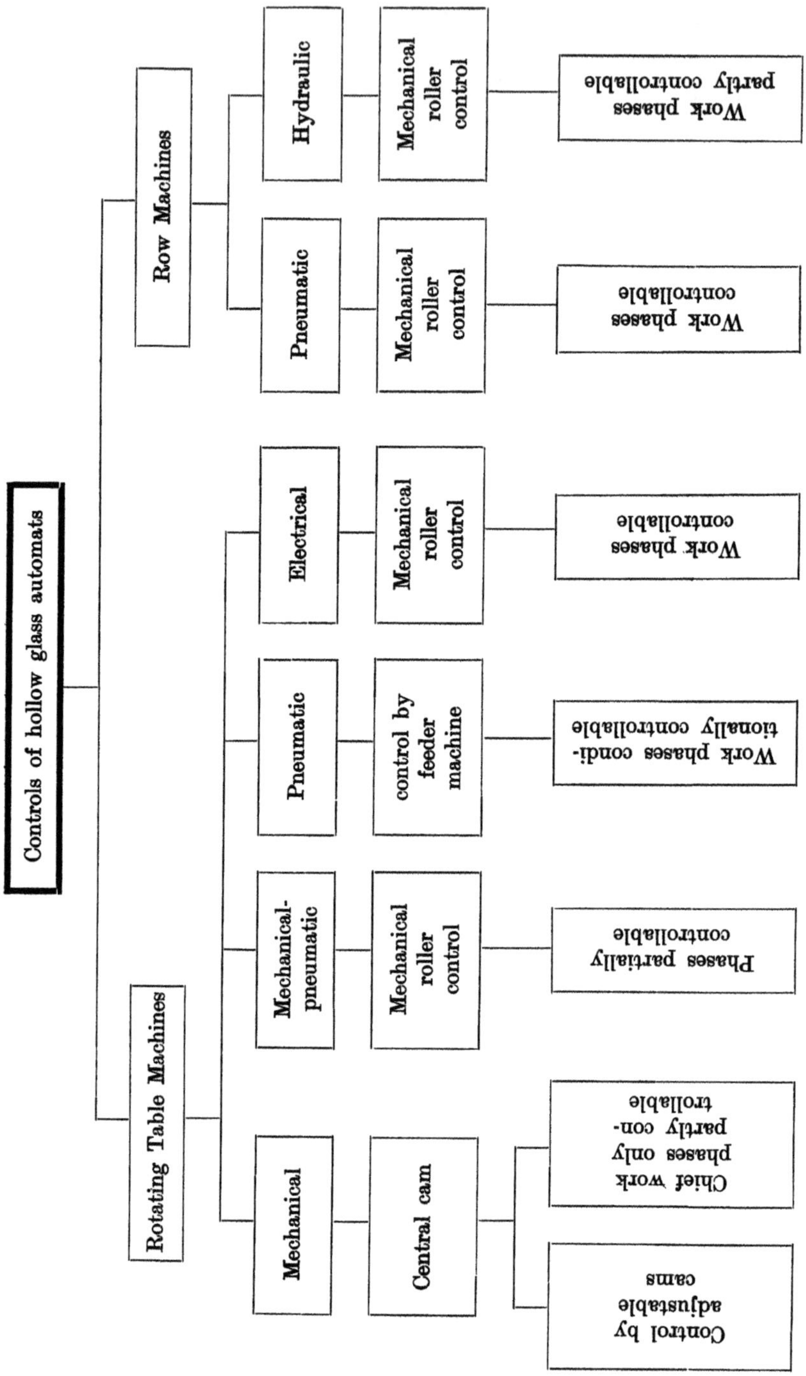

Fig. B 14. Survey of controls for different hollow glass machines.

2.1 Principles of Control; Function Times

Before considering the details of control installations, working phases and prescribed times of function should be considered more closely. Apart from the pure press process, press-blow and blow-blow processes require several finishing phases (Fig. B 13). The coordination of the finishing factors requires definite work steps and sequences with precise timing of each. The various properties of the glass in working, and the wide range of hollow ware of various sizes, weights and shapes, demand great adaptibility of the control system to meet the many requirements. A system which does not permit time regulation of at least the most important work steps is not flexible and machines utilizing such a system are limited as to their production program. Hollow glass machines which are to work functionally, covering a wide range of production, must permit a wide range of adjustments in the main finishing phases, particularly in timing, which demands — depending upon the construction of the machine — a more or less complicated control system.

In the framework of a review of control systems it is not decisive whether the timing of various machines can be reduced to a common denominator or exhibit definite relations. However it is always advantageous to know the timing of each function in a given production machine. If the supplier does not furnish this information, it must be established with precision and diagrammed correspondingly. The diagram must distinguish between work phases that are variable or invariable. Such diagrams, together with the proper consideration of the respective processes, make ist possible to examine more closely the finishing conditions of the machine.

The work flow of the individual functions of rotating machines with cam control may be well represented by circular diagrams (Fig. B 15). It is, however, equally possible to consider the disk unwound and to enter the control phases on a line diagram. In this case the run of the cam can be represented together with the time sequence. A schematic diagram of this sort is shown in Fig. B 16.

Pneumatic controls with mechanically driven control drums permit plotting of control times in a system of coordinates in which the horizontal coordinate represents work steps in proper sequence and the vertical coordinate the angle of the control drum. Data are simply lines. The control times and angles entered in Fig. B 17 represent an arbitrary example. Such a diagram, however, says nothing about possible changes in settings.

The timing of the various work steps is restricted by factors deriving from kinematics and from the production characteristics of glass as a material. Intersecting motions can only be controlled short of mechanical

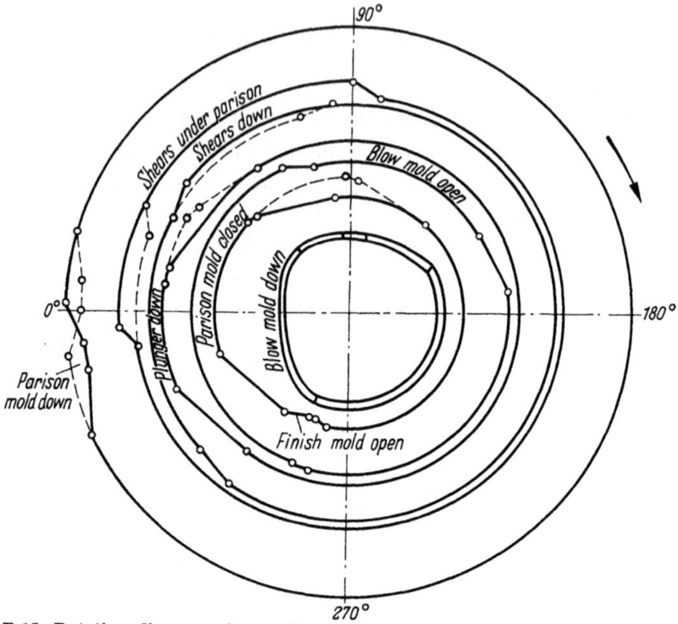

Fig. B 15. Rotating diagram of a suction process rotating table (Owens AR machine).

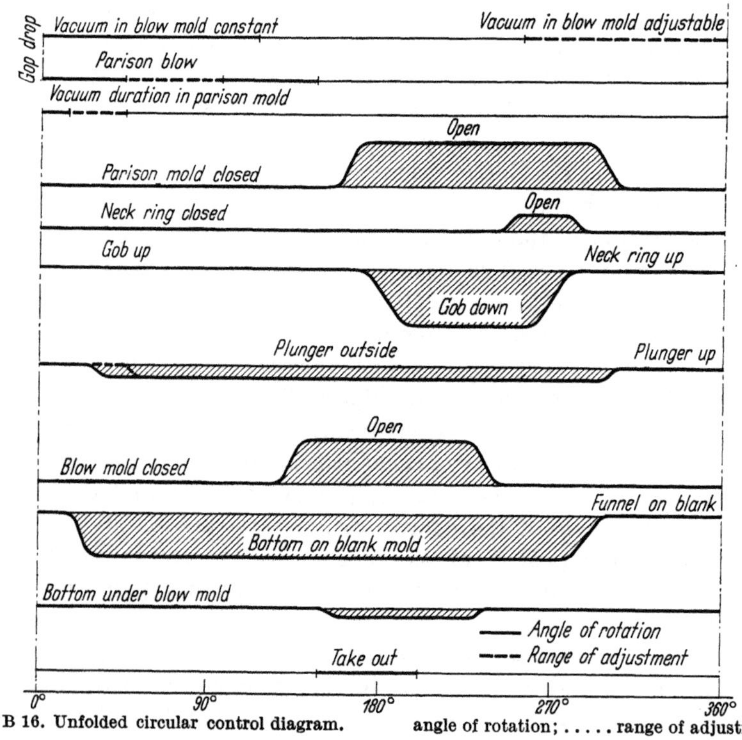

Fig. B 16. Unfolded circular control diagram. —— angle of rotation; . . . . . range of adjustment.

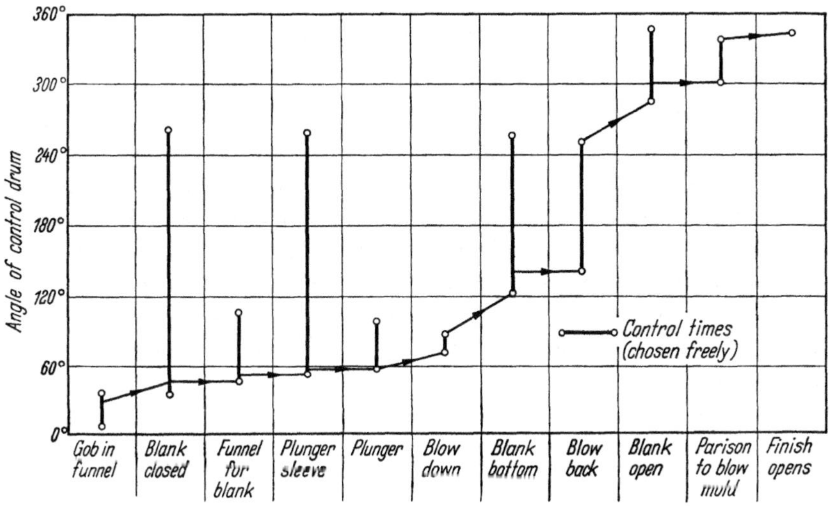

Fig. B 17. Control diagram of the blank part of an air driven machine with mechanical control drum.

collision. For full consideration of productivity it is often advantageous to examine exactly interdependent production phases. For this purpose one can establish a path-time diagram on a suitable yardstick, the vertical

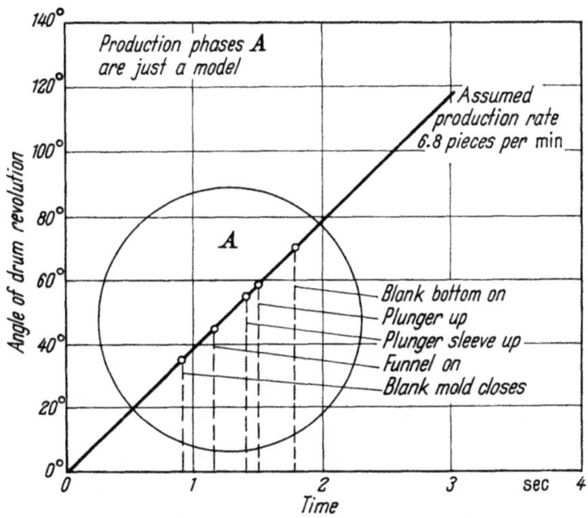

Fig. B 18. Diagram for study of production process.

coordinate of which is divided into radians of angle (representing curve or circumference of control drum). The inclination of the straight line passing the zero point is a measure of the desired production in terms of pieces/min. A diagram of this kind is given in Fig. B 18. Such investigations of different machine operations help to avoid many faulty settings which may cause breakage of machine parts or excessive mold wear.

2.2 Construction and Operation

2.2.1 Mechanical Control

The production steps in rotating or turntable machines as well as all desired tool motions are released and controlled by centrally fixed cams. The cams of larger machines are composed of two or more segments and held on a central fixed center column (or center tower). But there are also glass machines (one-arm machines) in which the cams rotate. To provide individual steps in this rigid cam various constructions permit the introduction of inserts of longer or shorter control time. However, this adjustment is usually possible only when the machine is at rest.

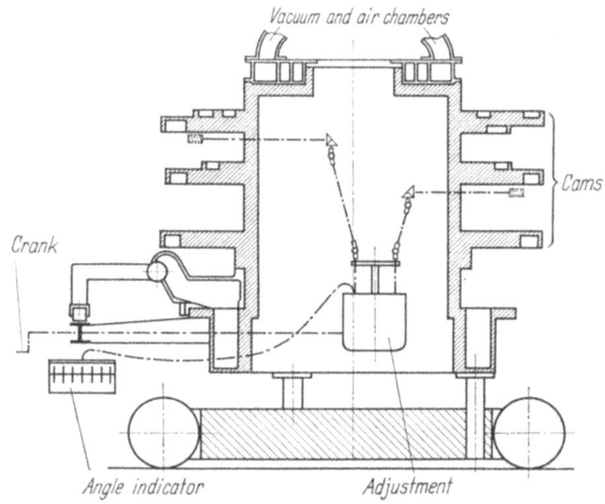

Fig. B 19. Construction and adjustment of cams on the Owens AR machine (schematic).

An exemplary solution permitting large changes in important steps is the construction of variable cams and appropriate control systems in the Owens AR and AQ machines. The change in cams (i.e. the desired change in control time), may be executed in the shortest time and, with the proper skill, even during the running of the machine. The following production phases of the Owens AR and AQ machines can be adjusted: suction, cut off, plunger action, neck ring mold action, and parison action. The cam change mechanism is in the fixed central tower and is activated by a shaft through a crank. One first selects the production phase to be changed, then by means of the crank the desired angle is selected. The exact angle of change can be read on an indicator which is part of this clever mechanism. Fig. B 19 shows the construction of the variable cams with the change mechanism. Fig. B 20 shows a picture of the change mechanism.

Fig. B 20. Cam adjustment in an Owens AR machine (exposed).

2.2.2 Mechanical-Pneumatic Control

In mechanical-pneumatic control some of the work phases are accomplished and controlled by the mechanical drive of the table, while the rest of the activations are pneumatic. The mechanical and pneumatic parts are linked by a mechanically driven control shaft which possesses a certain number of control roller cams which carry changeable valve tips to activate the respective control valves. These valves release the compressed air for the respective locations. Two systems may be distinguished. The air coming from the control block and released by the control valve may either

(a) directly activate the work cylinder, or

(b) activate a control cylinder with a cam slide control gate so that the separately introduced air drives the work cylinder.

The difference between the two systems consists in the time required for the air supply, which is long for the work phase but short for the cam slide. Method (a) provides air for all locations from an air chamber

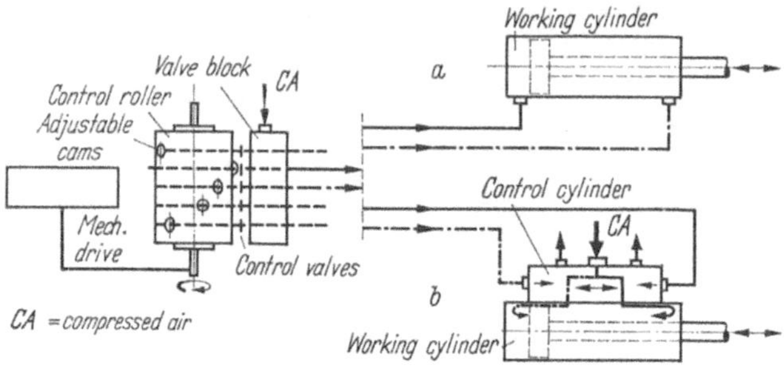

Fig. B 21. Mechanical-air control system.

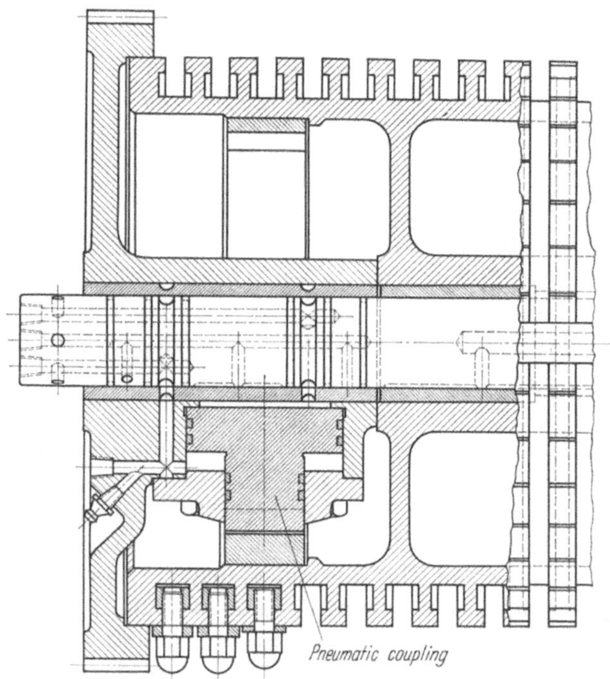

Fig. B 21a. Construction of a control roller (timing drum) with coupling for the Hartford IS machine.

housed in a complicated valve block, while (b) provides a short impulse with air release which is directly connected with the control or work cylinder.

Fig. B 21 illustrates the mechanical-pneumatic control combination chat is basic for many hollow-glass machines. The construction of a tontrol shaft is shown in Fig. B 21 a. The drum is connected with the drive wheel by means of a special coupling. A control valve used for this control shaft is shown in Fig. 21 b. In order to be able to influence the pneumatically activated work phases in their motions the network

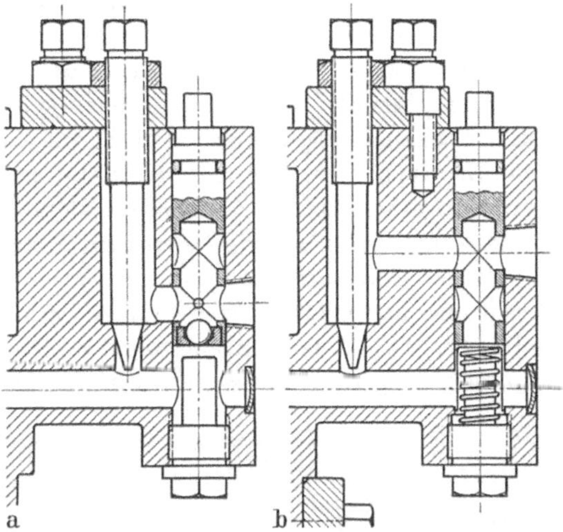

Fig. B 21b. Construction of a pneumatic drive control unit (a) with ball, (b) with spring.

also contains various accessory aggregates such as back pressure valves, regulating valves, and regulating needles which may cushion exhaust air etc. It is also necessary to work with different pressures which necessitates the insertion of some changeable and fixed pressure valves. Typical examples are for (a) the Hartford IS machine control and for (b) the Lynch press-blow machine control.

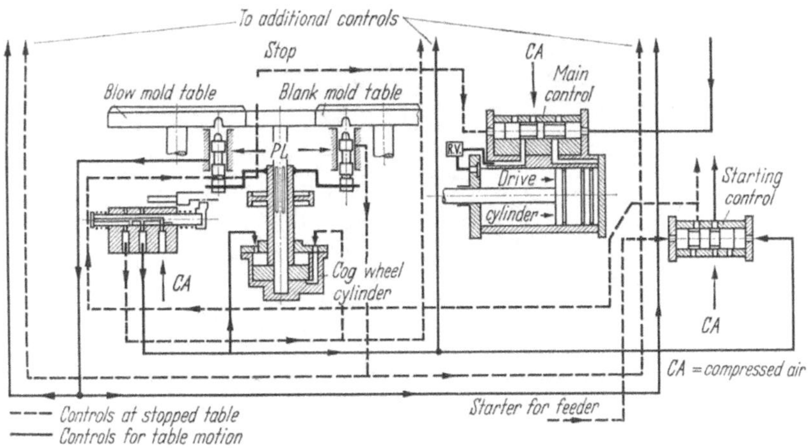

Fig. B 22. Schematic of the pneumatic control of a 2-table machine (control circuit incomplete).

2.2.3 Pneumatic Control

Purely pneumatic hollow-glass machines lack all mechanical drives and receive, instead, control impulses from the feeder mechanism. These impulses release in timed sequence the various motions and activations in the machine. Since this pneumatic "control chain" represents a closed circuit, the possibilities for influencing time are limited. In other words, the system is not flexible. Alterations in the impulses coming from the feeder mechanism change only the period of the entire control run i.e. the production number is changed, but no individual steps are influenced.

The accessory aggregates listed in 2.2.2 are absolutely necessary in the purely pneumatic machines in order to realize a stable run of the work steps in the case of pressure changes or variations in production rate. Nevertheless it may happen that after hours of perfect performance the machine suddenly jumps one step and must be re-regulated. The scheme of a pneumatic control chain is shown in Fig. B 22. For better

Fig. B 23. Hillmann machine with electric control.

orientation only some phases are drawn into the control cycle. Examples of purely pneumatically controlled production machines are the Lynch BB and L 10 machines.

2.2.4 Electrical Control

Even though there is at present no hollow-glass machine whose chief phases are controlled electrically, one machine should be mentioned here which had been provided with a very well conceived electrcial control 30 years ago. This was Hillmann's ten-arm rotating machine which worked according to the Owens principle. The head of the machine was a kind of control drum with continuous arms or rings. From these rings, electrical leads went to the electrical control board next to the machine on which control times could be varied (Fig. B 23).

2.2.5 Hydraulic Control

Glass machines with hydraulic drive or controls are still rare. The trend however is in this direction. Hydraulic control permits the transfer of large forces without mechanical links and precise mastery over force and velocity. Transfer of effect is chiefly by potential energy; such energy transport allows the spatial separation of pressure generation and work mechanisms. Transfer mechanisms have small dimensions. These properties all correspond with demands on glass machine controls and drives. It thus seems probable that in the future hydraulics will increasingly enter the construction of controls and drives in glass machines.

3 Mold Mounting

By W. Giegerich, Heilbronn, and G. Kiper, Munich

The two mold halves of blank and blow molds are inserted into special carriers which by means of various gear mechanisms allow the opening and closing of molds as well as maintenance of their closed condition against blow pressure. In the following section some characteristic examples are discussed in accordance with Kiper's [1] report.

The simplest mounting used in hand and semiautomatic operation is shown in Fig. B 24. It represents the bolting of the two halves to a closing jaw (mold carrier and clasp). By an appropriately selected lever ratio easy activation and secure bolting of the mold is achieved. A further development is the gear with the pneumatic drive represented in Fig. B 25. The mold carriers or links here are activated by means of compressed air through a piston via a coupling arm. The diagram shows the closing angle ψ and the transfer angle μ as functions of the piston part s. The arm is used for parison and finish molds. The range of motion is s_a to s_{eV} for parison and s_a to s_{eF} for finish molds. Closure is activated by the air

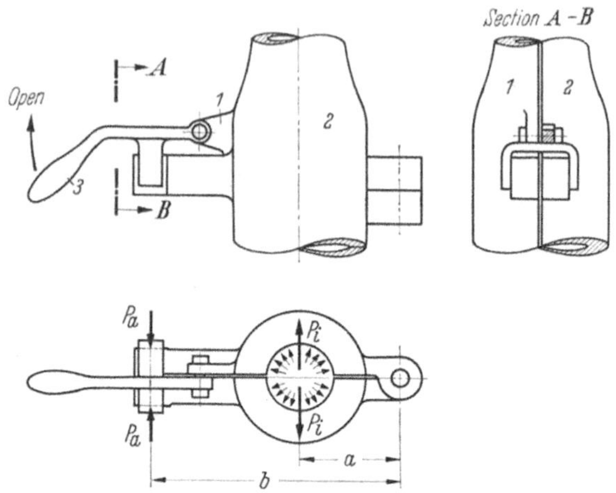

Fig. B 24. The closing of a blank mold on a semi-automatic bottle blowing machine.

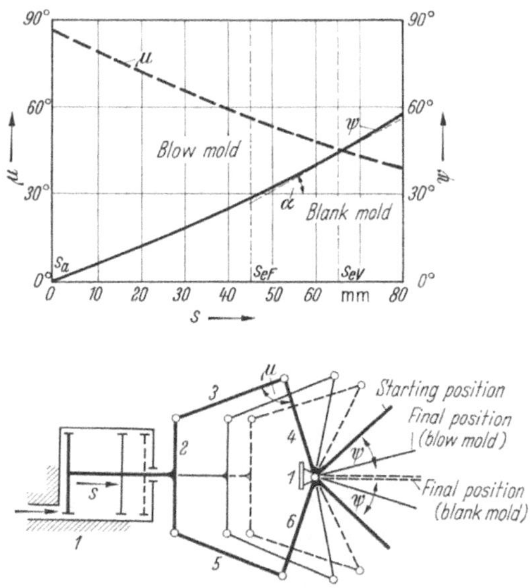

Fig. B 25. Pneumatic drive for a blank closing on a $^3/_4$ automatic machine and the motion diagram.

pressure upon the piston. Further examples for gears on mold closures of fully automatic bottle blow machines are shown in Figs. B 26 to B 30.

Fig. B 26 shows the links of the parison mold of the fully automatic Roirant B machine.

A rotating disc cam *2* is the driven member which periodically moves lever *4* via shaft *3*. This lever couples through roller *5* and arm *6* to give a double action. The coupling motion of link *5* is transferred into a swing of angle lever *8*. The control of shaft *7* in the link *5* plane can be regulated to influence the angle of *8* between its limit positions. By means of the transmission member *9* the swing of member *8* is transformed into the to-and-fro motion of slide *10*. The drive of the mold halves *12, 14* follows over slides *11* and *13*. Members *2* to *10* work in parallel planes turned 90° against the planes of motion of members *11*—*14*. Yielding members *11, 13* consist each of two other members which can work against each other

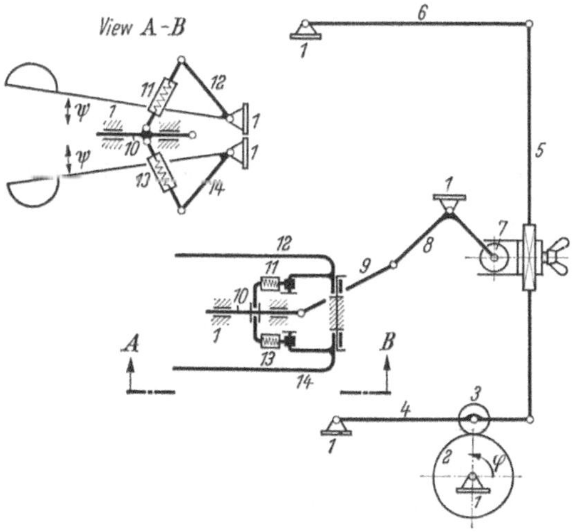

Fig. B 26. Drive for opening and closing of a blank mold in a Roirant B machine.

and through springs and stops of a certain length. This allows further motion of the gear parts in the moment of closure. At this momement the length is shortened while the spring tension increases. During blowing the two prestressed springs receive the moments resulting from the blowing pressure. These moments effect the closures *12, 14* from the "outside", so that the closed position is maintained. The entire gear train is constructed from two gears arranged in two different planes and contains 16 members and 21 pairs of elements. The mechanical drive is activated from one cam and is supported by the two springs in the „open" phase.

The gear used for the activation of the holders of the finish mold halves used in a fully automatic Roirant B machine is shown schematically in Fig. B 27. Here too the drive operates from a rotating cam *2* which forcefully activates swing lever *4* via shaft *3*. Members *4, 5, 6* furnish with drive member *1* a joint quadrangle which works as a double swing with the equally working quadrangle *1, 6, 7, 8*. Here too coupling member

5 is worked out as a yielding member, actually consisting of two members with a prestressed spring as an auxillary drive. The limiting positions of the swing angle of member *8* can be influenced by changes in the

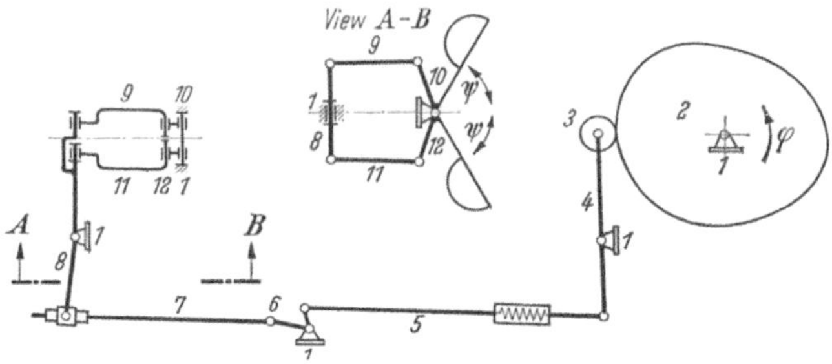

Fig. B 27. Drive for opening and closing of blow mold in a Roirant B machine.

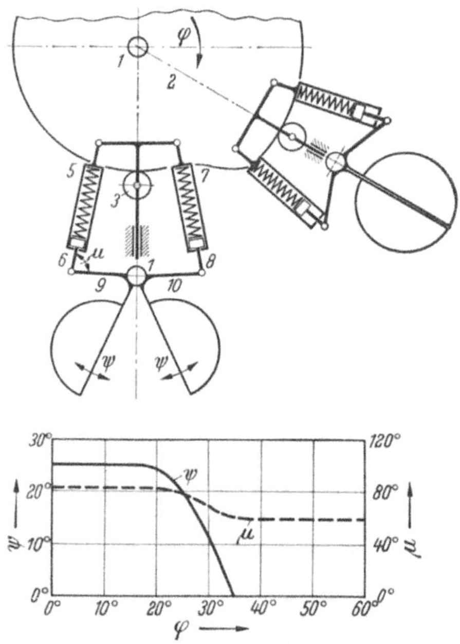

Fig. B. 28. Drive arrangement for blank and blow mold closing in a Roirant A 6 machine with motion diagram.

length of coupling *7*. The motion of member *8* goes to carriers of the finish mold halves *10, 12* through transmission members *9, 11*. The carriers can be rotated around a common axle and can move in planes turned 90° related to the motion planes of members *2* to *8*. Since the curves in which members *9* and *11* move are not planar, these transmission members describe space curves: the gear is a space gear. In this realization of the pair of elements forming rotating links, the gear theoretically represents a rigid construction. However, sufficient play in the menbers frequently provides the necessary mutual flexibility, so that the transfer of motion is assured. This arrangement will be satisfactory only for low rotational speeds; if, however, these speeds are increased and the transmitted forces are considerable, it is of doubtful value because of excessive wear. As in the gears driving parison mold halves, the yielding member *5* is shortened at the start of closure and the spring is stressed. This stress serves to maintain closure during blow-out. The gear train consists of 13 members and has 17 pairs of elements. The mechanical drive starts from a cam and is supported by the spring in the "open" phase.

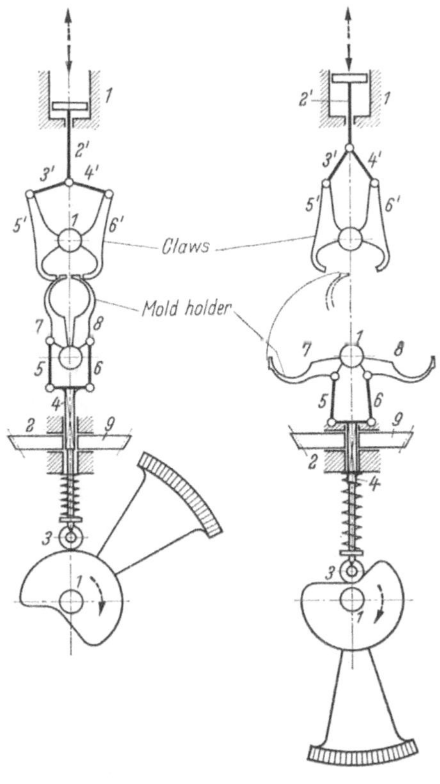

Fig. B 29. Drive arrangement for opening and closing of the blank mold on the Lynch B machine. Left: mold holder and claws closed; right: open.

The Roirant A 6 bottle machine works as a rotating unit with a turntable operation in steps and having six equal stations. Fig. B 28 shows schematically the gear construction for two stations as they are used for the activation of mold support or hangers, both for parison and blow molds. In one location the halves are open, in the other they are closed. The drive again starts from a disc *2* which is centrally located in the axle of the rotating turntable. All six locations receive their impulses from the common disc via follower *3*. The latter moves a radially constructed pivot *4* which drives coaxial swing levers *9, 10* via yielding effective member pairs *5, 6* and *7, 8*. The levers carry the mold halves. On closure a further motion of pivot *4* is received by the pairs of elastic transmission members *5, 6* and *7, 8* while the compression of the spring increases. This compressions maintains closure during blowout. Fig. B 28 also shows the dependence of the motions of the mold halves (angle ψ)

upon the relative angular rotation between table and cam angle (angle ϕ) between two rest positions of the table.

At $\psi = 0$, $\phi = 35°$ the motion of the halves stops at the moment of closure. The graph also shows the change of the transmission angle μ between members *6*, *9* and *8*, *10*. It can be seen that μ attains a minimum at closure but this is still satisfactory. The gear train contains 10 members and 12 pairs of elements. The mechanical drive starts from a disc and is supported by the two springs in the "open" phase.

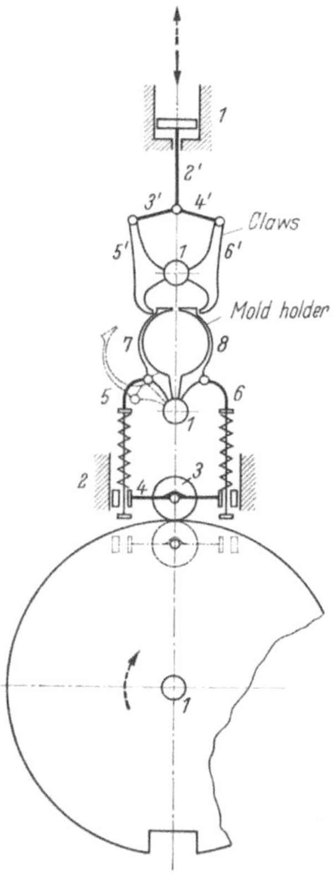

Fig. B 30. Drive arrangement for opening and closing of the blow mold on the Lynch B machine.

Fig. B 29 is a schematic of the gear construction assigned to all stations of the turntable of a Lynch B bottle machine for the control of opening and closure; it also illustrates the accessory gear which in the rest position during blowing drives the clamping claws (left: closed, right: open position). Disregarding the apparatus required for the opening of the mold halves by180°, the gear consists of the gate *4* carried by *2* and moved by the centrally fixed disc *1* via follower *3*. As in the 3/4 automat illustrated in Fig. B 25 this motion is transmitted to the coaxial turnable holders *7*, *8* via coupling members *5*, *6*. Gate *4* again is yielding so that a movement of the shaft after closure shortens the member, while the stress of the spring is increased. Here the stress can be quite small because on blowing the mold halves are closed positively by claws *5'*, *6'*. The six member pneumatic drive used for this purpose corresponds with the one used for the 3/4 automat (Fig. B 25), in which, however, the links of members *3'*, *4'* are not coaxial with drive piston *2'*. All in all the combined gear train for the opening and closing of parison halves has 14 gear members with 18 pairs of elements. In addition to a mechanical drive starting from a cam and, in the "open" phase, supported by a spring, there is found here a second drive, a pressure controlled piston. As Fig. B 30 shows, a similar pneumatic accessory installation is used to maintain closing during blow-out in the blow mold. Mold holders *7*, *8* are used for this purpose. These are put in motion by arm *4* via transmission members *5*, *6*, motion being transmitted from disc *1* to arm *4* via follower *3*. Clamping is secured by means of two springs.

A comparison of the various constructions and drives is given in Table B 3 [1]. An examination of these gears shows, for example, that the selection of proper member dimensions easily permits lower power requirements. Attention should be given to the power ratios in the bearings, to power

Table B 3. *Drives for closures in bottle blowing machines. Comparison of different constructions* [1]

Machine	Blank mold	Blow mold	Drive	Closure			Links	Pairs of elements	Number of drives
				Through	During blow				
					Aktive	Passive			
$^1/_2$-Automat	×	×	Manual	Elasticity of parts		×	3	3	(1)
$^3/_4$-Automat	×	×	Pneumatic	Air cushions	×		6	7	1
Roirant B	×		Mechanical	2 Springs	×		16	21	1+(2)
		×	Mechanical	Spring	×		13	17	1+(1)
Roirant A 6	×	×	Mechanical	2 Springs	×		10	12	1+(2)
Lynch B	×		Mechanical Pneumatic	Springs-/Cushion		×	14	18	1∼1∼(1)
		×	Mechanical Pneumatic	2 Springs/Cushion		×	13	17	1+1+(2)

requirements, and to the centering of mold halves. In the finishing operation faulty centering may be recognized in a disagreeable manner by forming defects such as offset seams. It is therefore important to know the possible causes of such defects which may arise even in symmetrical gear constructions. They are: unequal mold halves, off centering, unequal changes in the shapes of corresponding gear parts from yielding, plastic or thermal causes, differential wear of bearings, and unequal fatigue of springs in symmetrically placed transmission members [2].

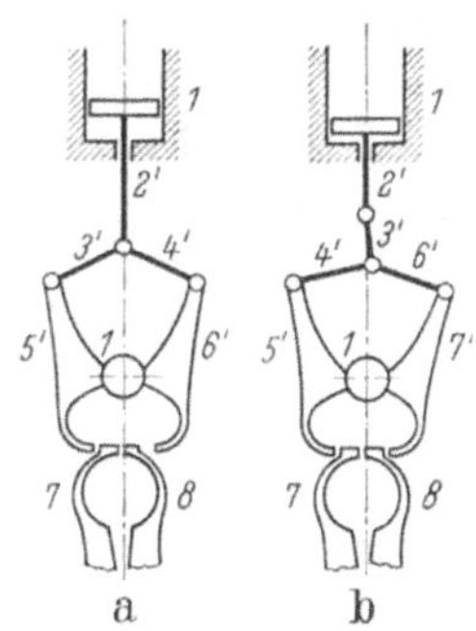

Fig. B 31. Drive arrangement for closing claws: (a) constrained, (b) unconstrained (asymmetrical touch).

The susceptibility of mold closing gears is different depending on whether gears are constrained, or conditionally constrained. Constrained gear arrangement is said to exist, if the relative motion of gear members is determined uniquely; in this case all mobility vanishes at closure. Examples are the gears of Figs. B 25, 27, 29, 31. Conditionally constrained are gears in Fig. B 26, 28, 30, in which relative motion depends upon the activating forces through the insertion of spring members. Frequently thy insertion of an intermediate member — as shown in Fig. B 31 — can alter an unconstrained gear to one which is constrained. This example illustrates the importance of a thorough examination in terms of gear technology.

References

[1] KIPER, G.: Die Getriebe für Formenzuhaltungen bei Flaschenblasmaschinen. Glastechn. Ber. 28 (1955) 475—482.

[2] MILLER, E.: Air pressure and springs insure safe operation. Machine Design 5 (1933) 13—16.

4 Molds

By H. RIEDE, Düsseldorf

4.1 General Information

Forming of hot, viscous glass is effected by special forming tools, usually called in short "molds". Molding of hollow glass occurs in two successive work steps: one distinguishes between blank (parison) and blow molds. In pressing frequently one step is enough.

Flat glass forming processes are of a different kind. The drawing process succeeds by the exact mastery over the glass temperature, and consequently its viscosity, in carrying out the formation of the sheet without specific molding tools. The rolling process generates the sheet between rolls rotating in opposite directions.

The most important papers found in the extensive literature on glass molds are listed at the end of this chapter (general [1—10], design [1—22], manufacture [24—33], cleaning [34—45], maintenance [46—47], lubrication [49—60]).

Fig. B 32 represents molding tools for suction machines. Parison and blow molds are supported by special hangers or they can be held by

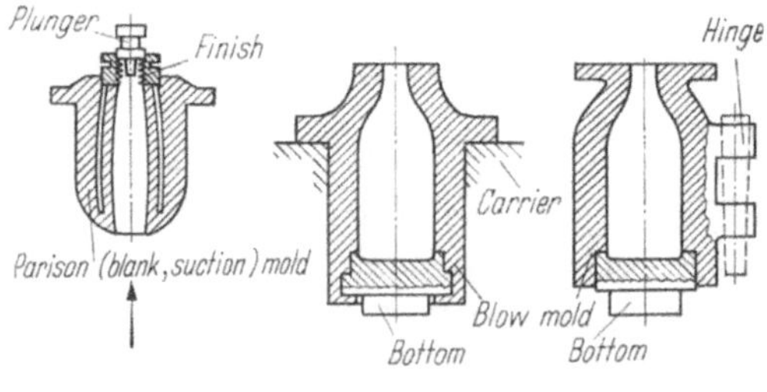

Fig. B 32. Mold implements for vacuum machine.

cast-on hinges with a conical sleeve. In feeder machines one distinguishes the blow-blow process (for narrow-neck containers such as bottles) and the press-blow process (for wide-neck containers). The required tools are shown in Figs. B 33a and b. In the blow-blow process a funnel sits

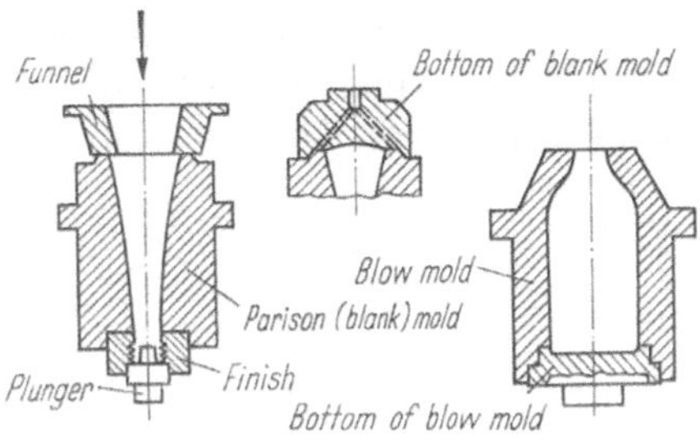

Fig. B 33a. Mold implements for blow-blow machine.

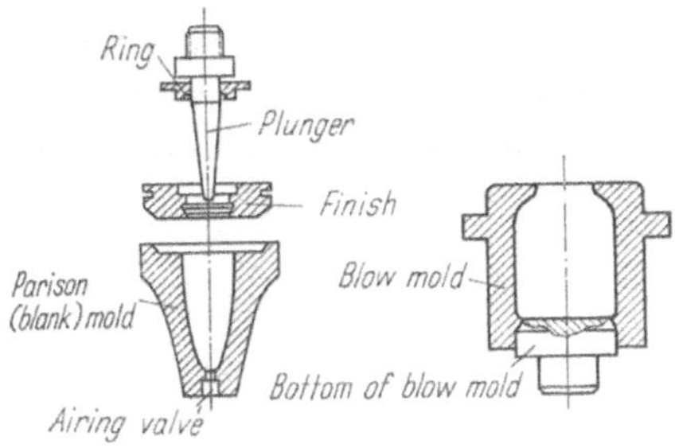

Fig. B 33b. Mold implements for press-blow machine.

above the parison during gobbing. The funnel is removed before preblowing and replaced by the baffle mold. The funnel serves to facilitate the dropping of the gob into the parison mold. In the press-blow process the finish mold is placed on top, similar to the suction-blow process. After the introduction of the gob a plunger with ring is introduced into the parison mold from above. Fig. B 34 permits a view of the construc-

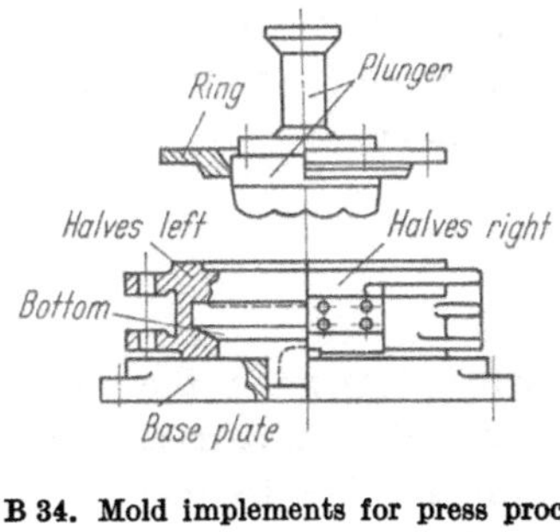

Fig. B 34. Mold implements for press process.

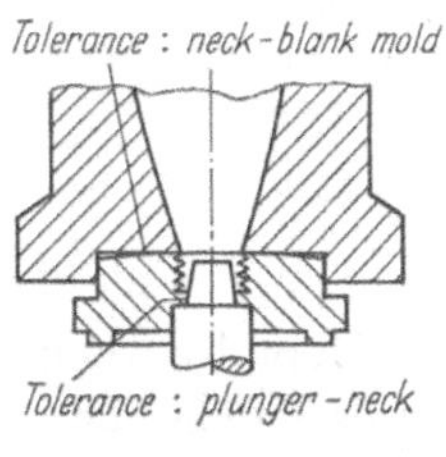

Fig. B 35. Neck ring detail of a feeder mold blank.

tion of press molding tools (half of a glass brick). Pressed ware such as bowls, tumblers, etc. have one-part molds.

The high operating temperatures of mold parts (500—600 °C) and the temperature conditions in the machine control definite dimensions and tolerances of interlocking parts. The interlocking must occur without excessive friction or jamming, while too much play causes undesirable features in the ware, such as broad seams. In Fig. B 35 an example illustrates the many places where dimensional tolerances must be observed.

When the glass is introduced into the parison mold and when the parison is blown in the blow mold, the air must be capable of escaping from the molds. If the air can not escape fast enough production troubles and defects occur. The chief air escape from the molds occurs through the seams, some part of the air also escaping between other interacting parts. In certain processes (press-blow) the parison mold contains special air escape vents. Heavy engravings in molds are supplied with small drilled holes, to obtain fast escape of air in these zones. These measures

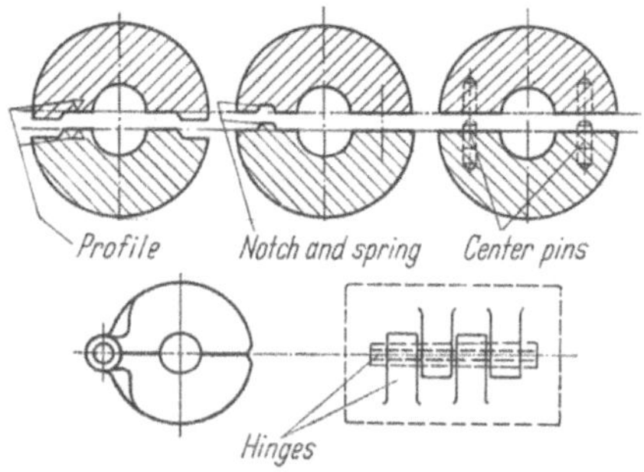

Fig. B 36. Centering of mold halves.

are all limited by the appearance of marks and seams in the finished ware. Experience teaches that it is difficult to give generally valid tolerances, since they differ too much for every process or machine. Temperature and composition of glass also play a part in this problem. It is important that in processes demanding 2-part molds, mold parts can not move against each other when closed. Some possibilities for the centering of mold halves are shown in Fig. B 36.

4.2 Mold Design

4.2.1 External Shape

Mold holders by their dimensions partly determine the external dimensions of molds; but in some machines different carrier sizes are used for small, medium and larger containers. The IS machine, for example, has six carrier sizes differing in diameter. The user of the IS machine is supplied with diagrams based upon experience from which holder size can be determined as a function of glass weight, container size, etc. Fig. B 37 shows the fastening of a blow mold on a holder.

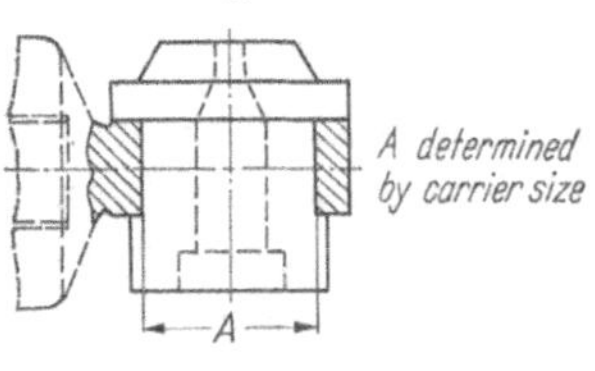

Fig. B 37. Mold mounts through carrier.

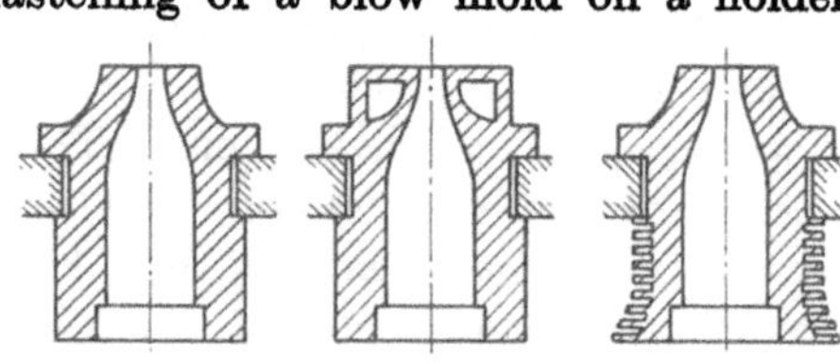

Fig. B 38. Examples of blow mold construction.

While the external dimensions of the mold near the holder do not permit much size variation, the parts above and below can be adapted to the demands of heat transfer or the shape of the container or of the gob or parison (Fig. B 38). In the Owens blow molds the external shape is ready-cast, and only hinge bores must be worked.

4.2.2 Design of Hollow Glass Molds [1—4]

The shaping of hollow glass container molds is, in most cases, a compromise between the necessities of glass manufacture and the desires of the customer as to appearance and end use of the ware. The chief dimensions of height and diameter, content and weight represent the basis of the design. Content and principal external dimensions are usually specified by the customer. They are subject to laws on measures and weights and to the recommendations of standard specifications.

The moldmaker in the glass factory must now start the not always easy task of combining and reconciling the third factor (weight of the glass, Table B 4 and B 5), with the specified dimensions. The following relation will serve as the first approximation to such a calculation: Capacity of container = capacity of blow mold — volume of gob. Fig. B 39 illustrates this relationship.

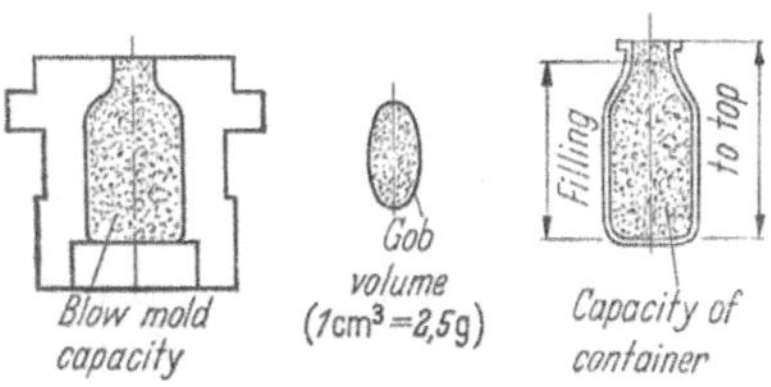

Fig. B 39. Relationship between the overall capacity of blow mold and glass.

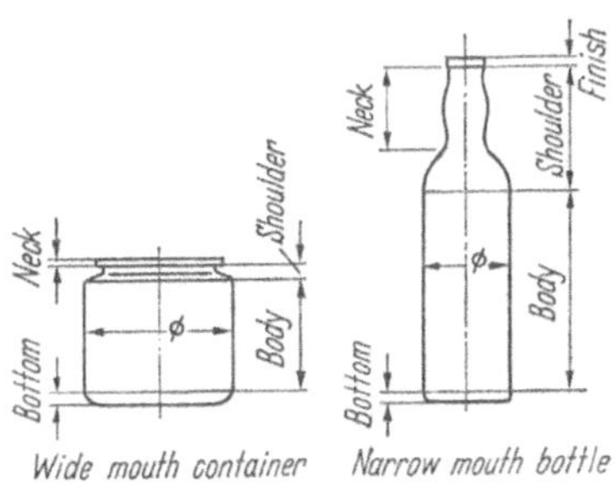

Fig. B 40. Notations for hollow ware.

Hollow-ware may be divided into wide-neck and narrow-neck containers. Designations used are represented in Fig. B 40. Wide-neck containers have a relatively short shoulder, with a short neck or no neck at all. The narrow-neck containers are characterized by the narrow mouth (finish), the shape of the neck and shoulder, as well as by the varying diameter. As a rule the neck and shoulder are characteristic of the particular kind of container such as water, beer, wine, medicine, cordial, etc. The curved parts making up the neck and shoulder are conic sections such as circular arcs, ellipses, parabolas, hyperbolas, or combinations of these curves. Calculations and actual mold construction

Table B 4. *Bottle weights* (*g*)

Capacity					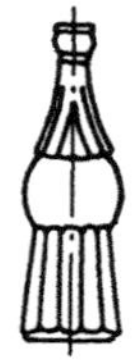			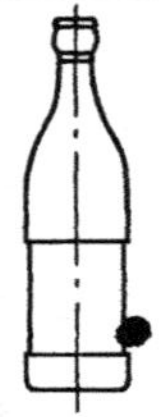			Containers and prescription bottles	
											Round or oval bottles cross section	Special shapes with high weight
30 cm³		55 ± 10							40 ± 5		50 ± 10	60 ± 10
50 cm³		80 ± 20	90 ± 20						60 ± 10		70 ± 10	75 ± 10
0,1 l		150 ± 30	170 ± 30					140 ± 20	95 ± 10	155 ± 20	120 ± 20	140 ± 20
0,15 l		230 ± 30	230 ± 30						125 ± 15	200 ± 30	175 ± 25	190 ± 25
0,2 l	$260 \pm {}^{100}_{30}$	270 ± 30	250 ± 30		380 ± 40		240 ± 30		170 ± 20	260 ± 30	190 ± 25	220 ± 30
0,25 l	$300 \pm {}^{100}_{50}$	300 ± 50	300 ± 50	300 ± 40	380 ± 40	340 ± 40	320 ± 40	270 ± 30	210 ± 30	290 ± 30	250 ± 30	280 ± 30
0,3/0,37 l	380 ± 70	400 ± 70	440 ± 70	440 ± 50	430 ± 50	440 ± 50			300 ± 40		280 ± 30	340 ± 40
0,5 l	490 ± 70	530 ± 70	500 ± 70	520 ± 60		530 ± 60	500 ± 60	460 ± 50	360 ± 40		390 ± 50	450 ± 50
0,6/0,66 l		610 ± 70	550 ± 70	600 ± 60		600 ± 60						
0,7 l	610 ± 80	610 ± 80	600 ± 80	660 ± 60		680 ± 60			480 ± 40			
1 l	720 ± 80	720 ± 80	720 ± 80	750 ± 60			760 ± 60	750 ± 60	590 ± 40		700 ± 60	760 ± 60

Weights are average. Indicated tolerances depend on bottle use, machine type and weight of finish.

Table B 5. *Weights of wide mouth containers* (g)

 Round, low thread or vacuum finish		 Angular or oval Low thread or vacuum finish		 Jelly or marmelade Round cylinder with normal thread			 Household jars		
Cap. (cm³)	Wt. (g)	Cap. (cm³)	Wt. (g)	Designation (kg)	Cap. (cm³)	Wt. (g)	Nominal cap.-lit.	Cap. (cm³)	Wt. (g)
50	55 ± 15								
				$^1/_{16}$	60	80 ± 15			
100	110 ± 20								
				$^1/_8$	110	110 ± 20			
				$^1/_4$	210	140 ± 20			
235	140 ± 20	235	170 ± 30						
410	200 ± 30	410	225 ± 50	$^1/_2$	410	200 ± 30			
510	240 ± 40								
							$^1/_2$	600	340 ± 40
				1	820	360 ± 50			
							$^3/_4$	850	380 ± 50
863	350 ± 50	863	400 ± 50						
							1	1100	450 ± 50
				$1^1/_2$	1220	570 ± 50			
							$1^1/_2$	1600	650 ± 70
				2	1610	700 ± 70			
							2	2100	700 ± 70
				3	2400	950 ± 100			

Weights are average. Indicated tolerances depend on glass use, machine type and weight of finish.

must agree. Contents are calculated from well known relations. Irregular shapes are sometimes divided into smaller sections, 5 to 10 mm apart, each of which can easily be determined geometrically. Since the calculation of each design is time consuming work, time saving tools have been developed. The design of containers of the some shape, but of different capacity, is done by proportional enlargement or diminuition. Irregular cross sections are measured with a planimeter using empirical corrections. Hot glass shrinkage is accounted for by an assumed 1—2%. There are tables using calculated and empirical values for the conventional mouths and bottoms. A further means to determine volume is to use a model together with hydrostatic scales.

In designing hollow glassware the mold maker should pay attention to the following guidelines: the bottle should have a usable shape and a low center of gravity, with due consideration of conveyor, bottling and capping processes. Sharp corners and edges may cause breakage in handling or transport. They must be avoided to a large extent. Sharp edges in molds, including press molds, cause crizzles or checks. These defects are deleterious to the strength of the glass ware. Hollow glassware that is exposed to strong internal pressure demands very uniform wall thickness. A circular cross section is the best warranty for uniform glass distribution. Cornered bottle shapes introduce the problem of bringing enough glass into the corners; therefore corners should be rounded as much as possible. A minimum bottom size is needed for stability, and certain weight to capacity ratios should be maintained. The respective empirical values are represented to advantage by nomograms including some additional factors. Often it is found necessary to redesign the shape of the parison, for example, for better distribution of glass.

4.2.3 Mold Design for Pressed Glass Articles

Just as for bottle molds the chief dimensions, capacity, and glass weight are basic in the design of pressed glass molds. In addition there are specific requirements such as the following:

(a) For pressed containers closed by metal or plastic caps the exterior mouth diameter is the starting point. The capacity is specified and glass weight is assumed from experience, with particular consideration of wall thickness.

(b) In the design of metal covers the internal diameter is the starting point. That is, the plunger diameter is fixed. Glass weight and wall thickness are taken from empirical data.

(c) Building blocks must not only have fixed external dimensions, but must also tolerate specified loads. This is why wall thicknesses also are determining factors in their design. Glass weight is thus given within definite limits. Table B 6 summarizes all factors which influence the design for pressed glass ware depending upon the kind.

Table B 6. *Factors affecting mold design*

Article	Outside diam Finish diam.	Inside diam.	Wall thickness	Cap.	Wt.	Load	Shape
Containers, bottles	×		×	×	E		×
Covers		×	×		E		
Plates	×		×		E		×
Glass blocks	×		×		E	×	×

E = from experience, in part depending on other factors

4.2.4 Design of Parison, Parison Mold, and Plunger for Glass Containers

While the production process has little or no influence upon the design of blow molds, it becomes the determining factor for the construction of parison molds. This is clearly seen in the comparison of the cross sections of parison mold (and parison) for suction and feeder processes. (Fig. B 41).

In the suction process the capacity of the parison mold, including its finish, corresponds with the capacity of the container (1 cc glass = 2.5 g). The design of the parison shape must account for the fact that the glass is sucked from below. This is why a parison mold design for the suction process should never tolerate sharp transitions or large bulges which make suction difficult. Some suction mold parison cross sections are shown in Fig. B 42. Parison sections (parison mold contours) for the feeder process are shown in Figs. B 43 and B 47.

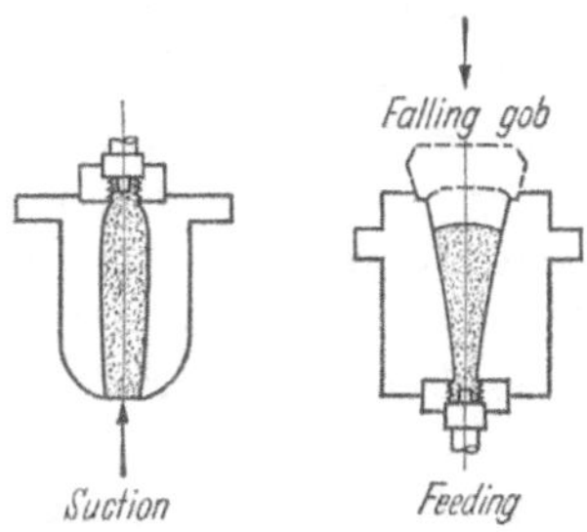

Fig. B 41. Blank mold sections.

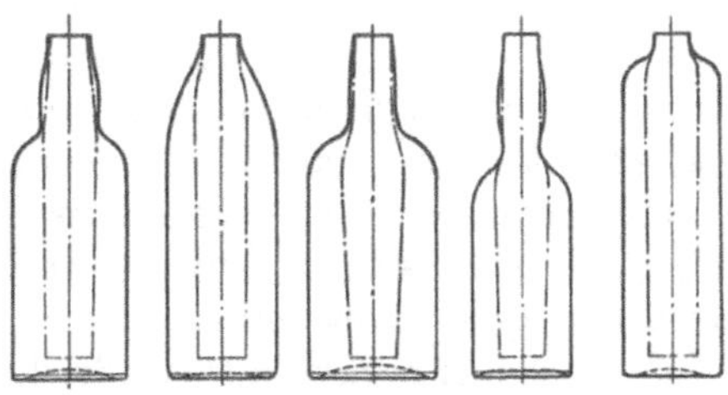

Fig. B 42. Comparison of parison and bottle shapes (vacuum process).

In the feeder process the capacity of the parison mold is larger than that of the required volume of glass. The relation between these two capacities is defined by the term "overcapacity." This term plays a basic part in the calculation and design of parison molds. The "overcapacity" in % is the ratio of:

(Total volume of parison − glass volume) × 100/glass volume. The total parison volume = capacity of mouth + capacity of parison mold + parison bottom part. The total parison volume $= G_H \left(1 + \frac{\%\,\text{overcapacity}}{100}\right) \times 0.461$.

Factors influencing the value of the overcapacity are: mouth size, height below mouth, diameter, shape and the ratio of the glass weight to the capacity. Depending on these factors the overcapacity may be said to be between 20 and 50%.

For irregular shapes the shape of the parison (blank) mold is similar to the shape of the blow mold to assure favorable glass distribution. In such cases the overcapacity serves more as the control than the basis for shaping the parison mold.

The principle, valid for the suction process, to put a minimum of resistance into the path of the glass, is equally valid for feeder machines. The perfect introduction of the dropping gob into the blank mold is important for good production. The parison mold should be free of internal protrusions. Protruding edges and corners prevent free and frictionless drop. Abrupt shoulders should be avoided (Fig. B 43).

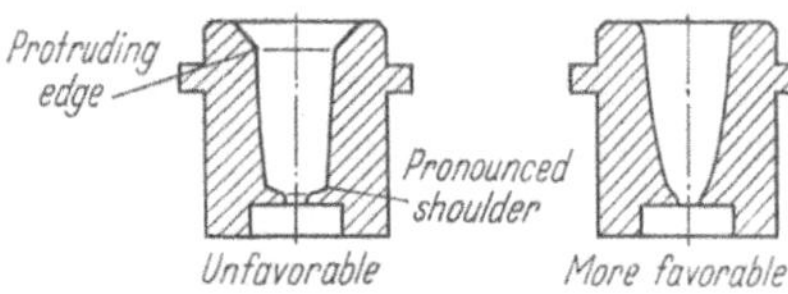

Fig. B 43. Design of feeder blank molds.

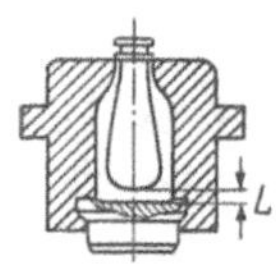

Fig. B 44. Length difference *L* between parison and blow mold.

The parison mold, and thus the parison, are always shorter than the finished bottle (by L in Fig. B 44). This permits elongation of the parison on transfer to the finish mold. No general values are given for this elongation; each hollow-glass has its empirical data. Roughly, it can be estimated that for narrow-neck containers elongations of 4—25 mm, for wide-neck containers elongations of 10—30 mm are possible, depending upon glass weight, height of the container, glass composition and color, and type of machine.

The design of parison molds is facilitated by empirical data. Table B 7 shows, for instance, the correlation of glass weight, bottle content,

Table B 7. *Relationship between weight, capacity and parison diameter of bottles*

Ratio wt./cap.	Parison diam. in % of bot. diam.	Ratio wt./cap.	Parison diam. in % of bot. diam.
1:4	24	1.5:1	70
1:2	50	2:1	75
3:4	60	2.5:1	80
1:1	65		

parison diameter and bottle diameter. Starting from the glass weight, tables give information about the associated gob dimensions (Table B 8). This table also gives indications about funnel and parison mold diameters.

Table B 8. *Comparison of gob dimensions and weight*

Weight (g)	15	20	30	50	75	100	150	200	250	300	400	500	600	700	800	900
Length (mm)	40	48	53	68	75	80	95	105	118	120	130	142	150	158	165	170
Diameter (mm)	15	16	18	21	24	27	30	34	37	40	43	46	48	50	52	54

The gob is rounded at both ends, falls vertically and should, if possible, not touch the upper third of the parison mold. For this reason the diameter of the funnel is always made a little smaller than the mean diameter of the parison mold.

The shape of the plunger is also important. The task of the plunger is to assist in the quick formation of the finish. Plungers which are too long impede the fast penetration of the glass into the finish. Plunger points should be rounded. For narrow-neck containers the shape and size of the plunger is determined by the nominal mouth diameter. For wide-neck containers the shape of the plunger determines the glass distribution in the parison.

Some important dimensions in the design of parison and plunger for wide-neck containers are shown in Fig. B 45. In the pressing operation the plungers get in intimate contact with hot glass and are therefore hollow, with water or air cooling. The open part of the blank mold is closed by the baffle. Sharp edges between parison and baffle must be avoided or marks and weak spots in the container result. Fig. B 46 schematically shows three types of baffles. Parison cross-sections for narrow- and wide-neck containers are shown in Fig. B 47.

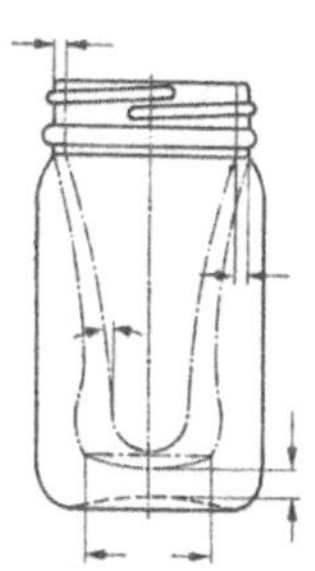

Fig. B 45. Dimensions important for the design of plunger and blank mold for wide mouth containers.

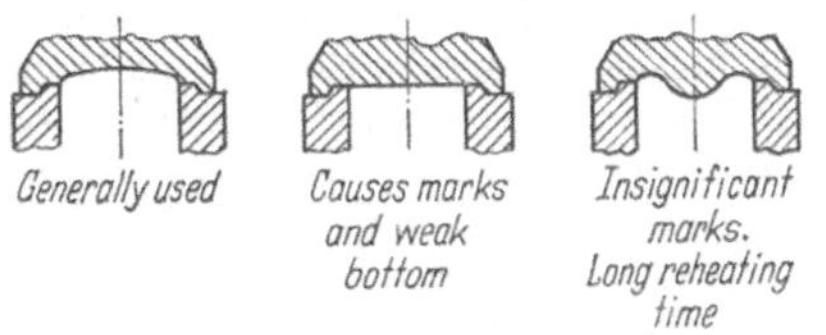

Fig. B 46. Construction of a blank bottom plate.

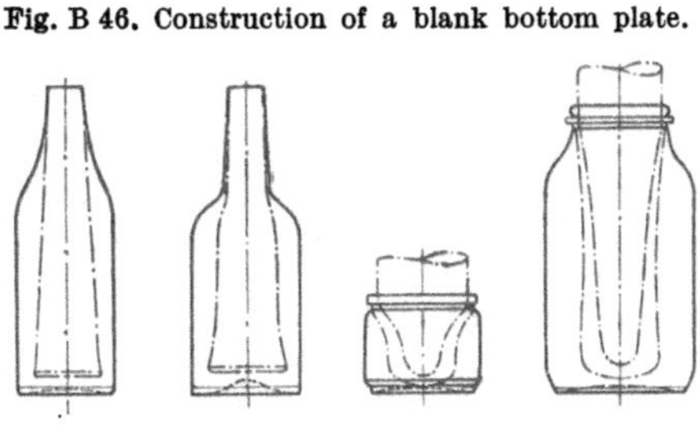

Fig. B 47. Parison sections (feeder process).

4.3 Moldmaking

The fabrication of molds is quite demanding. They may consist of one or of several parts and, since some of these may be interchangeable, close dimensional tolerances are necessary. Glass molds can be fabricated by simple machine tools and clamping devices. A large mold shop will, however, utilize modern machine tools and fabrication methods. Under most circumstances it will establish special production lines for the making of specific parts such as blank molds, blow molds, neck rings, plungers and accessories (baffles, funnels etc.). To limit the change of clamping devices to a minimum, the conveyors can be subdivided into different ranges of diameters. Almost every hollow glass plant makes some of its molds, and almost always repairs them. Therefore, it must be intimately concerned with the problems of moldmaking.

A few hints are given here which experience shows are preliminary conditions for economical manufacture. The raw castings should be made in such a way as to minimize further machining. Excess metal must be limited to what is absolutely necessary — this requires perfect metal patterns. Only efficient machines having hard alloy tools can work at high cutting speeds. The manufacturing operations are advantageously divided up into partial operations and distributed to the proper machine tools. Careful preparation and maintenance of the tools is necessary, as for cutter, hard alloy cutting knife heads, drills, and tool bits. Wear due to improper treatment, particularly of hard metals, is large and becomes expensive. The manufacture of bottle molds is explained below. The blank and blow molds of the Hartford IS machine are used as an example:

(a) Machining of mold halves

In divided molds the first operation involves working each half mold. This may have to be preceded by machining the holding or clamping devices. The machining of the halves is done by milling, planing, grinding, or combinations of these processes. The process selected depends upon the kind and number of available machine tools. The surface quality must be very good and hand finishing is usually necessary. In the case of blank molds for the suction process the vacuum slots are recessed into the surfaces.

(b) Turning of halves

The next operation is the machining of the outer mold faces which fit the mold holders. Turning lathes with multiple cutters or special turret lathes are used to advantage (Fig. B 48).

(c) Machining of neck ring and neck section

For this operation tools are used with special chucking devices as in section (b) above. Multiple cutter lathes or special turret lathes are again used. A special device for this purpose is shown in Fig. B 49.

(d) Machining of bottom plate and connector

The same types of machine are used here (Fig. B 50).

(e) For the working of the shape proper in the blank and blow molds copying lathes are used. Special copying lathes are used for both oval and angular shapes, as indicated in Fig. B 51. Very complex, irregular shapes can be worked mechanically on multi-spindle hydraulic copying automatics. Depending upon the particular copying system used the basis for shaping is either a steel template or a complete model piece in 1:1 scale.

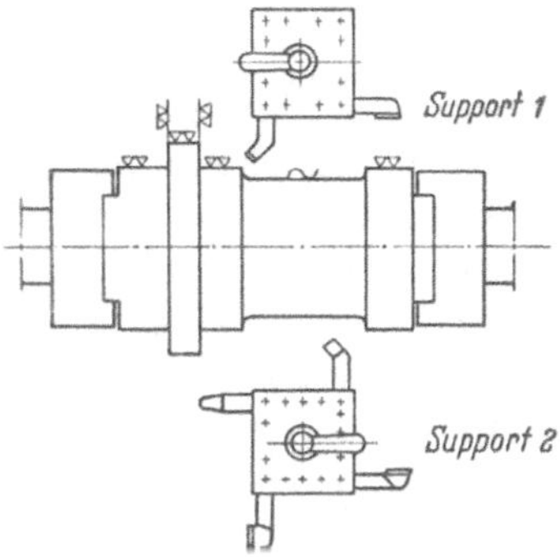

Fig. B 48. Machining of mold exteriors on turning lathes with multiple cutters.

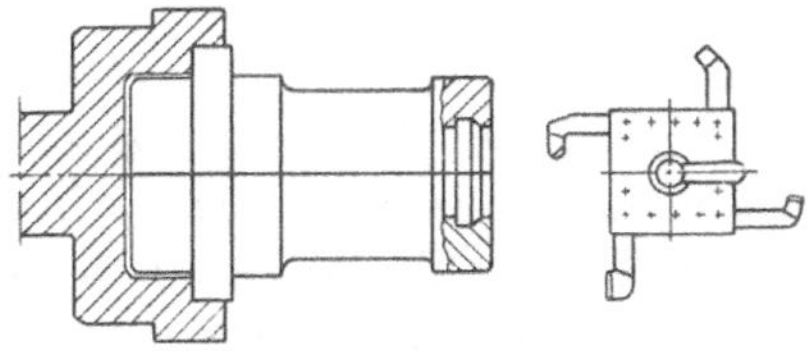

Fig. B 50. Machining of bottom plate and connector.

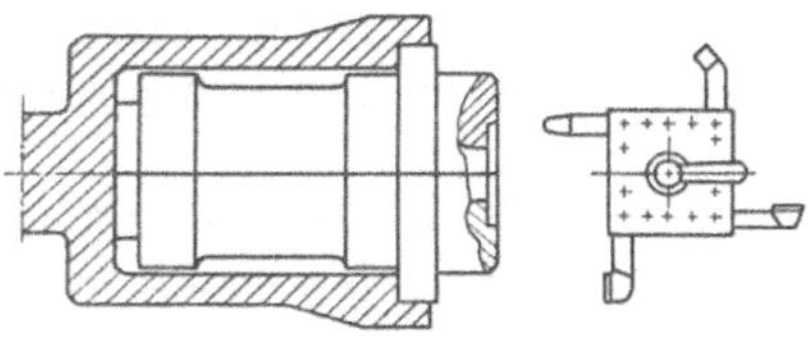

Fig. B 49. Special holding device for the machining of neck ring and neck section.

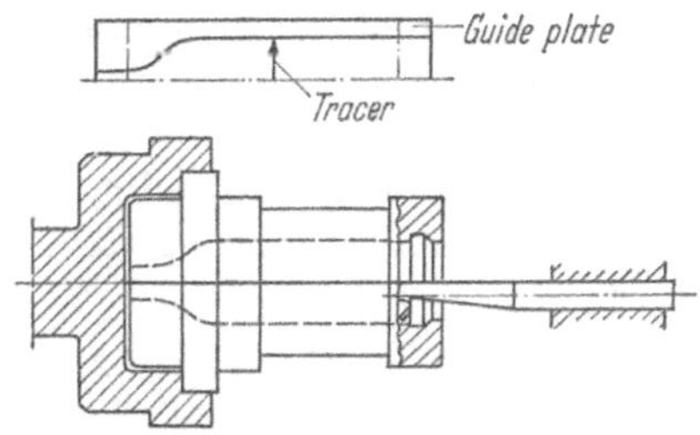

Fig. B 51. Turning of blank and blow molds on a copying lathe.

(f) Refining of the surface

The surface obtained by mechanical working of the sections which will contact the glass is not smooth enough. It is known that every roughness will show on the glass container. For this reason all faces that will touch glass are sanded and polished. Angular, oval and decorated molds are reworked by hand by chiselers and engravers.

(g) Mold inspection

After completion of all fabrication operations and manual afterworking mold parts must be inspected before they are released for use. In this inspection the parts are assembled and checked for tolerances and play. Certain inspection tools are used (special calipers, plug gauges, caliper gauges, etc.). Special inspection stations are used which reproduce the conditions prevailing in the run a of particular machine. Mold volume is tested by filling with water and weighing it. There also are procedures in which special hot materials are cast into the mold which solidify rapidly. The volume of the solidified model is weighed on displacement scales. The resulting value is corrected by means of a shrinkage factor to obtain the actual content.

The large consumption of neck rings, plungers, bottoms and funnels caused by thermal abuse stimulated their economical production on automatic machines characterized by short fabrication times. The prerequisite for their economic use is a certain minimum number of parts to be produced, otherwise set-up times have an unfavorable effect upon the economics of the situation.

4.4 Cleaning and Repair of Mold Parts

The strong thermal abuse of molds, scaling and crusting by oil refuse, and mechanical damage necessitate continuous cleaning and repair of molds.

4.4.1 Cleaning

Scaling on glass exposed faces, supported by burned oil refuse, becomes gradually so heavy that the finished ware shows marks and becomes unusable. When this is observed, the mold must be disassembled and cleaned. Various cleaning methods are used. The older method, still found today, is manual emery sanding. For round sections the piece is rotated while the sander is manually held under some pressure. The electrically driven grinding stand must be strong since the rotating mold represents a heavy mass and cannot be precisely centered. For this reason safety precautions must be taken. Emery sanding proper is done by coarse emery cloth; polishing is done with fine grained emery cloth. Molds for angular, oval and other containers must be cleaned and polished with manual grinding equipment. While this latter work must be done by skilled labor, unskilled labor can be used for grinding stand work.

Another process is cleaning by dry or wet sandblast. Depending upon the required surface quality this process too may call for subsequent polishing. Sandblasting is also termed jet lapping. The cleaning material consists of loose grinding particles (quartz sand, corundum, glass beads) which are mixed with water. The mixture is circulated by a pump system. In a special blast nozzle the lapping mixture is accelerated by compressed air and ejected toward the parts to be cleaned (Fig. B 52).

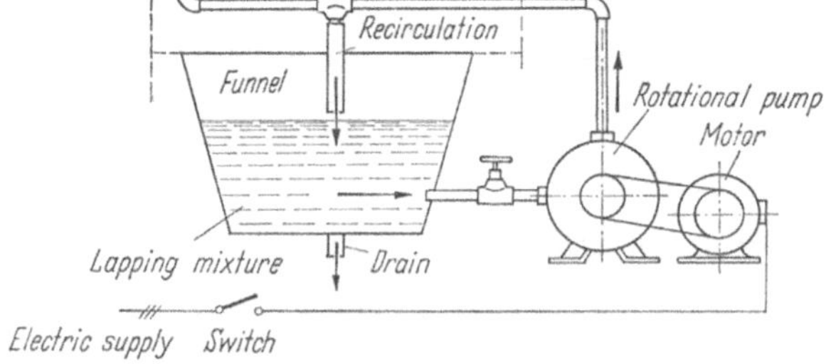

Fig. B 52. Scheme of a lapping device.

Both processes require some manual skill, attention and experience, since inexpert procedure easily causes damage to edges and faces resulting in premature wear.

Small parts such as neckrings or plungers may be cleaned in a special salt bath. In the electrically or gas heated open salt bath the parts to be cleaned stay for 30—40 minutes at temperatures from 400—450 °C. Subsequently the mold parts are taken from the bath, air cooled to about 200 °C, and rinsed in clean water.

The newest development is electrolytic mold cleaning. Details of the installations required are shown in Figs. B 52a and b. An alkali bath effects descaling and derusting as well as moderate degreasing. In this cleaning process the base metal is not attacked. In contrast, repeated sandblasting is found to cause removal of carbon ("graphite nests")

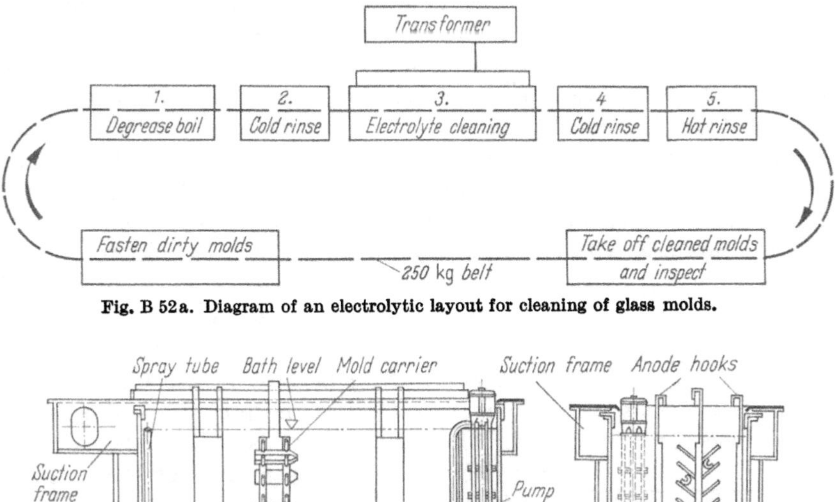

Fig. B 52a. Diagram of an electrolytic layout for cleaning of glass molds.

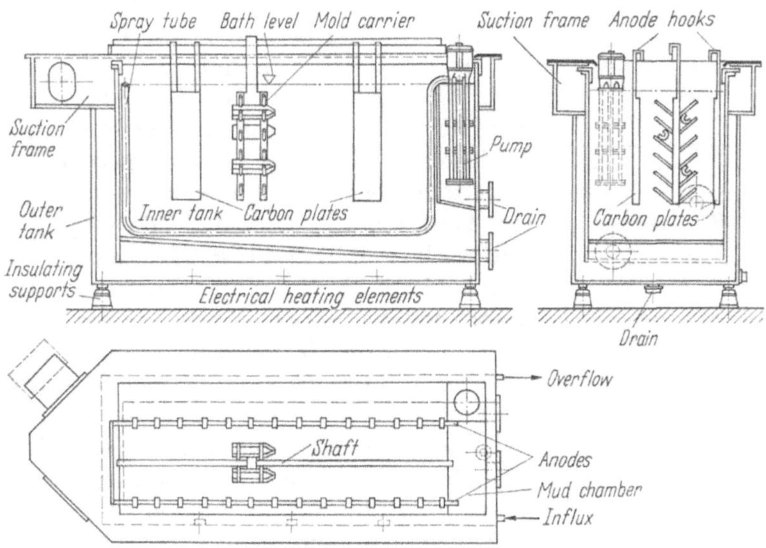

Fig. B 52b. Construction of an electrolytic cleaning bath.

at the surface. To limit the electrolytic cleaning process to a minimum in time extra dirty molds should be precleaned by boiling in a degreaser followed by a rinse in clear water. After electrolytic cleaning the cleaned parts emerge with a polished surface so that emery polishing can be omitted or minimized [45).

4.4.2 Repairs

The most frequent damage to molds is at edges or corners. The damaged parts leave marks on the ware. The damage may be caused by faulty setting of the machine (incorrect control times), by abnormally high wear of machine parts, or by improper fit of matching parts. In most cases, however, edges and corners are damaged by glass jammed between halves at start of a mold change. The usual type of repair is by inserting nickel or steel pins followed by afterworking. Pinholes are advantageously drilled by light, high speed tools. A valuable aid for mold repair has for some time been provided by autogenous and electrical welding. If molds are repaired by the welding process a basic requirement is that the cooled welded spots can be filed and are non-porous. For this purpose, in the autogenous process, the mold or mold half must be brought to a certain uniform temperature and must be cooled slowly after welding. In electrical welding with suitable electrodes good results can be obtained without heat treatment. Much used and frequently repaired parts are not suited for welding. In most cases welding spots then turn out to be too hard and too porous. In all cases welding results on molds will depend very much on the skill and experience of the operator. If larger areas have been welded (e. g. neck or bottom areas) mechanical afterworking is necessary to return to the prescribed dimensions within specified tolerances. Effective repair also demands suitable machine tools for proper chucking of the mold. A fringe area of welding is metal spraying, used on certain mold parts, Since sprayed alloys are very hard, finishing can only be done by special steel working or by grinding. Of course new mold tools also can be made more resistant to wear by metal spraying.

An important factor in mold repair is the finishing of the mold faces. Temperature differences in production cause more or less pronounced faults in the mold halves particularly of new molds. The molds then must be disassembled and manually afterworked or touched up the faces. This time consuming manual labor can also be executed by a special guided grinding machine.

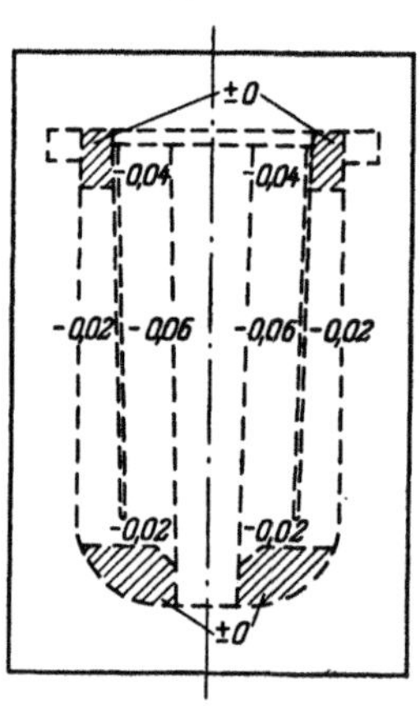

Fig. B 53. Example of a template for an Owens blank mold.

The contours that are to be ground are transferred by a special template (Fig. B 53). This process has the advantage that all molds become uniform. Here, too, all cleaned and repaired molds are carefully inspected before reuse. Every part is checked for dimensions, tolerances, and volume, using, as on new molds, the proper calipers, gauges and special tools. To avoid disturbance of the production process standby parts should be readied waiting for the repairs which inevitably will be required.

References

[1] Tooley, F. V.: Handbook of Glass Manufacture: Holscher, H. H.: Feeding and Forming, New York: Ogden Publishing Company 1953.
[2] Anon.: 20 years of progress in glassware moulds. Glass Ind. 21 (1940) 108/81.
[3] Scafe, R. M.: Glass molds present many problems. The Foundry 68 (1940) No. 8, pp. 30/31, 103—105; No. 9, pp. 36, 101—103.
[4] Mejer, L.: Czesci automatow butelkowych formujace szklo. Szklo i Ceram. 6 (1955) 168—172.
[5] Zumpe, K. A.: Betrachtungen über die Herstellung von Glaskörpern und Glasformen. Die Glashütte 75 (1949) 225—228.
[6] Pol, A. R.: Die Abnutzung der Teile von Glas-Einformmaschinen und ihre Bestimmung (from the Russian). Stekoln. prom. 15 (1939) No. 12, pp. 18—21.
[7] Schuk, A. W.: Formen zum Pressen von Spezialgläsern (from the Russian). Steklo i Keram. 5 (1948) No. 11, pp. 20/21. [Abstr. J. Amer. ceram. Soc. (Ceram. Abstr.) 34 (1951) 108.]
[8] Giegerich, W.: Wärmetechnische Probleme der Gestaltung von Glasformen. Glastechn. Ber. 23 (1950) 161—169.
[9] Zumpe, K. A.: Maschinen und Geräte zur Glasverarbeitung, Leipzig: Fachbuch-Verlag 1953.
[10] Anon.: Symposium über Formen für die Glasindustrie (Society of Glass Technology. A Symposium on moulds for the glass industry). Glass 38 (1961) 15—19.

Mold Design

[11] Steinke, G.: Betrachtungen über die geometrische Gestalt von Glasbehältern, insbesondere von Flaschen. Die Glashütte 69 (1939) 148—153.
[12] Holl, F.: Inhaltsbestimmung von Drehkörpergefäßen ohne Rechnen. Sprechsaal 88 (1955) 223—227.
[13] Smith, P. A.: Planning and Developing the Glass Container. Glass Packer 33 (1954) No. 12, p. 25.
[14] Langer, G. B.: Mold Design and Machine Operation. The Glass Industry 19 1938) 215—219.
[15] Wardley, T., and C. V. Pearson: Mould Design for Automatic Machines. J. Soc. Glass Technol. 33 (1949) 59—74.
[16] Eckert, F.: Gestaltung der Vorform (Abstract). Glastechn. Ber. 11 (1933) 99.
[17] Rühl, A. Wyss, and H. Severin: Gestaltung der Vorform von Glasblasemaschinen (Abstract). Glastechn. Ber. 9 (1931) 443.
[18] English, S.: Notes in the design of parison moulds. Glass 4 (1927) 518—520.

[19] Anon.: Formen zum Massenpressen von optischem Glas. Feingerätetechn. 1 (1952) 357. [Original in Czech: Zlepšovatel a vynálezce (Prague) 1/2 (1951) 313.]
[20] Schleifer, P.: Form zum Pressen von Gläsern für Wasserstandsanzeiger Typ „Barex". Feingerätetechn. 2 (1953) 278/79.
[21] Greeg, C. V., and W. R. B. Frank: Methods and instruments for the calculation of mould capacity. J. Soc. Glass Technol. 36 (1952) 300—326.
[22] Hampton, M. M.: Measurement of the volume of solids of revolution. J. Soc. Glass Technol. 36 (1952) 327—329.

Manufacturing of Molds

[24] Dickore, G.: Strahlläppen und sein Anwendungsgebiet. Techn. Mitteilungen 1957, No. 1, pp. 12/13.
[25] Baker, G. W.: Making patterns for cast figure molds. Ceram. Ind. 58 (1952 No. 4, pp. 119/20.
[26] Baker, G. W.: How to make a cast figure mold. Ceram. Ind. 58 (1952) No. 6 pp. 93/94.
[27] Anon.: Speed engraving glass models. Ceram. Ind. 52 (1949) No. 4, pp. 104/05.
[28] Prosser, W.: Making workable molds for glass containers. Ceram. Ind. 30 (1938) No. 2, pp. 35/36.
[29] Hannich, W.: Das Härten der Glaspreßformen. Glashütte 60 (1930) 774/75.
[30] Swain, R. E.: The construction of common types of press molds. Glass Ind. 10 (1929) 7—12.
[31] Armytage, F. W.: Mould production by copying methods: Automatic and semiautomatic reproduction of mould and blank shapes from three-dimensional masters or two-dimensional templates. J. Soc. Glass Technol. 38 (1954) 3N—12N.
[32] Anon.: Machine tools for the manufacture of moulds. Glass 9 (1932) 420 to 425, 434.
[33] Scheidweiler, R.: Die Bearbeitung der Glasformen mit Kraftwerkzeugen. Die Glashütte 59 (1929) 569/70.

Cleaning of Molds

[34] Anon.: New liquid honing process for cleaning glass moulds. Glass 26 (1949) 235, 236, 240.
[35] Ganss, H.: Elektrolytische Formenreinigung (Abstract). Glastechn. Ber. 31 (1958) 217.
[36] Anon.: How they clean glass molds without damage. Ceram. Ind. 65 (1955) 59.
[37] Cheney, A.: New mould treatment techniques. Glass 29 (1952) No. 5, p. 188.
[38] Evans, N. L.: The sodium hydride process and its application to the cleaning of moulds for the glass industry. J. Soc. Glass Technol. 34 (1950) 55—62.
[39] Anderson, F.: How wet blasting cuts mold cleaning costs. Ceram. Ind. 60 (1953) No. 6, pp. 104/05.
[40] Gooding, E. C.: Mould cleaning by the vapour blast process. J. Soc. Glass Technol. 34 (1950) 101—107.
[41] Spring, G. E.: Glass mold cleaning by liquid honing. Bull. Amer. Ceram. Soc. 30 (1951) 264.
[42] Anon.: Increased mold life may result from newly developed cleaning process. Glass Ind. 28 (1947) 357, 384.
[43] Anon.: Glass mold cleaning device saves money. Ceram. Ind. 57 (1951) No. 6, p. 81.

[44] Anon.: Chemical cleaning of molds. Glass Ind. 17 (1936) 47/48.
[45] Ganss, H.: Reinigung von Glasformen. Fachausschußbericht No. 60 der Dtsch. Glastechn. Ges. (1961) 599–607.

Maintenance of Molds

[46] Anon.: Armstrong uses fusion-bonded alloy casting on glass blank molds. Glass Ind. 36 (1955) 145, 169.
[47] Anon.: Molds and mold parts salvaged. Glass Ind. 35 (1954) 81.

Lubricating of Molds

[49] Richard, A.: Les emplois des silicones dans l'industrie du verre. Verr. et Réfract. 11 (1957) 218–222.
[50] Stapp, J. A.: Why glass plants are using more silicones. Ceram. Ind. 68 (1957) No. 2, pp. 66/67.
[51] Singer, S. J.: Silicones open new era in glass molding. Ceram. Ind. 67 (1956) No. 5, pp. 82/83, 117.
[52] Stander, B. B.: The value of correct lubrication in the glass industry. Glass Ind. 33 (1952) 126/27.
[53] Murphy, J. R.: Colloidal graphite treatment of molds. Ceram. Ind. 54 (1950) No. 2, p. 59.
[54] Murphy, J. R.: Colloidal graphite treatment of mold surfaces. Glass Ind. 31 (1950) 250–253.
[55] Kurdjumow, K. P.: Spritzpistole zur Schmierung der Formen bei der Bearbeitung von Glas (from the Russian). Steklo i Keram. 13 (1956) No. 9, p. 29.
[56] Chmielenski, J.: Zastosowanie grafitu koloidalnego w szklarstwie. Szklo i Ceram. 5 (1954) 136/37.
[57] Cheney, A.: Le graphite dans l'industrie du verre. Nouvelles techniques de traitement des moules. Silic. industr. 17 (1952) 169–171.
[58] Anon.: Mold-smoking speeds glass bottle output. Ceram. Ind. 55 (1950) No. 5, p. 76.
[59] Marshall, J. B., and G. W. Mackay: Some notes on the application of colloidal graphite to glass bottle moulds. J. Soc. Glass Technol. 22 (1938) 214 to 217.
[60] Botvinkin, O. K., and V. T. Marinina: Pouziti koloidniho grafitoveho mazadla ve sklarskem prumyslu. Ceskosl. Sklar. 2 (1952) 199, 214–219, 233–237; 3 (1953) 13–17, 31/32.

5 Characteristics of Mold Materials

By F. Oberlies, Würzburg

Because of the highly variable demands of glass molding there is no universal mold material. Heat resistance is a primary requirement since mold materials are exposed to hot, viscous glass in the forming process. Next in importance are requirements on wear, corrosion, working and polishing behavior. Operational conditions (wall strength, external cooling, lubrication, timing etc.) put the stress on one or the other property, while some differences can be compensated.

Table B 9 surveys the most used mold materials. Fig. B 54 gives the permitted temperature ranges for these materials. Among the large number of possibilities grey cast iron, unalloyed or with small additions, still dominates because of low cost and good workability. This is true even though density, volume constancy, and suitability for polishing often no longer best satisfy modern demands. For more pronounced heat and mechanical resistance, as well as to obtain smoother glass surfaces, heat resistant steel is used, frequently at the expense of heat conductivity.

Table B 9. *Mold materials*

Material	Applications
Cast iron without or with small additions of Ni, Cr, Cu, Mo, Ti, Al	Press and hollow ware molds, bottle glass and rolled glass
Heat resistant steels: Ni-Cr-Fe; Ni-Cr-Wo-Cu-Fe; Cr-Co-Mn-Mo-Si-Fe; V5M;	Hollow and press ware (stirrers, plungers, needles, rings, inserts)
Vanadium steels Cr-Fe; Cr-Ni-Fe; Cr-Ni-Ti-Fe; Cr-Al-Ti	Jets, bending rollers, table rollers
Bronze with or without addition of Al	Suction molds, plungers
Nickel-Chromium	Molds for flacons
Monel (Ni-Cu)	Small molds
Nickel	Puncture molds
Copper	Puncture molds
Platinum, Platinum-Rhodium	Nozzles
Hard and artificial carbon	Rotating molds
Wood	Mouth blown hollow ware

For working temperatures between 500 and 700 °C ordinary unalloyed cast iron is used, while nickel and chromium alloys or pure nickel and chromium are used for higher temperatures up to about 1000 °C.

During the past decade various experimental attempts were made to survey the most suitable materials [14a, 53, 67, 69, 82, 82a, 90]. The wear of glass molds is most unusual, when compared to that of other machine parts, because one has to do with an intermittent heating and cooling which is difficult to reproduce even in simulation tests.

The properties of iron and steel types are determined essentially by their composition and processing (fusion, casting, cooling, heat treatment) [9a, 71, 74d]. Processing involves metallurgical problems which cannot be discussed here. The heat resistant steels used by the glass industry are so well known and standardized that reference can be made to the literature as well as to standard lists [30, 32].

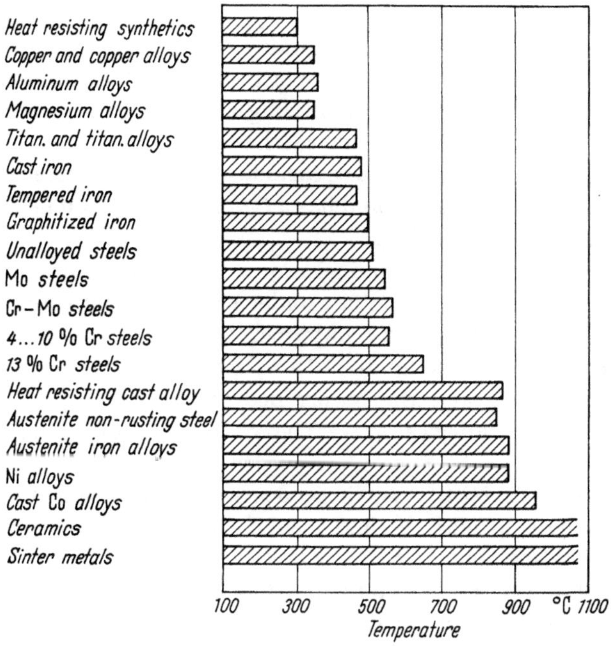

Fig. B 54. Operating temperature of high temperature mold materials. Temperatures given are the upper temperature for operation under extended life [8].

5.1 Chemical Composition and Structure

5.1.1 Cast Iron

The chemical composition of the mold materials used by the glass industry [19, 53] lies within the following limits:

Usually contained in iron	Alloying additives
C 2.7—3.8% (of which graphite 2.2—3.5%)	Cr 0.06—0.75%
	Ni 0.13—2.0%
Si 1.2 —2.8%	Cu 0.1 —0.54%
Mn 0.4 —0.9%	Mo 0.06—0.6%
P 0.14—0.73%	Ti 0.1 —0.25%
S 0.05—1.61%	V 0.15%

In all cases a heat resistant grey cast iron is desired. It is characterized by a relatively high carbon content (largely in graphitic form). Graphitizing elements are thus important.

Carbon primarily determines the properties of cast iron. If the eutectoid content rises above 0.86% the strength decreases [71, p. 191].

Graphite. A decrease in the strength of cast iron is caused by the notch effects of graphite inclusions; this situation can be improved by suitable development of graphite in special or discontinuous lamellar shape. Graphite also decreases scaling.

Silicon, which next to carbon exerts the greatest influence on the properties of cast iron, has a graphitizing effect up to a content of 3.5%, also forms ferrites, and so decreases hardness.

Manganese forms and stabilizes carbide above 0.5%, increases strength and hardness, bonds sulphur, and has a favorable effect on heat resistance and on resistance to scaling.

Phosphorus improves wear resistance, but also increases brittleness and therefore cracking, as well as, under certain circumstances, sticking of glass [19, 53a, 74a—c] by improper release.

Sulphur is innocuous up to 0.15%, favors carbide formation, but increases brittleness and the occurrence of hard spots.

The alloying additives are effective in the following directions [71, 74d]:

Al to 4.5%, Ti, Ni to 4%, Cu to 3% are graphitizing; Cr is an outstanding carbide former; V, and also Ni, Mo, Ti maintain carbide or pearlite and refine them. Cr, Ni, Mo, V, and also Cu improve strength, hardness, scaling resistance and constancy of volume. By addition of Al the scaling resistance is also increased, but below 0.6%, and in the absence of Cr, volume constancy is impaired and the metal is hard to machine. In general, improvement of mechanical properties goes with improvement of heat resistance.

The development of proper structure is decisive for the properties of a grey cast iron. The structure is influenced not only by composition but also by the cooling rate, the nucleation of the melt and other factors. The Fe—C phase diagram (Fig. B 55) [40] shows the separation relationships. The diagram has fundamental importance: main precipitation of carbon follows the stable system, while the formation of basic structure follows the metastable system. Since every cast iron also contains accessory constituents (Si, Mn, P, S) the temperatures and concentrations shown in Fig. B 55 are shifted. The structure of a grey cast thus consists of two components differing chemically and structurally: a metallic mass and embedded graphite. The chief structural components of cast iron and steel are summarized in Table B 10 and can be discussed using illustrations of structures which are also given. The basic structure then may consist of ferrite or cementite, or of pearlite, or, depending on the carbon content, on additives and on quenching conditions, of ledeburite, austenite, martensite or sorbite. The shape of graphite formation is also important for the properties of cast iron, that is, whether in large veins, lamellae, nodules or as graphitic carbon. Furthermore the cast is characterized by what is termed its degree of saturation (Table B 11). As to metallographic studies reference is made to the literature.

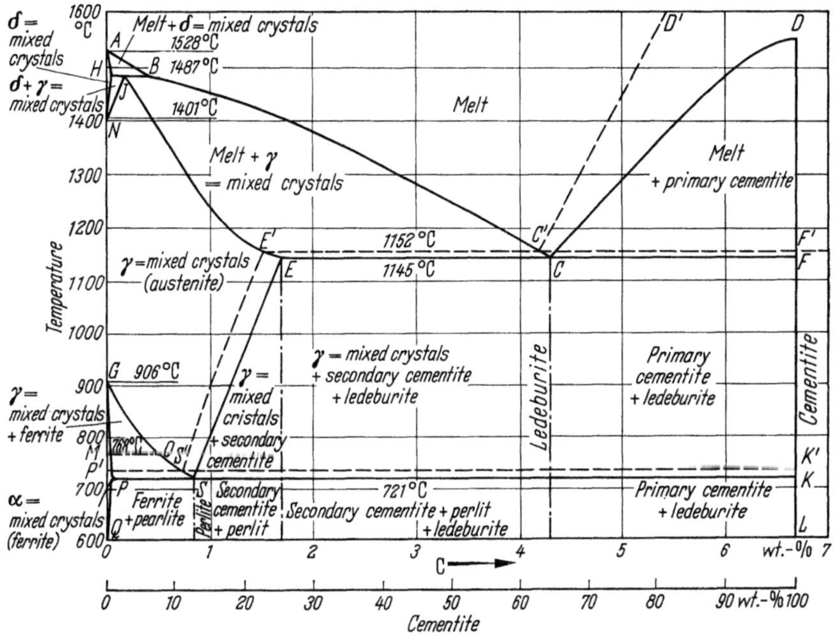

Fig. B 55. The phase diagram iron-carbon (Rep. 180, Werkstoffausschuß VDEh).

Table B 10. *Microstructure and texture*

Compostion of structure		Habit
Graphite		Eutectically fine, lamellar, spherical
Ferrite	α-mixed cr. (Fe + very Little C)	Bright white base
Cementite	Iron carbide (Fe_3C)	Bright white planes
Pearlite	Eutectic mixture of cementite and ferrite	Lamellar or granular
Steadite	Ternary phosphide eutectic	Heterogeneous planes, often narrow
Manganese sulfide		Sharply cornered planes
Ledeburite	Eutectic mixture of cementite and saturated undecomposed Fe-C mixed cr.	Like cementite with dark stripes and dots
Austenite	γ-mixed cr. decomposed into ferrite and pearlite under certain conditions	Dark bright plates
Martensite	From quenching above mixed cr.	Needles
Sorbite	Particularly fine pearlite structure with varying C content	Plates

Table B 11. *Texture of carbon in cast iron*

Carbon

- Bound
  metastable
  System: Fe—Fe_3C
  (cementite or iron carbide)
  = white cast
- Free
  stable (in metallic base)
  System: Fe-C
  - Graphite
    - Cast
  - Tempered carbon
    - Tempered cast
      (all Fe_3C decomposes into Fe + C (tempered)

$$\text{Saturation } s_c = \frac{\%\text{C}}{4.23 - \frac{\%\text{ Si}}{3.2}}$$

$s_c < 1$	$s_c = 1.0$	$s_c > 1$
Subeutectic γ-mixed cr.: pine structure clearly dendritic. Graphite nests, finer inside, coarser outside. Steadite net around graphite. Manganese sulfide parallel to dendrites	Eutectic	Supereutectic No pronounced pine structure, graphite nests: inside also coarse, steadite and manganese sulfide uniformly distributed

Iron carbide, Fe_3C, can be preserved only by rapid quenching. On slow solidification, and in the presence of graphitizing elements, which in a sense also decreases the cooling rates, the transition from the metastable to the stable system (ferrite, graphite) occurs easily. For mechanically superior cast irons, however, one aims during the subsequent transformation in the eutectoid area to obtain the metastable carbide phase because of the concomitant improvement of mechanical properties, (for instance in the case of grey cast iron one desires a pearlitic mass). In this case heat effects in operation may cause structural changes because of the decomposition of pearlite at 720 °C [38]. If heat effects are prolonged, as in glass molds, it may even take place at much lower temperatures, from about 400 °C. Depending upon the relative importance of mechanical or thermal properties one may prefer ferritic or pearlitic grey casts. It is also possible to stabilize pearlite.

The most used mold materials are usually ingot cast, and so have a 1—2 mm quench layer with a particularly fine and dense structure on the face that will be in contact with the glass. Unfortunately this fine structure is often removed too much in afterworking [19]. In thick walled casts, such as glass molds, the structure becomes coarser through the cross section as shown in Fig. B 56. The nearly eutectic cast represented ($s_c = 0.97$) has, on the glass side, a ferritic base mass in dendritic arrangements with extremely fine graphite, (a). In the center of the mold the cast is pearlitic with coarser graphite lamellae and remainders of ferrite (b); towards the outer edge, caused by slower cooling in the sand bed, the structure has become coarse with a ferritic base mass and remainders of pearlite (c) (dark areas). Differences which are too great through the cross section of a thickwalled cast may cause strains and start heat cracks in use. The main causes for strains are differential cool-

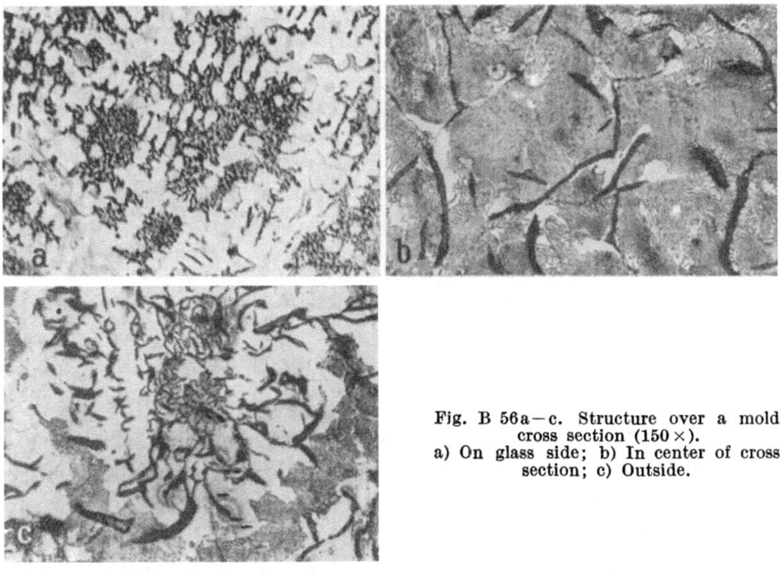

Fig. B 56a—c. Structure over a mold cross section (150 ×).
a) On glass side; b) In center of cross section; c) Outside.

ing rates and shrinkage in certain zones. Strain formation is favored by decreasing degree of saturation, increasing hardness, increasing phosphorus content, and the appearance of quench layers. Strain can be supressed by compositional measures, shaping, and by mold and cast techniques. Existing casting strain is removed by subsequent heat treatment at 500—650 °C; to cause the decomposition of pearlite the casts are heated at 700—760 °C, which results in ferritization, in whole or in part, depending upon wall thickness and treating time.

Fig. B 57a and B 57b show the structures on the glass side of what was known to be a good casting. (a) shows a ferritic cast, (b) a fine pearlite cast with steadite spots. These structures are contrasted with the coarse structure of a poor cast (c).

Recently low addition alloy casts have been introduced. In Table B 12 the chemical compositions of casts found in practice are listed. The structure of these casts, which greatly differ in hardness, is partly purely

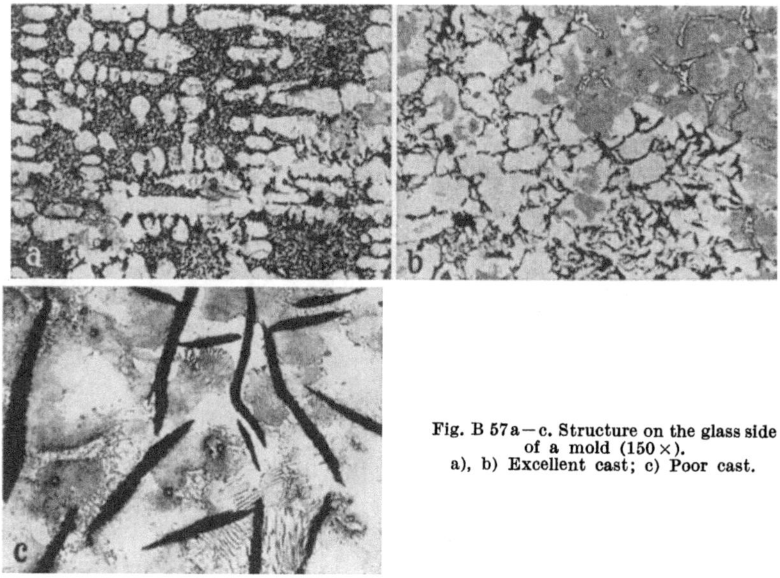

Fig. B 57a—c. Structure on the glass side of a mold (150×). a), b) Excellent cast; c) Poor cast.

ferritic (B 58a), or pearlitic with brighter ferrite spots (B 58b), or also ledeburitic (B 58c) with cementite plates in the pearlite. The Meehanite cast (B 58d) [52a], also showing ledeburite, also has a ray-like quenching layer with cementite plates. It is notable that alloying additions have stabilized ledeburite to such as extent that the beginning of transformation can only be observed after 8 hours heat treatment at 750 °C.

Generally the highly resistant alloys are hard to machine. The so-called spherocast or spherographite castings also fall into this category [11b, 17, 67, 69]. It is usually produced with a magnesium base; some nickel gets into the cast, since the magnesium addition is done by means of a magnesium-nickel alloy. The spherographite cast reputedly has a particularly high heat resistance so that it is considered a possible replacement for the more expensive chromium containing casts. While various spherographite casts have been proven in press mold rings, blank molds from this material showed bad heat cracks after short operation times (Fig. B 59a, b). In this cast the spherolites are surrounded by a wealth of secondary graphite so that the heated cast becomes ferritic. The compact structure of the spherographite cast permits the development of heat strains more readily than a normal grey cast iron which permits the decay of strain though graphite lamellae. Also, when spherographite casts are used, there must be careful coordination of composition, casting conditions and alloying additives.

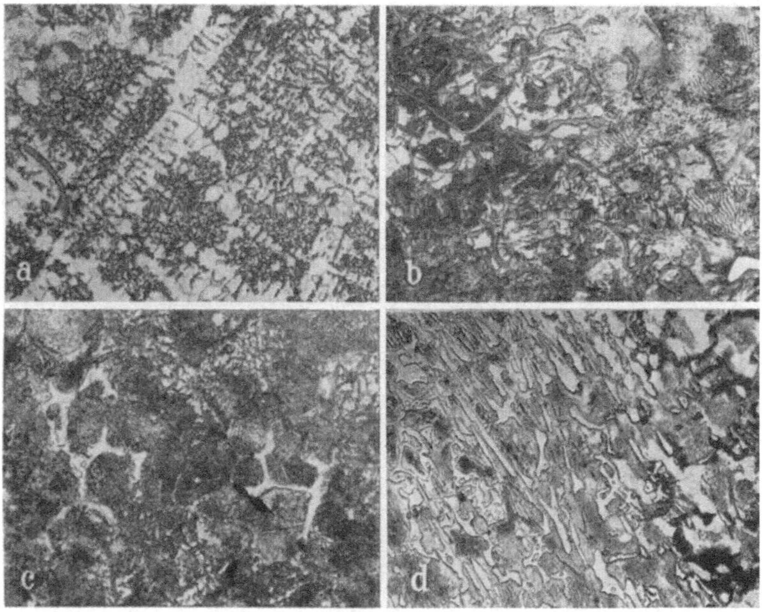

Fig. B 58a–d. Structure of the glass side of a mold of low alloy cast iron.
a) Blank mold (*II*) ferrite (140 ×); b) Same, used (*III*) pearlitic (140 ×); c) Same, used (*IV*) ledeburitic (140 ×); d) Blow mold ledeburitic (Meehanite *VI*) (150 ×).

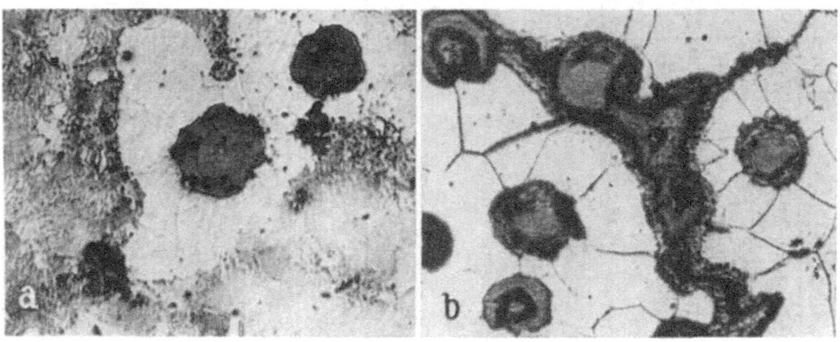

Fig. B 59a and b. Globular graphite or spherical cast iron.
a) Ring used (*VIII*) (150 ×); b) Blank mold used (*IX*) (150 ×).

Table B 12. *Chemical composition of low alloy mold materials* (%)

	C	Graphite	Si	Mn	P	S	Cr	Ni	Cu	Mo	Ti	Al	V	Ce	Mg	s_c	HB (kp/mm²)
I. a) Finish	3.68	3.67	2.21	0.56	0.26	0.116	0.10	0.20	0.69	0.061	0.052	0.023	—	—	—	1.04	89
b) Finish	3.42	3.31	2.29	0.41	0.23	0.060	0.26	0.20	0.78	0.42	0.056	0.042	—	—	—	0.97	95
II. Blank mold	3.64	3.46	2.18	0.52	0.050	0.077	0.30	0.25	0.36	0.030	0.050	0.047	—	—	—	1.02	163
III. Blank mold	3.60	2.93	1.80	0.60	0.18	0.060	0.36	0.65	0.32	0.061	0.130	0.020	—	—	—	0.98	170
IV. Blank mold	3.31	2.78	2.49	0.56	0.18	0.77	0.69	0.14	0.84	0.061	0.054	0.034	—	—	—	0.96	187
V. Blow mold	3.49	3.31	1.96	0.57	0.070	0.074	0.13	0.055	0.096	0.030	0.100	0.039	—	—	—	0.97	103
VI. Blow mold (Meehanite)	3.09	2.21	1.73	0.72	0.22	0.086	0.25	0.13	0.54	—	0.25	—	—	—	—	0.84	190
Spherolite cast																	
VII. Pressed ring	1.94	—	2.86	0.41	0.060	0.008	0.016	1.86	0.12	0.061	0.14	—	0.033	0.12	0.18	0.60	156
VIII. Pressed ring	2.20	—	2.68	0.34	0.030	0.016	0.062	0.126	0.22	—	0.070	—	—	—	—	0.65	156
IX. Blank mold used	2.88	—	3.27	0.38	0.032	0.012	0.07	0.10	0.11	0.61	0.02	—	0.16	—	0.035	0.90	162

Table B 13. *Chemical composition of heat resistant steels and metals* (%). (*After* HESSENBRUCH [30])

	Cr	Ni	Fe	Mn	Si	C	Cu	Al	Zn
Cr-Fe alloys									
Martensitic 6% Cr	6	—	Rem.	0.5	0.5—2	<0.2	—	—	—
13% Cr	13.5	0.5	Rem.	0.5	0.3—1	<0.15	—	—	—
Ferritic 18% Cr	17—19	<0.5	Rem.	0.5	0.5—2	<0.15	—	—	—
30% Cr	28—30	<0.5	Rem.	0.5—1	0.5—1.5	<0.15	—	—	—
Cr-Ni-Fe alloys									
Austenitic 18/8	18—21	8—10	Rem.	0.5	0.5	0.1	—	—	—
25/15	20	10—15	67—62	0.5	2.0	0.35	—	—	—
20/35	20	27—35	41—42	1—2	0.5—1	<0.15	—	—	—
15/60	15—18	60—65	22—18	1—2	0.5—1.5	<0.15	—	—	—
Cr-Ni alloys									
Austenitic 20/80	19—21	76—79	0.5	0—3	0.5—1.5	0.15	—	—	—
Al-Bronze alloys									
Minox A	—	—	3	3	—	—	80	7	5
Minox DV	—	22	—	—	—	—	62	7	7

5.1.2 Heat Resistant Steels

Table B 13 lists the chemical compositions of some heat resistant steels used in the glass industry. Examples of steels resistant to 1000 °C are:

a) a ferritic steel with nitride precipitation used for cast molds: 0.12% C, 1.5% Si, 0.9% Mn, 1.4% Al, 24.0% Cr; hardness 190 HB (Fig. B 60),

b) a pearlitic steel for pressed glass molds with tempered structure and finely dispersed carbides (0.22% C, 17.5% Cr, 1.4% Ni) hardness 210 HB (Fig. B 61),

c) a used pressed glass mold, austenite with fine precipitations of the alloying elements at the grain boundaries, hardness 180 HB (Fig. B 62),

d) a used blow pipe: nitrogen absorbed in use caused the formation of needle shaped nitrides in the austenite structure (Fig. B 63).

For bending and table rollers there are alloys with 13—27% Cr, 9—60% Ni, some with Al (about 1%) or Ti (about 0.2%).

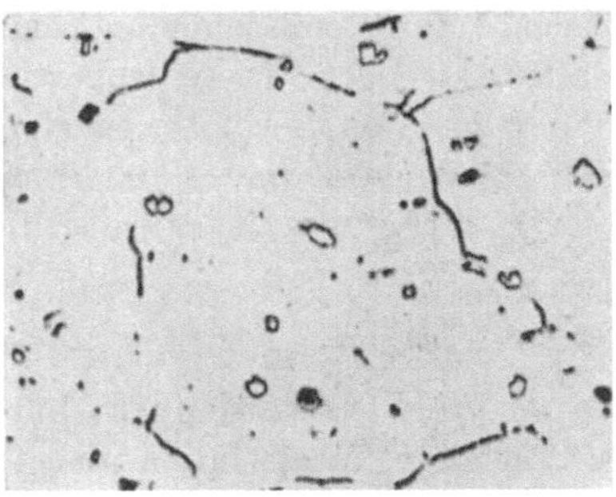

Fig. B 60. Ferritic steel for casting molds (150 ×).

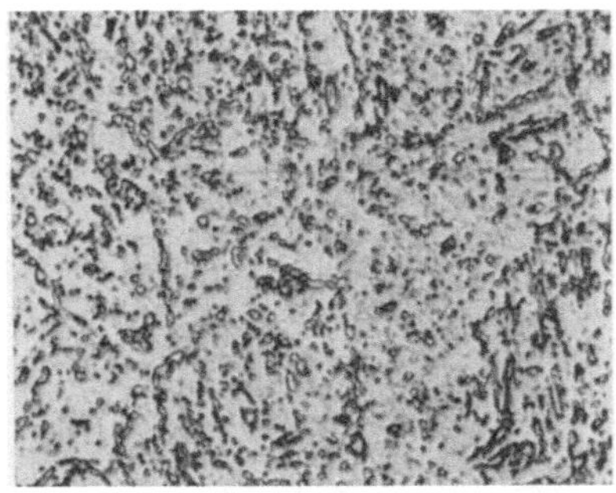

Fig. B 61. Pearlitic steel for pressed molds (600 ×).

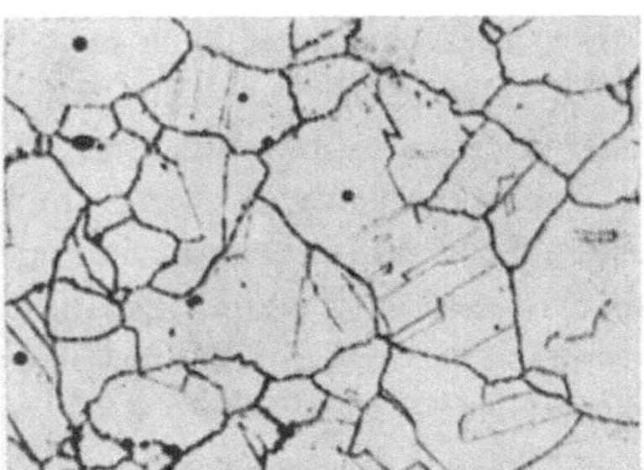

Fig. B 62. Austenitic steel. Used press mold (150 ×).

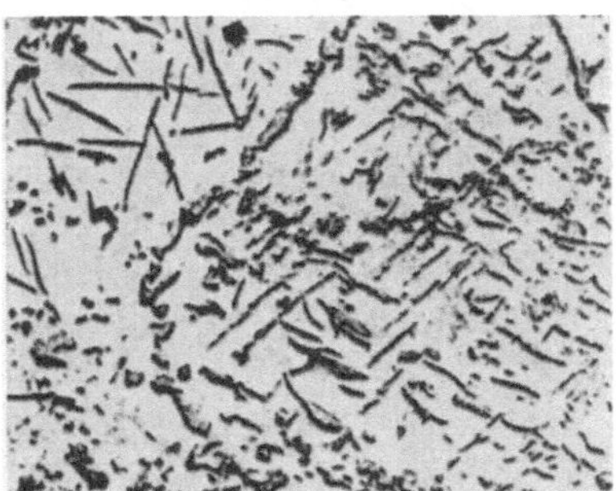

Fig. B 63. Boundary structure of a used glass blowpipe made of steel containing 25% Cr and 20% Ni (0.14% N_2 in the total cross-section) [32].

It has been reported that Cr—Ni-steels are not resistant to sulphur bearing gases and that high Al alloyed steels fail, in spite of their high scaling resistance, because they polish badly [10].

If such steels are used it is often necessary to consider not only heat resistance but high quality of resulting glass surfaces or specific conditions such as the attack of fluorine containing glasses on molds. In general, austenitic structures are more heat resistant than ferritic structures. Heat cracks depend more upon hardness, strain formation, and heat resistance than on expansivity. Steels with lower carbon content and inferior hardness as well as those containing special carbides are a little less inclined to have heat cracks. In these steels, too, scaling resistance is important. It should also be pointed out that water vapor is quite corrosive [32].

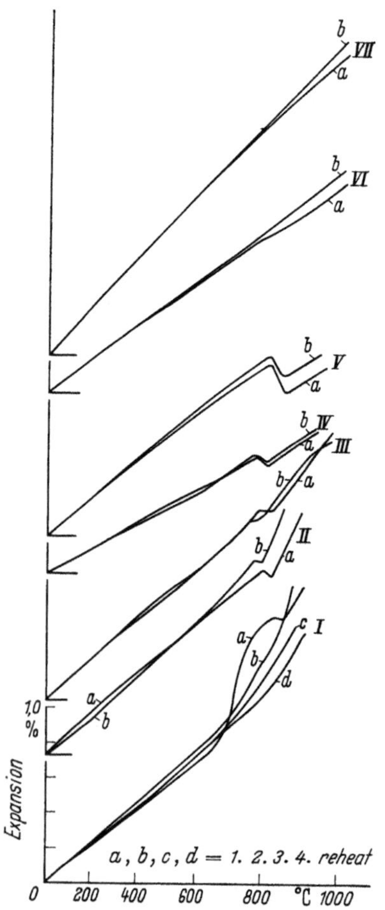

Fig. B 64. Expansion characteristics of different mold materials: *I* Pearlitic (Fig. B 56); *II* Ferritic; *III* Ledeburitic; *IV* Ledeburitic and *V* Pearlitic alloyed casts; *VI* Ferritic; *VII* Austenitic steels.

5.2 Heat Resistance

Heat resistance includes scaling resistance, volume constancy and high temperature strength. These properties in turn depend upon thermal expansion, thermal shock resistance, oxidation behavior, heat conductivity and specific heat.

5.2.1 Thermal Expansion

The thermal expansion of pure metals is usually linear, in contrast to that of alloys. In the case of grey cast iron one has to consider that graphite has a smaller coefficient of expansion (8×10^{-6}) than the metallic base (e. g. Fe, with 12.5×10^{-6}). This causes cracks and flaws on heating. Fig. B 64 shows the expansion of some mold materials. Irreversible volume changes on repeated heating, which also occur in alloyed steels, are important. These are caused by:

a) the temperature and time dependent decomposition of iron carbide $Fe_3C \rightarrow 3\,Fe + C$ (crack formation because of the pearlite transformation at 720 °C, or on prolonged or alternating heating from 400 °C). The precipitated carbon diffuses through the base mass and is deposited as secondary graphite at existing graphite locations.

b) Oxidation of the metallic base mass Fe + O → FeO, with graphite lamellae giving access to oxidizing gases.

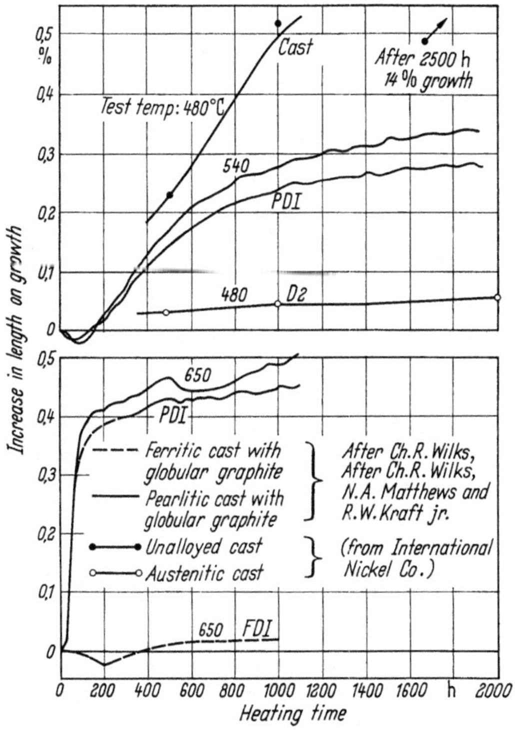

Fig. B 65. Growth of cast iron at high temperature [85].

The volume change caused by cementite decomposition amounts to only a few percent, but the expansion caused by internal oxidation may exceed 60% depending upon the composition, hence the need for denser structures. In Fig. B 65 the growth of cast iron at elevated temperatures is shown. Volume constancy can be improved by finest spheroidal or temper carbon graphite formation, by a stable or alloy-stabilized non-transformable structure, or by the bonding of carbon in the form of special carbides.

5.2.2 Oxidation

At temperatures above 500 °C almost all mold materials will form oxide or scale layers which disturb glass shaping and mold life. All consideration regarding the polish of the glass side of the mold face must refer to operational temperatures and materials must be as scale resistant as possible. The attack of oxygen rich gases forms a more or less connected, dense and adhesive layer of scale which gives some protection against further oxidation. The slower oxidation of cast iron as compared

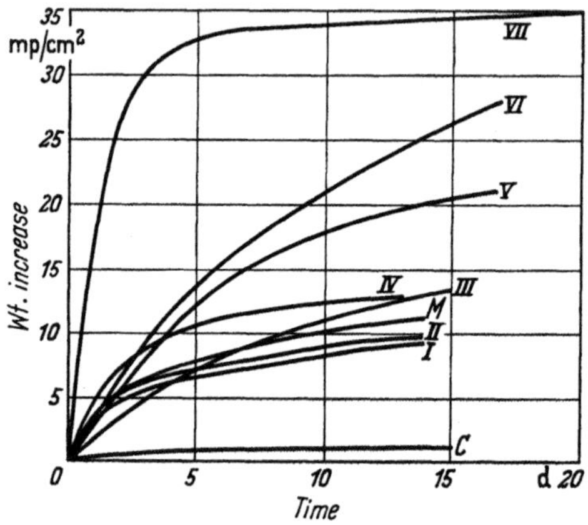

Fig. B 66. Scaling of different mold materials at 600 °C. *C* = Nickel free Cr. cast; *I* Pearlitic; *II* Ferritic; *M* Meehanite; *IV* Pearlitic; *III*, *V*, *VI*, *VII* Coarse ferritic or pearlitic casts.

to steel is explained by this dense, adhesive layer of scale which permits the use of cast pieces between 550 and 650 °C under normal conditions.

From the interference color of oxidic layers compared to equivalent air layers [29, 19] the following descending order for the tendency to oxidize was determined:

carbon steel; cast iron; chromium nickel steel with 35% Ni; alloyed steel with Cr, and small amounts of Mo, Si, Al.

Table B 14. *Thermal conductivity of metals excepting cast iron*

Material	Thermal Conductivity (kcal/m h degree)	Temperature (°C)
Iron	60	
Cr-Fe alloys	17.3—25.2	
Cr-Ni-Fe alloys	10.8—14.4	
Nickel	37.5—45	300—500
Chromium	54	
Magnesium	135	100
Red cast	60	
Tin	56	
Brass	92—101	
Zinc	95	
Aluminum	151.2—175; 235—274	300—400
Copper	330—340	200
Gold	256	24
Silver	360	100
Wood	0.32—0.31 ∥ 0.07—0.19 ⊥	

It is reported that Cr-Fe alloys are oxidation resistant in air between 700 and 1180 °C, Cr-Ni-Fe alloys between 1050 and 1200 °C. Pure chromium shows no oxidation and therefore no crack formation at high temperatures. The scaling behavior of some mold casts can be seen in Fig. B 66; a nickel free chromium cast is shown for comparison. Of the various structural constituents ferrite oxidizes before carbide; steadite oxidizes most easily; its oxidation layers foliate and cause pitted glass surfaces or sticking [48, 51, 53b, 74a, a, c]. The skin of the cast provides some natural protection against corrosion.

5.2.3 Heat Conductivity

Table B 14 gives the coefficients of thermal conductivites of various metals. Cast iron is not included in this table. The coefficient of thermal

Table B 15. *Thermal conductivity of different cast irons as a function of temperature*

No.	Chemical composition					Thermal conductivity (kcal/m h degree)			
	C	Si	Mn	P	S	100 °C	200 °C	300 °C	400 °C
1	2,89	1.87	0.32	0.27	0.046	40.3	39.6	38.5	37.8
2	3.87	2.81	0.28	0.28	0.045	37.8	37.1	36.3	35.2
3	3.03	4.20	0.28	0.30	0.043	34.9	34.2	33.8	33.1
4	3.34	1.90	0.76	0.18	0.065	42.1	41.0	40.3	39.6
5	3.40	1.90	0.92	0.59	0.060	41.3	40.7	39.6	38.6
6	3.30	2.00	1.00	0.95	0.050	40.0	39.2	38.1	37.1

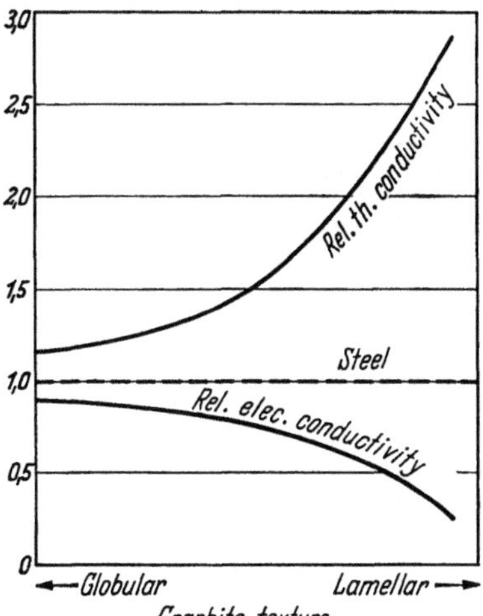

Fig. B 67. Thermal conductivity and electrical conductivity of cast iron compared to that of steel [43].

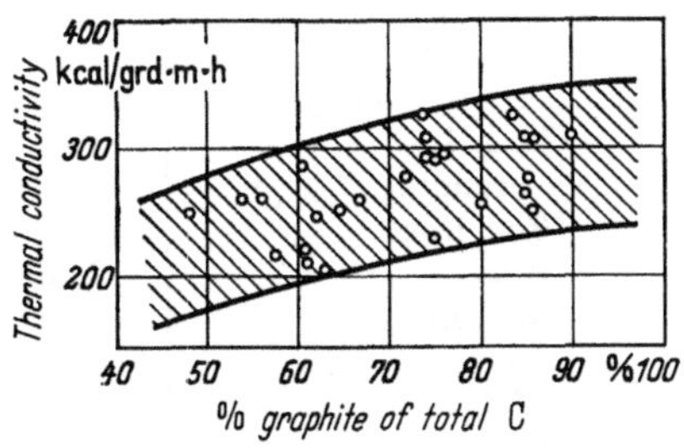

Fig. B 68. Effect of the percentage of graphite of the total carbon content on conductivity [83].

conductivity of cast iron at room temperature is between 36 and 54 kcal/m h degree. Alloying decreases it. The heat conductivity of cast iron decreases linearly with increasing temperature (Table B 15).

Heat conductivity is much influenced by structure, especially by the formation of carbon. Bound carbon causes more of a decrease than free carbon, and, as in steels, particularly as a sorbitic or martensitic formation is approached. Figs. B 67 and B 68 show the heat conductivity of cast iron in comparison to that of steel and also show the influence of the amount of graphite. It can be understood that the inferior conductivity of globular graphite in the cast may cause heat accumulation leading to heat cracks (Fig. B 59).

5.2.4 *Thermal Diffusivity*

While heat conductivity represents a measure of the heat flux in a body of constant temperature gradient, thermal diffusivity measures how fast temperature variations per unit time propagate in a body. Thermal diffusivity is very important in glass molds characterized by periodic temperature fluctuations on the inside; so is the penetration depth of these fluctuations. These values are now used to characterize mold materials [17a, 87].

5.2.5 *Specific Heat*

Table B 16 shows the specific heat of mold materials and shows that it generally increases with temperature. The true specific heat also increases with the carbon content of cast iron [80].

Table B 16. *Mean specific heat of mold materials*

Material	Mean specific heat (kcal/kg degree)	Temperature (°C)
Cast iron	0.16	500
Heat resting steels		
Cr-Fe alloy 1	0.11	
Cr-Ni-Fe alloys	0.12	
Nickel	0.124	15—630
Chromium	0.113	0—100
Red iron	0.0937	100
Brass	0.0917—0.099	
Aluminium	0.237	18—500
Gold	0.031	20—100
Silver	0.0581	20—100
Wood	0.32	0—100
Copper	0.0994	
Graphite	0.21	0—100

5.2.6 *Strength*

Occasionally the specific weight (Table B 17) and the tensile strength of mold materials are of interest. This is particularly true for the high temperature strengths such as those shown in Fig. B 69. The first noticeable changes in strength occur above 400 °C. Alloy additions to stabilize the iron carbide are important above this temperature. Globular graphite casts are classified between steel, on the one hand, and cast iron having lamellar graphite, on the other hand. The small temperature sensivity of austenitic structures shoult be mentioned.

Table B 17. *Specific weight of mold materials*

Material	Specific weight (kg/m^3)
Cast iron	6980—7300
Ferrite	7850—7920
Graphite	2250—2550
Cementite	7660
Pearlite	7740
Cr-Fe alloys	7600—7700
Cr-Ni-Fe alloys	7750—8400
Nickel	8850
Chromium	7100
Red cast	8850
Brass	8400—8700
Aluminum	2560
Copper	8900
Gold	19300
Silver	10420—10600
Wood	320—1030

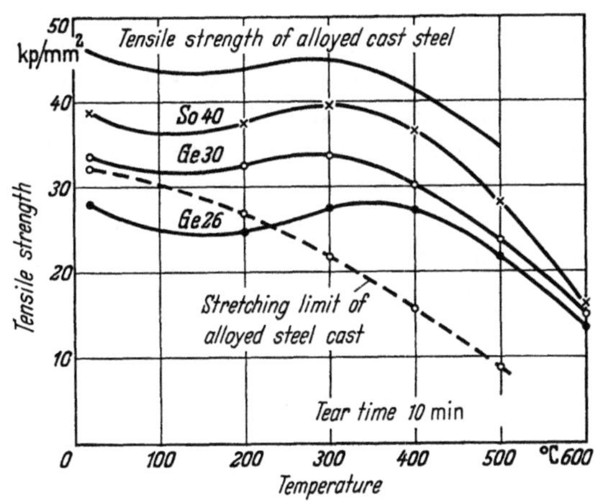

Fig. B 69. High temperature strength of high grade cast irons in relation to that of cast steel [46].

5.3 Hardness and Machinability

Between the tensile strength σ and the Brinell hardness HB in kp/mm² the following direct relations exist:

carbon steels: $\sigma_B = 0.36$ HB alloyed steels: $\sigma_B = 0.34$ HB

For grey cast iron the relative hardness RH i. e. that related to tensile strength, varies within wide limits (0.8—1.3). If the casts are similar, equal Brinell hardnesses indicate equal tensile strengths. Table B 18 gives the Brinell hardness of various materials. It depends essentially upon the structure of the basic matrix, i. e. upon the content of bound

Table B 18. *Hardness of mold materials-BHN*

Material	Brinell hardness (kp/mm²)
Pure iron	50—70
Cast iron, very soft, ferritic	100—140
Cast iron, pearlitic	175—225
Cast iron, martensitic	360—500
Mold cast iron	103—187
Heat resisting steels Cr-Fe alloys	180—230
Heat resisting steels Cr-Ni-Fe alloys	140—180
Nickel, soft	80
hard	200
Red cast	50—100
Brass	50—100
Aluminum	15— 25
Gold	18.5
Silver	25.0
Wood, according to kind	0.8—2.5
Graphite	~30

carbon, and the distribution of pearlitic carbides, and upon the interlamellar distance of pearlite lamellae. Hardness decreases with increasing temperature, noticeably above 400 °C. The hardness of cast iron at 550 °C has decreased to 23—35% of its value at room temperature. The hardness of the fine structure is measured by the Vickers method (microhardness). Hardness is considered a measure of machineability, including planing, grinding, turning, engraving, polishing, and repairing. The degree of the desired polish depends upon fine structure and the hardness of the various structural components.

5.4 Application of Coatings

To improve mold life, correctness of size, and the surface quality of the glass, coatings are sometimes applied to the inside surfaces [2, 16a, 37, 92a]. Frequently the interior of glass molds is electroplated with nickel or chromium. Poorly applied coatings tend to peel. Nickel or copper coating prior to chrome plating is reported beneficial on steel molds. The thickness of such coatings is about:

0.01 – 0.2 mm for chromium
0.025 mm for nickel
0.01 mm for copper

After some time chromium coatings must be renewed because of loss by diffusion.

In what is called "Inchroming" the diffusion of chromium is used to coat edges of steel parts with 0.1 mm of a scale-proof layer of high chromium content. Steel and cast steel are easily inchromed. Grey cast

Fig. B 70a. "Colmonoy C 20"-alloyed cast iron interface.

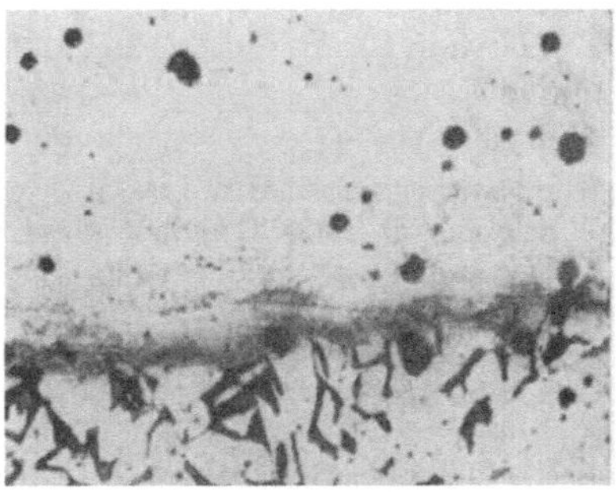

Fig. B 70b. "34 C" on a plunger of structural steel (150 ×).

iron is hard to treat by this process. Since afterworking is hardly possible, and since the parts warp easily at the high temperatures necessary for inchroming the method is rarely used for glass molds.

Heating at certain temperatures in suitable atmospheres permits the production of a hard coat of Fe_3O_4 at the surface (0.075–0.085 mm thick). There is, however, no data available in the glass industry on "calorizing", the spraying on of an aluminum layer followed by heat treatment. "Alitizing", the diffusion of aluminum by heating in a metal powder, has only rarely been successful, and certainly is not in glass molds where the hard surface is veined by fine cracks.

Recently the so-called "Colmonoy" coatings [17, 40a, 52d, 64, 65] have been applied successfully. A hard chromium-nickel-boron alloy is sprayed on and subsequently sintered with a welding torch. These layers are up to 1.5 mm thick and adhere very well, since alloying and interdiffusion take place between base material and coating, as shown in Fig. B 70a. Depending upon their compositions these coats possess hardnesses from 230 to 320 kp/mm² HB. A similar material, also suited for such coatings, is the metal powder "34 C" which is also sprayed on and burned in to produce a surface hardness of approximately 180 kp/mm^2 HB (Fig. B 70b). Both materials can be repaired.

5.5 Sticking of Glass

Hot glass sticks to metal when the latter exceeds a certain temperature, the "sticking temperature". In Table B 19 sticking limits are given for two kinds of glasses and several materials. The very complicated

Table B 19. *Sticking limits in °C of various types of glass and mold materials*

	Lead glass	Soda lime glass
After THOMAS on:		
Cast iron	555	605
Steels	515–530	525–615
Cr plated iron	560	625
After SARJANT and MIDDLEHAM		
Sticking at 1200 °C of hot glass on:		
Cast iron	<400	
13% Cr-steel	500	
30% Cr-steel	600	
Special steel	>700	

sticking process is influenced [14, 37, 47, 48, 50, 53a, 74a, b, 94] by: temperature, viscosity, surface tension, glass composition, mold surface condition, and specific heat as well as heat conductivity of the mold material. The sticking tendency of glass is decreased by thin, adhesive oxide layers and the use of lubricants which prevent the direct contact of the glass with the surface of the mold. According to recent experiments the sticking temperature increases with heat conductivity [19]. In practice, however, steel molds, having 3 to 4 times lower conductivities than cast iron have sticking temperatures 100–300 degrees above those of cast iron, probably because of their smoother surfaces. Sticking at low temperatures is often caused by the netlike arrangements of steadite in high phosphorus (about 0.9%) materials. The phosphorus, because it is a very hard constituent, protrudes because of lesser wear on polishing and favors sticking (B 71). Finally, Table B 20 summarizes characteristic properties and conditions for metallic mold materials.

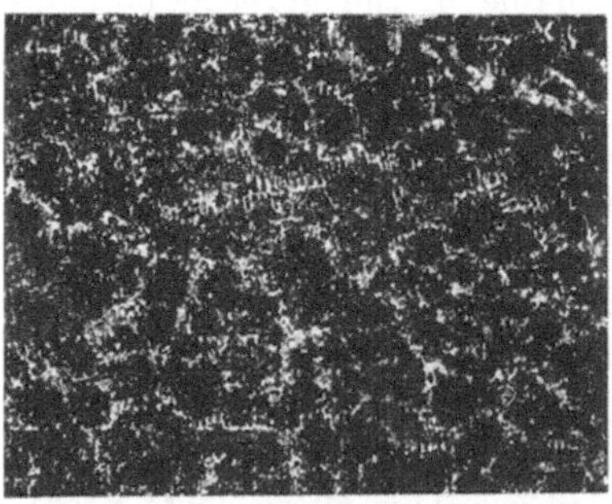

Fig. B 71. Structure of cast iron, showing appearances of sticking, steadite network (10 ×).

Table B 20. *Properties of mold materials*

Property	Requirement
General	Fine dense structure; in cast: eutectic graphite
Volume stability	Globular graphite, ferritic or alloy-stabilized structure, special carbides in steels
Small oxidation tendency	Thin, adhering oxide layers, enhanced by Mn, Al; globular graphite, incoherent graphite lamellae Cast iron 550— 650 °C Cast iron alloys 600— 900 °C Cr-Fe alloys 700—1180 °C Cr-Ni-Fe alloys 1050—1200 °C
Thermal expansion	Uniformly low up to working temperature, no irreversible volume changes by structural transformation as in pearlite; little oxidation
Thermal shock resistance	Ferritic structure with coarse graphite veins
High temperature strength	Austenitic structure better than ferritic; decreases in order steel—globular graphite cast—lamellar graphite cast
Thermal conductivity	High for cast with lamellar graphite; decreases with globular graphite and alloying additions (carbides)
Specific heat	Increases with C
High strength and elasticity limits at working temperature	Fine grain, ferritic cast with additions of Ni, Cr, Mo; steel
Good workability	Ferritic; low hardness, about 130 kp/mm² HB
Polishing behavior	Fine structure, alloyed cast, steel
Wear resistance	Pearlitic (or ledeburitic, austenitic) structure
High sticking temperature	Fine dense structure (steel); in cast fine graphite formation, no steadite ($s_c \geqq 1$), good heat conductivity
Cracking, growth spalling, warping	Caused by structural transformations, excessive oxidation (particularly interior), strain, low thermal conductivity; alloyed cast tends to tear, as does globular graphite cast

References

[1] Avery, H. S., and C. R. Wilks: Cast heat-resisting nickel-chromium-iron alloys. Trans. Amer. Soc. Metals 1948, Nickel Bull. 20 (1947) No. 11, p. 165.

[2] Balderson, S. C.: Chromium plating of moulds for glassware. Glass 17 (1940) 193.

[3] Bauret, P., A. Portevin, and P. Chevenard: Alliage devant résister aux effets d'un gradient alternatif de températures et en particulier, alliages, pour moules de verrerie fine. Compt. rend. 184 (1927) 1655/56.

[4] BELL, G. S.: Austenitic alloy irons and their applications in glass industry. Glass 9 (1932) 474—477.

[5] BORNHOFEN, O.: Grauguß als Werkstoff für Glasformen. Glastechn. Ber. 12 (1934) 339—345.

[6] BRUCKNER, W. H., L. R. KOVAK, and A. B. WILDER: Metallurgical study of cast iron glass molds. Bull. Amer. Ceram. Soc. 19 (1940) 58—61.

[7] BRUCKNER, W. H., and H. CZYSEWSKI: Metallurgical study of cast iron for glass moulds. J. Amer. Ceram. Soc. 25 (1942) 381—388.

[8] CLAUSNER, H. R.: Materials and Methods 39 (1954) No. 4, pp. 117—132.

[9] COENEN, M.: Wärme- und Temperaturleitfähigkeitsmessungen an Gußformmetallen (Abstracts). Glastechn. Ber. 33 (1960) 35, 199, 482.

[9a] CUMBERLAND, J.: Production of die cast-iron bar. Glass Technol. 2 (1961) 114—116.

[10] DAMMER, A.: Stahllegierungen für die Tafel- und Formglasherstellung. Werkstoffe und Korrosion 2 (1951) 409—414.

[11] DAVIS, N.: (a) New cast iron increases glass mold life. Ceram. Ind. 60 (1953) No. 4, pp. 133, 162/163.
(b) Mold irons for the glass industry. Glass Ind. 35 (1954) 19/20.

[12] DIETZEL, A., and F. OBERLIES: Untersuchung über Eigenschaften von Werkstoffen für Glasformen und deren Verhalten im Gebrauch (Abstracts). Glastechn. Ber. 15 (1937) 125; 16 (1938) 151, 283; 17 (1939) 166.

[13] DIXON, F. W.: Leistungsfähigkeit von Metallformen. Ceram. Ind. 30 (1938) No. 6, pp. 50—52.

[14] DOWLING, W. C., H. V. FAIRBANKS, and W. A. KOEHLER: A study of the effect of lubricants on the adherence of molten glass to heated metals. J. Amer. Ceram. Soc. 33 (1950) 269—273.

[14a] EDGINGTON, J. H.: (a) Mould materials. A summary of information from Emhart Manufact. Comp., Overmeyer Mould Comp. Inc., The Toledo Mould Comp. Glass Technol. 2 (1961) 116—119.
(b) Surface treatment of moulds. Glass Technol. 2 (1961) 140/41.

[15] Techn. Staff Emhart Manufact. Comp. Hartford: Investigation of materials for glass container molds. Glass Ind. 40 (1959) 21—23, 42, 44; 80—81, 96, 98; 131, 154, 155.

[16] EVANS, N. L., W. GOACHER, and J. E. HURST: Cast iron for the manufacture of glass bottle moulds. J. Soc. Glass Techn. 20 (1936) 45—53, 63—71.

[16a] FAWCETT, J. A.: Metallizing of glass machinery and moulds. Glass Technol. 2 (1961) 142—145.

[17] FLATTER, E. R.: Mold irons and the glass mold situation. Bull. Amer. Ceram. Soc. 33 (1954) 101—103.

[17a] FRANK, W. R. B.: A review of recent work on the working temperatures of glassware moulds in various countries. Glass Technol. 2 (1961) 107—110.

[18] GIEGERICH, W.: (a) Wärmetechnische Probleme der Gestaltung von Glasformen. Glastechn. Ber. 23 (1950) 161—169.
(b) Thermische Eigenschaften der Werkstoffe für Glasformen (Abstract). Glastechn. Ber. 24 (1951) 142.
(c) Die Oberfläche der Formenwand und ihr Einfluß auf die Strömung des Glases. Glastechn. Ber. 26 (1953) 333—341. Eng. Transl.: Proc. Internat. Comm. Glass 2 (1955) 143—153. Fr. Transl.: Proc. Internat. Comm. Glass 2 (1955) 154—162. Fachausschußbericht No. 50 of the Dtsch. Glastechn. Ges. (1953).

[19] GIEGERICH, W., and F. OBERLIES: Formenwerkstoffe und Vorgänge in Glas-Formen. Part I: Eigenschaften der Werkstoffe und Anforderungen im Betrieb.

Glastechn. Ber. 26 (1953) 363—378; Fachausschußbericht No. 50 of the Dtsch. Glastechn. Ges. (1953).

[20] GOERENS, P.: Einführung in die Metallographie, 7th and 8th Ed., Halle: Knapp 1948.

[21] GOW, J. T.: A survey of hot-gas corrosion of iron-nickel-chromium alloys. Corrosion 3 (1947) 311—324, 383—403; Nickel Bull. 20 (1947) 177.

[22] GRODSKY, V. A.: Bronze glass molds replacing cast iron molds. J. Opt. Soc. Amer. 40 (1950) 177.

[23] GUILLAMON, J.: Fontes pour moules de verrerie. Fonderie (1951) 2373/74.

[24] GUSS, S.: Glasformen. Keram. Rdsch. 34 (1926) 380—382.

[25] HANEL, R.: Amerikanische und französische Erfahrungen bei der Verwendung von Gußeisen für Glasformen (Abstract). Glastechn. Ber. 11 (1933) 267.

[26] HANEMANN, H., and A. SCHRADER: Atlas metallographicus, Vol. 1 and 2, Berlin 1936.

[27] HANNICH, W.: Das Härten der Glaspreßformen. Glashütte 60 (1930) 774/75.

[28] HENGSTENBERG, O.: Das Kleben des Glases an der Form (Abstract). Glastechn. Ber. 10 (1932) 464.

[29] HENGSTENBERG, O., H. MÜLLER, and F. STÄBLEIN: Über Glasformen aus Sonderstahl sowie den Wärmefluß in solchen und in Glasformen überhaupt. Fachausschußbericht No. 31 of the Dtsch. Glastechn. Ges. (1934).

[30] HESSENBRUCH, W.: Metalle und Legierungen für hohe Temperaturen, Part 1: Zunderfeste Legierungen, Berlin: Springer 1940.

[30a] HILL, D. K.: Investigations of the influence of thermal treatments on the performance of mould irons. Glass Technol. 2 (1961) 110—113.

[31] HOOPER, G., and C. E. WILLIAMS: Zusammensetzung und Eignung der Glasformen. Glashütte 61 (1931) 521/22.

[32] HOUDREMONT, E.: Handbuch der Sonderstahlkunde, Vol. 1, 3rd Ed., Berlin/Göttingen/Heidelberg: Springer; Düsseldorf: Verlag Stahleisen 1956.

[33] HOWES, H. W.: Some general considerations of glass-making moulds. J. Soc. Glass Technol. 17 (1933) 68—84.

[34] ILLIG, K.: Elektrolytische Verchromung von Glasformen und Walzenmaterialien. Glastechn. Ber. 5 (1927/28) 241—252.

[35] ILLIG, K., and W. BIRETT: Chromium plating in the blown, pressed glass industry. Ceram. Ind. 13 (1929) 170—176.

[35a] JOHNSON, J. W., H. SKELAND, and E. R. FLATTER: Bottle and mould design in the glass industry. Glass Technol. 2 (1961) 103—106.

[36] KALPERS, H.: (a) Warmfeste Stähle für die Glasindustrie. Glas-Email-Keramo-Technik 1 (1950) 8/9.
(b) Über hitzebeständige Stähle für die Glasindustrie. Sprechsaal 85 (1952) 53/54.

[37] KAPNICKY, J. A., H. V. FAIRBANKS, and W. A. KOEHLER: Study of adherence of molten glass to heated metals. J. Amer. Ceram. Soc. 32 (1949) 305—308.

[38] KLEIN, E. H.: Der zeitliche Verlauf des Zementitzerfalls im Gußeisen. Stahl u. Eisen 54 (1934) 827—830.

[39] KORECKY, J.: Chromovani sklarskych forem. Skarské Rozhledy 10 (1933) 23/24.

[40] KÖRBER, F., W. OELSEN, and coauthors: Das Zustandsschaubild Eisen–Kohlenstoff. Bericht No. 180 des Werkstoffausschusses des Vereins Deutscher Eisenhüttenleute, Düsseldorf, 1949.

[40a] LEACH, T. C.: (a) Increasing life of moulds and mould equipment. Glass Technol. 2 (1961) 135—139.
(b) The colmonoy sprayweld process applied to Roirant suction machine components. Glass 38 (1961) 454—456.

[41] LE GRAND, R.: Proves machinability depends on microstructure. Amer. Mach. 94 (1951) 1077.
[42] LIFSCHITZ, M. A.: Preßformen aus Sonderstählen. Steklo i Keram. 13 (1937) 29—32.
[43] LÖHBERG, K., and J. MOTZ: Gießerei 44 (1957) 305—308.
[44] Anon.: „Lunz"-Eisen, ein neues Metall für Glasformen. Glass 10 (1933) 198.
[45] MAFFEI, C. A.: I prodotti siderurgici nell industria vetraria. Vetro e Silic. 1 (1956) 15—18.
[46] MAILÄNDER, R., and H. JUNGBLUTH: Techn. Mitt. Krupp I (1933) 83—91.
[47] MARBOE, E. C., W. D. SMILEY, and W. A. WEYL: Some adhesion phenomena involving glasses, metals and fused salts and their interpretation. J. Soc. Glass Technol. 32 (1948) 264—280.
[48] MARININA, W. T.: Untersuchung über das Ankleben der Glasmasse beim Verformen des Glases auf der RW-Maschine (from the Russian). Steklo i Keram. 12 (1955) 9—11.
[49] MCGRATH, K., D. P. FORBES, and F. W. DIXON: What constitutes good glass house iron. Glass Ind. 31 (1950) 81—83, 106.
[50] MÜLLER, H.: Kleben des Glases an der Form (Abstract). Glastechn. Ber. 11 (1939) 99.
[51] NEHMITZ, A.: Beitrag zum Zusammenhang von Preßglas-Oberflächen und Gußgefüge der Formen (Abstract). Glastechn. Ber. 16 (1938) 283.
[52] OBERLIES, F.: (a) Untersuchungen an dem Spezial-Formenguß „Meehanite". Glastechn. Ber. 26 (1953) 263—271.
(b) Untersuchungen über das Verhalten von Formenflächen aus Grauguß (Abstract). Glastechn. Ber. 16 (1938) 151.
(c) Weitere Versuche über das Formenproblem und über den Abdruck der Formenoberfläche im Glase (Abstract). Glastechn. Ber. 17 (1939) 166.
(d) Untersuchungen über amerikanische Schweißmaterialien für Glasformen (Abstract). Glastechn. Ber. 31 (1958) 218.
(e) Untersuchung ausländischer Formengußsorten; weitere Ergebnisse des Formenuntersuchungsprogrammes (Abstract). Glastechn. Ber. 29 (1956) 344.
[53] OBERLIES, F., and A. DIETZEL: (a) Metallische Werkstoffe für die Glasverarbeitung und Untersuchungen an Formenguß. Glastechn. Ber. 24 (1951) 79—102.
(b) Verzunderung von Gußeisenoberflächen und deren Abdruck im Glas (Abstract). Glastechn. Ber. 16 (1938) 218.
[54] Anon.: Emilite, a new metal for glass molds. Nat. Glass Budget 50 (1934) 3, 7, 13.
[55] Anon.: Materials for glass moulds. Glass Ind. 16 (1935) 180/81.
[56] Anon.: Properties of various alloy metals used in the glass industry. Glass Ind. 17 (1936) 14.
[57] Anon.: Some applications of Nickel alloys in the glass industry. Nickel Bull. 9 (1936) 229. — Glass Ind. 18 (1937) 340—342.
[58] Anon.: Gußeisen für die Herstellung von Flaschenformen. Metallurgie 13 (1936) 141/42.
[59] Anon.: Applications des alliages de nickel en verrerie. Verre et Silic. Ind. 9 (1938) 122—126.
[60] Anon.: Les metaux pour moules a travailler le verre. Verre et Silic. Ind. 9 (1938) 94.
[61] Anon.: Formen zum Massenpressen von optischem Glas. Feingerätetechnik 1 (1952) 357. [Original in Czech: Zlepšovatel a vynalezce 1/2 (1951) 313.]
[62] Anon.: Chromium diffusion of glass molds. Glass 30 (1953) 466/67.

[63] Anon.: Overmyer holds sixth iron clinic. Glass Ind. 35 (1954) 549/50, 578/79.
[64] Anon.: Armstrong uses fusion-bonded alloy castings on glass blank molds. Glass Ind. 36 (1955) 145, 169.
[65] Anon.: Nickel-base alloy insert increases thimble life. Glass Ind. 36 (1955) 472/73.
[66] Anon.: Manufacture of glass moulds. Glass 32 (1955) 606.
[67] Anon.: Eighth Overmyer Iron clinic. Glass Ind. 39 (1958) 380/81, 394/95; 1960 Overmyer Iron clinic. Glass Ind. 41 (1960) 395, 414/15.
[68] Anon.: Glass molds. Glass Ind. 40 (1959) 596–599.
[69] PADNI, P., and K. D. SHARMA: (a) Glass moulds. Bull. Central Glass & Ceram. Res. Inst. 6 (1959) 23–28.
(b) An investigation on glass moulds: Part I. Materials for moulds. Bull. Central Glass & Ceram. Res. Inst. 8 (1961) 3–14; Part II. Technique of casting. Bull. Central Glass & Ceram. Res. Inst. 8 (1961) 73–78.
[70] PARKIN, M., and W. E. S. TURNER: The use of special alloys in the glass industry. J. Soc. Glass Technol. 21 (1937) 247–262.
[71] PIWOWARSKY, E.: Hochwertiges Gußeisen, 2nd Ed., Berlin/Göttingen/Heidelberg: Springer 1951 (see pp. 735–780, 1022/23).
[72] POL, A. R.: Die Abnutzung der Teile von Glas-Einformmaschinen und ihre Bestimmung (from the Russian). Stekoln. prom. 15 (1939) 18–21.
[72a] RENN, H. v.: Iron for glass bottle moulds. Glass 37 (1960) 485–490, 544 to 546.
[73] RICHARDSON, B. F.: High-temperature alloys for the glass industry. Glass Ind. 40 (1959) 421/22, 445, 473–75, 498, 530/31, 560, 562.
[74] ROLL, F.: (a) Ursachen des Klebens des Glases an der Form (Abstract). Glastechn. Ber. 12 (1934) 39.
(b) Kleben von Glas an Glasformen (Abstract). Glastechn. Ber. 16 (1938) 284.
(c) Entstehung einer narbigen Oberfläche im Preßglas. Glastechn. Ber. 15 (1937) 427/28.
(d) Handbuch der Gießerei-Technik, Vol. I, Part 1 and 2, Berlin/Göttingen/Heidelberg: Springer 1959/60.
[75] SANTOS, L. A. L.: Chilled cast iron for glass molds. ABM, Bol. Assoc. Brasil. metais 11 (1955) 381–387.
[76] SARJANT, R. J., and T. H. MIDDLEHAM: Heat resisting steels for the glass industry. J. Soc. Glass Technol. 20 (1936) 685–716.
[77] SCAFE, R. M.: Glass molds present many problems. The Foundry 68 (1940) No. 8, pp. 30/31, 103–105; No. 9, pp. 36, 101–103.
[78] SCHLEIFER, P.: Form zum Pressen von Gläsern für Wasserstandsanzeiger Typ „Barex". Feingerätetechnik 2 (1953) 278/79.
[79] SCHUK, A. W.: Formen zum Pressen von Spezialgläsern (from the Russian). Steklo i Keram. 5 (1948) 20/21.
[80] SCHWARZ, C.: Zusammenstellung wichtiger spezifischer Wärmen für metallurgische Berechnungen. Arch. Eisenhüttenw. 7 (1933) 281–292.
[81] SIMON, R.: Werkstoffe für Glasformvorrichtungen. Fachausschußbericht No. 11 der Dtsch. Glastechn. Ges. (1928).
[82] SMITH, R. D.: Metallurgical needs of the glass industry. Metals and Alloys 7 (1936) 217–220, 263–268. — Foundry Trade J. 55 (1936) 501–504.
[82a] Society of Glass Technology: (a) A symposium on moulds for the glass industry. Glass 38 (1961) 15–19.
(b) Meeting of glass molds. Nat. Glass Budget 77 (1961) 3, 11, 12.
[83] SÖHNCHEN, E.: Arch. Eisenhüttenw. 8 (1934/35) 223.
[84] STAHL, C. J.: Chromium-plated glass-forming molds. Glass Ind. 13 (1929) 191.

[85] STELZENMÜLLER, H.: VDI-Berichte 27 (1958) 61.
[86] STRAUSS, B.: Werkstoffe für Glasformen. Glastechn. Ber. 3 (1925/26) 407 to 412.
[87] TRIER, W.: Temperaturmessungen in Glasformen. Glastechn. Ber. 28 (1955) 336–351.
[88] TSCHERNJAK, M. H., and M. S. ASSLANOWA: Gußformen aus Metall und deren spezielle Oberflächeneigenschaften beim Glasblasen unter hohem Druck. Steklo i Keram. 12 (1936) 26–32.
[89] UXA, V.: Gußeiserne Glasformen (Sklarske litinove tvarnice). Sklar a Keram. 5 (1955) 52–56.
[90] VANICK, J. S.: (a) Metallurgical analysis of glass mold cast iron. Glass Ind 19 (1938) 263–266, 308–313.
(b) Gütegrade und Auswahl von gußeisernen Glasformen. Ceram. Ind. 31 (1938) No. 1, pp. 36/37, 48–50.
[91] WALKER, H. L.: Cast iron molds for glass ware. Ceram. Ind. 34 (1940) No. 6, p. 45.
[92] WARDLEY, TH.: Glass forming moulds: material and design. J. Soc. Glass Technol. 17 (1933) 50–67.
[92a] WEATHERLEY, E. G.: The diffusion of metals in relation to glass mould equipment. Glass Technol. 2 (1061) 146–149.
[93] WENDLER, A.: (a) Zur Frage der Verbesserung des Materials für Glasformen. Sprechsaal 58 (1925) 76/77.
(b) Grauguß-Gattierungen für Glasformen (Abstract). Glastechn. Ber. 10 (1932) 468.
[94] WEYL, W. A.: Adhesion to glass. Glass Ind. 26 (1945) 557–560, 574.
[95] WILLIAMS, C. E.: Effect of temperature on metals used in the glass industry. Ceram. Ind. 18 (1932) 225–227.
[96] ZMIHORSKI, E.: Die Auswahl des Formenmaterials für das Pressen von Glasgegenständen. Szklo i Ceram. 13 (1962) 229–238.

6 Mold Cooling

By W. GIEGERICH, Heilbronn, and R. WILLE, Berlin

The glass mass which is to be shaped in a viscous condition has to be cooled by the forming elements — mold, plunger, finish, bottom-plate and other — to the point where shape is preserved when it leaves the machine. The large amount of heat that must be removed during the molding process is dissipated in various ways:

1. Natural air cooling. Limited heat loss of the metal mold permits only moderate production rates. This method ist used in hand shops, especially for blowing of hollow glass.

2. Artificial cooling by periodic dipping in water. Wood or graphite molds are used "cold": a water vapor film between mold and glass serves as a separating and gliding layer.

3. Forced air cooling by jets (one or more) impinging on the mold wall vertically or at an angle. This is the most widespread method in hollow glass manufacture since it is relatively simple.

Forced air cooling will be treated in more detail. Depending upon the place to be cooled two methods are used: continuous cooling of external walls and periodic direct cooling of the interior faces of molds; or both methods are used jointly.

Cooling by other means has been proposed such as evaporation cooling and reflux cooling with sodium but no practical use has been reported.

Air cooling is preferred in spite of the low specific heat of air. The supply is practically unlimited, and it can be brought to any place and can be regulated easily. It does require considerable energy. The blowers for mold cooling are the largest energy consumers in a hollow glass plant. In the manufacture of hollow glass the contact of glass and hot mold wall is intermittent since filling and withdrawal occur in the same mold. The total time (period between two subsequent work phases) can therefore be divided into a glass or heat-up time and a cooling time. The periodic change of temperature in the mold wall is illustrated for inner and outer cooling by the scheme represented in Fig. B 72. While on the inside of the wall cool air is blown against the wall only during the "cooling time", on the outside air cooling may continue all the time or during a chosen period, depending upon the construction of the machine. In Fig. B 72, blowing time is marked as total time for external cooling. From a production viewpoint one will always run a mold close

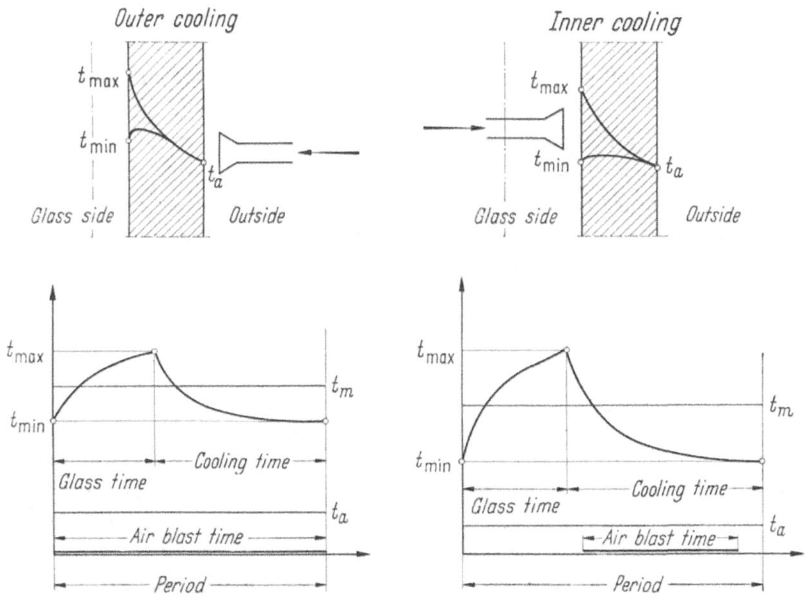

Fig. B 72. Outer and inner cooling of glass molds.

to the sticking limit. The position of the minimum temperature at the inner wall, t_{min}, depends upon the relation of blowing to total time, assuming equal heat load. Since, then, the maximum temperature is limited and since the sum of the heats lost at both wall surfaces is fixed for a constant machine output, the external cooling of a mold can be conceived in a simplified manner as a heat exchanger without storage; and the internal cooling as one with heat storage. Through the time temperature changes of the mold, which repeat in the rhythm of the machine periodically, one has in the mold a quasistationary temperature profile.

Heat transfer from the glass to the surroundings [1] occurs in three characteristic phases: heat transfer from glass to mold [2, 7]; heat

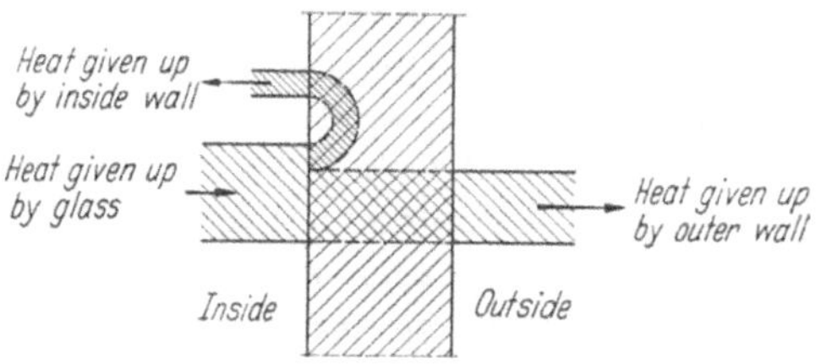

Fig. B 73
Diagram of heat flow in a mold wall.

storage and heat conduction through the wall [3]; and heat transfer from mold face to air [4, 5, 6]. These phases are represented schematically in Fig. B 73.

The problems of the heat transfer from glass to wall have been discussed in Sec. A 2. Schreck [3] analytically investigated the heat flux in mold walls and developed an experimental method for complicated cases. It works by means of an electrolytic trough in which under certain assumptions an analogy exists between electrical potential and temperature as well as between current and heat content. Schreck beautifully demonstrated the influence of local heat sinks at exterior mold surfaces on the temperature distribution at the inner face. Thick walled molds greatly facilitate the maintenance of a uniform temperature on the inner face when the external cooling varies considerably.

The task of removing heat from the external wall to the air has been treated by Wille in several studies [4—6). The influences of nozzle shape, diameter, and inclination of smooth sufaces and of surfaces with various projecting pins were investigated systematically. He used constant temperature model molds of dimensions, heat loads, and temperatures corresponding to those of molds used in production. The results of these measurements are summarized in Table B 21, where for various experimental conditions the measured air flows are given. The corresponding cooling devices are shown in Fig. B 74. The most important practical conclusions are summarized in the following Sec. B 6.1 to B 6.7.

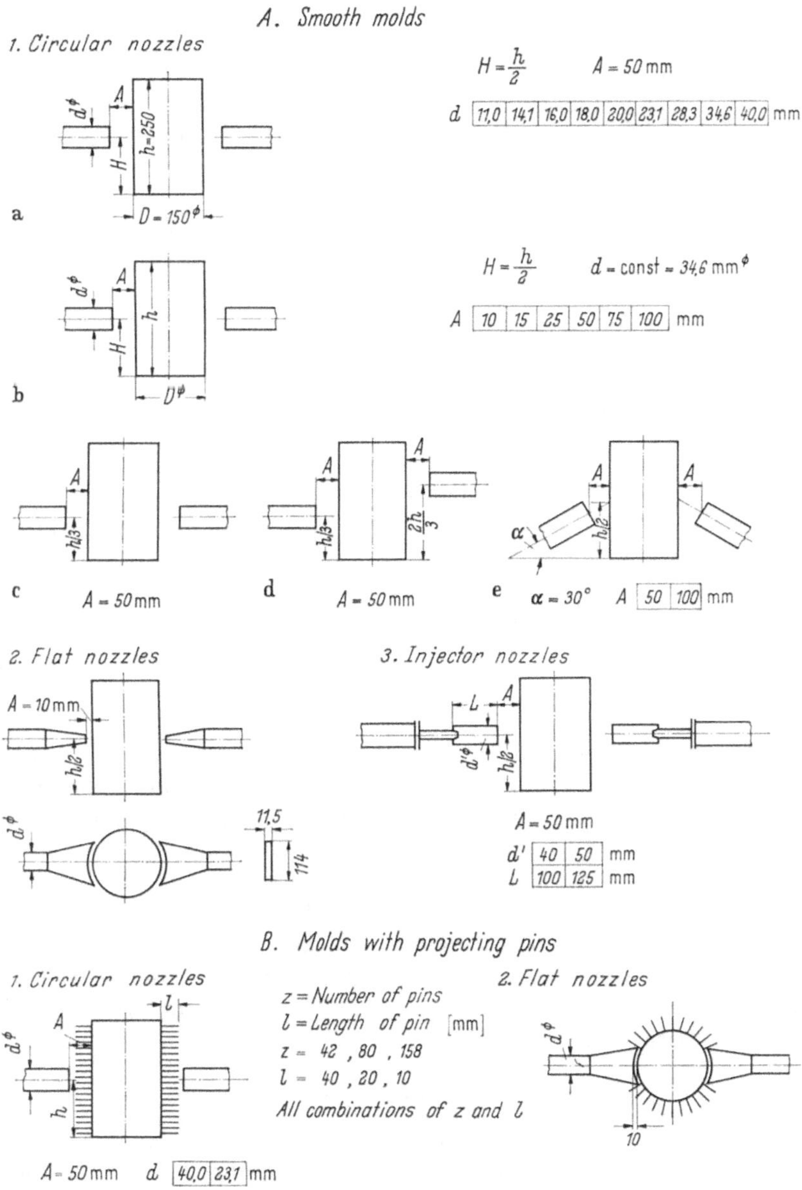

Fig. B 74a–e. Summary of investigated nozzles.

Table B 21. *Output of air jet in mkp/sec in different cooling systems*[1]

		Mean inner wall temperature (°C)			
		375	400	425	450
Smooth surface					
Circular nozzle, diam.	14 mm	—	17	10.7	6.7
	20 mm	13.6	9.2	6.1	4.1
	40 mm	12.9	8.7	5.9	4.1
Distance A	10 mm	11.3	7.7	5.2	3.65
(nozzle diam. 34.6 mm)	100 mm	11.8	7.7	5.3	3.65
Angle of inclination α	0°	11.5	7.7	5.5	3.7
(nozzle diam. 34.6 mm)	30°	—	12.1	8.3	5.7
Flat nozzle		6.3	4.5	3.2	2.2
Circular nozzle, diam.	40 mm	10.8	7.6	5.3	3.5
Injector nozzle		—	14.4	8.6	5.5
Circular nozzle, diam.	40 mm	13.0	8.6	5.7	3.9
Surface with z projecting pins					
Circular nozzle, diam. 40 mm					
$z = 42$	$l = 40$ mm	8.2	5.5	3.8	2.4
	$l = 10$ mm	10.6	6.8	4.6	3.1
$z = 158$	$l = 40$ mm	4.5	2.8	1.8	1.1
	$l = 10$ mm	7.6	5.1	3.4	2.2
Flat nozzle $z = 38$	$l = 40$ mm	5.1	3.5	2.4	1.6
	$l = 20$ mm	5.5	3.8	2.6	1.7
$z = 152$	$l = 40$ mm	3.2	1.94	1.2	0.073
	$l = 20$ mm	4.1	2.7	1.8	1.2

6.1 Inside Wall Temperature

Measurements were made at mean inside wall temperatures of 375, 400, 425, 450 °C. Air outputs for these temperatures are about in the relation 3.1 : 2.3 : 1.5 : 1.0, for equal heat loss. Energy requirements thus increases considerably for decreasing wall temperature.

6.2 Nozzle Diameter

The three diameters investigated (14, 20, 40 mm) gave, for 400 °C, air outputs in the ratio 1.94 : 1.05 : 1.0. For 450 °C internal wall temperature the ratios were: 1.64 : 1.0 : 1.0. The air output required thus increases with decreasing diameter.

6.3 Nozzle Shape

Flat nozzles require, for equal effect, only 58 to 62% of the energy of circular nozzles, while the injector nozzles require 1.4 to 1.6 as much as circular nozzles.

[1] The experimental mold used to obtain these data had an ext. diam. (without pins) of 150 mm, inner diam. of 70 mm, height 250 mm and consisted of cast iron.

6.4 Spacing of the Nozzles

No significant differences in air output required were found for the two investigated distances: 10 and 100 mm.

6.5 Inclination of the Nozzles

At 30° 1.5 to 1.7 more output was required than for the 90° positon.

6.6 Projecting Pins and Circular Nozzles

If the outer mold surface is provided with projecting pins and if the circular nozzle is placed at the favorable distance of 40 mm, the output required decreases to 64 to 57% for 42 pins, and to 35 to 26% for 158 pins, compared with impact on smooth surfaces.

6.7 Projecting Pins and Flat Nozzles

A flat nozzle in connection with projecting pins gave the smallest energy requirement at a 40 mm nozzle distance with 38 pins (37—39%); and with 152 pins (20—18%), compared to the use of a circular nozzle on smooth molds.

The experiments also showed that a considerable decrease in air requirement can be achieved without the use of projecting pins which are difficult to put on. In all considerations regarding cooling techniques it is important to know the relationship between air output and heat loss. This is the only way to estimate the effect of changes in air output on heat loss and, consequently on the performance of the machines. Wille and collaborators [4, 5, 6] have made pertinent investigations and have determined this relation. Between the air output N and the heat transfer experessed by the Nusselt number Nu, the general relation

$$N = K\, Nu^n \qquad (1)$$

can be established. The exponent n = 4.5 was derived from the experimental results. The constant K depends upon shape and position of the nozzle, and shape of the surface (outside diameter D_a and heigt h of the mold) and therefore is a fixed quantity for a given equipment. K might, for instance, depend on the outside diameter and the heigth as follows:

$$K = (D_a\, h)^{-0.515}.$$

If one designates by

Q the heat transferred to the mold
F the cylinder surface of the model mold $D_a\, h\, \pi$
λ_L the thermal conductivity of air
Θ_{aw} the average temperature of the mold surface
Θ_L the average air temperature

and if one substitutes for Nu in Eq. (1)

$$Nu = \frac{Q\,D_a}{F\,\lambda_L(\Theta_{aw} - \Theta_L)} \tag{2}$$

the following conclusions may be drawn:

1. Air output N will be decreased by diminution of Nu.

(a) N will be decreased by a decrease of external wall temperature Θ_{aw} by the use of high conductivity mold materials (copper, cast iron, hollow molds filled with sodium, thin molds, etc.). (An example for the influence of the wall thickness is: $D_a = 150$, $N = 100\%$; $D_a = 70$, $N = 40\%$.)

(b) N is decreased by a decrease of the heat accepted by the mold, for example by a reduction in glass weight.

2. N is decreased by an increase in the mold surface as by use of projecting pins.

3. From an energy viewpoint ist makes no difference whether the cooling effect is provided by a small air volume at high speed or by a large volume at low speed. An operational decrease in the air required is only attainable through the simultaneous and clever application of various constructional and dynamic flow considerations. For details reference should be made to the cited publications.

6.8 Arrangement of Cooling Nozzles in Practice

The effective cooling of molds in practice is complicated by construction features — mold holders, operating mechanisms, the opening and closing of the molds, and the motion of the machine itself. A few examples follow, all taken from popular machines.

6.8.1 Cooling Nozzles Connected with the Mold

Owens, in his suction-blow machine (Sec. E. 3.2.1), has made the mold holders hollow and has utilized them as cooling air ducts. This solution is not too simple, particularly rotating in a machine, but it is feasible, as Owens proved. The great advantage of this construction is that the distance between the nozzle and the mold surface remains constant during the entire working cycle of the machine and can be made small. The details of this feature can be seen in Fig. B 75, which shows the holder arm of an Owens machine with cooling canals.

Cooling is much facilitated in the molds of machines with fixed molds, e. g., the IS (Sec. E 3.3.3) or Lynch 44 (Sec. E 3.3.4) machine. Here the introduction and distribution of air to the rotating molds is not required. If the cooling nozzles are arranged on the machine frame, one has to tolerate for divided molds a large distance from cooling nozzles to mold surface, since the mold must open and close. In this construction it is not possible to direct the air stream close to the mold surface, since effective baffles can not be used. An instructive example of the skillful solution of the cooling problem under such conditions is the arrangement

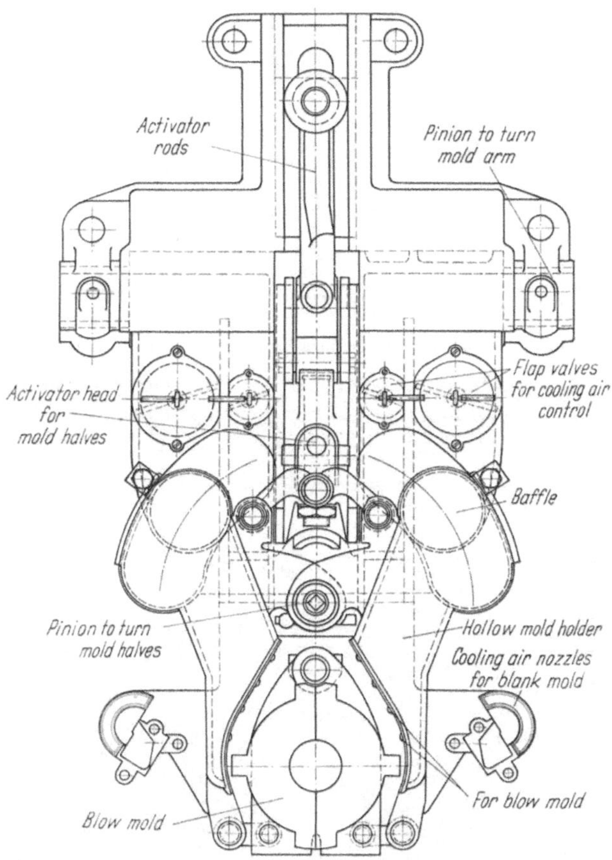

Fig. B 75. Blow mold holder arm of Owens machine with cooling canals [11].

and the construction of the cooling nozzles of the IS machine. The cooling nozzles are piled up, on both sides of the mold, like towers. Nozzles and nozzle holders are constructed in a manner permitting the adaptation to technological conditions by combinations of various nozzles. The air is introduced from below.

Mold cooling can be made particularly intensive if blank molds are solid as in the press-blow process. Here the nozzles can be easily arranged around the now—usually ribbed—mold at the correct distance. An example

Fig. B 76. Cooling of solid blank molds on a Lynch 44 machine (Courtesy Lynch Corp.).

Fig. B 77. Mold cooling in the Roirant R7 machine. Example of nozzles rotating with machine base.

is the cooling of the blank mold in the Lynch 44 machine (Fig. B 76). Considering the small distance and the symmetrical arrangements of the nozzles ist was possible to dispense with baffles without any loss in cooling efficiency. This facilitates mold changes.

For cooling molds in rotating machines cooling air ducts with their nozzles may rotate with frame and molds. This is examplified in the

Fig. B 78. Mold cooling in the L 10 machine. Example of fixed jets in intermittent machine.

Roirant R 7 machine (Sec. E 3.3.2) (Fig. B 77). One may also work with fixed nozzles (Fig. B 78). Fixed nozzles are recommended particuarly for intermittent machines such as the Lynch 10 machine (Sec. E 3.3.1). Rotating nozzles are recommended for continuous machines.

References

[1] GIEGERICH, W.: Wärmetechnische Probleme der Gestaltung von Glasformen. Glastechn. Ber. 23 (1950) 161–169.
[2] TRIER, W.: Zähigkeitsverteilung und Fließen des Glases in Vorformen. Glastechn. Ber. 32K (1959) II/17–II/25.
[3] SCHRECK, C.: Beitrag zur Ermittlung des Wärmeflusses in den Wänden von Glasformen. Glastechn. Ber. 28 (1955) 351–359.
[4] WILLE, R.: Methoden der Formenkühlung an Glasverarbeitungsmaschinen. Glastechn. Ber. 28 (1955) 351–359.
[5] WILLE, R.: Luftkühlung von Glasformen. Glastechn. Ber. 32K (1959) II/26 to II/32.
[6] BHATTACHARYA, A.: Der Einfluß von Nadeln auf die Kühlung von kurzen Kreiszylindern bei Freistrahlanblasung. Glastechn. Ber. 32 (1959) 397–401.
[7] TRIER, W.: Temperaturmessungen in Glasformen. Glastechn. Ber. 28 (1955) 336–359.
[8] SEVERIN, H.: Erfahrungen und Vorschläge zur Kühlung der Vorformen an Flaschenblasmaschinen (Abstract). Glastechn. Ber. 13 (1935) 257.
[9] KRUGOW, G. I.: Kühlen der Formen von WSch-Halbautomaten mit Wasser und Luft (from the Russian). Stekoln. Prom. 16 (1940) 3/4.
[10] STEFAN, Z.: Das Kühlen der Formen bei halbautomatischer Erzeugung auf Saugblasmaschinen. Ceskol Sklar. Keram. 4 (1954) 179.
[11] WENDLER, A.: Maschinelle Glasverarbeitung, Leipzig: Akad. Verlagsges. 1929.

C. The Historical Development of Glass Forming Machines

By L. MUSCHALEK, Bad Godesberg

Historical observations of the gradual evolvement of, first, the mechanical, and then the fully automatic forming of glass is of interest for several reasons. It may depict the technical history of a specialty and through this of the intellectual aptitude and the creative power of engineers. It can also teach and stimulate the contemporary engineer, since many elements of present-day complex automatic machines are found in older tools and in early equipment and machines. Frequently the differences of modern processes simply reflect various combinations of older basic ideas. The study of original manual processes permits particular insight into single forming operation and suggests instructive comparisons.

The history of machines still used today—treated in Section E — will not be discussed in this section, since it represents the history of definite types such as of the Owens machine, while the purpose here is the elucidation of the basic technical development of a process.

1 Development of Processes Leading to the Machine Age

The history of glass production has been investigated and is known, at least in its broad features. The exploration of detail which may be significant to present day engineers still requires thorough research. Often reliable sources and contemporary relics such as artisans' tools are not readily available. With respect to the development of artisans' tools and the mechanization of glass manufacture the history of glass production may be divided into the following periods (with the annotation that the boundaries must be considered flexible since after a new invention, the older and newer processes frequently are used concurrently):

1. Manual processes, from the beginning of glass production to the present,

(a) Before the invention of glass blowing: pressing, squeezing and casting; and for hollow ware what is called the sand and clay core technique,

(b) Invention of the glass blow pipe around 30 B. C. and introduction of the glass blow-pipe to the Western Mediterranean. This invention started a mode of production which became typical for a long time and is being replaced gradually by mechanical processes only since the end of the 19th century. Forming through blowing is a process typical for the material glass.

2. Mechanical processes,

(a) Mechanical processes in the hollow glass field were first developed in the 19th century, but became effective only in the 20th century.

(b) Fully automatic processes were invented in the 20th century.

The historical development of artisans' tools may be conceived as a prestage of modern mechanization. For this reason the following survey is given.

The true beginning of glass production is not known so that even in antiquity myth filled this gap in our knowledge. This is not the place to consider various reports about accidents which may have resulted in glass formation, nor to weigh claims about the location of the cradle of glass production. This allocation has already been made to a majority of the antique civilizations without a really final decision.

On the basis of their knowledge of Egyptian and Mesopotamian conditions archaeologists suspect [31] that glazes — thin glassy coats on other materials, particularly steatite, — represent the beginning of the history of glass. This development is dated as in the 5th or 4th millenium B. C. but no detail can be given about techniques or tools. Soon thereafter a material was used which consisted largely of sintered quartz flour and replaced clay materials. These quartz frits are known unter the misleading designations of "Egyptian faience" or "Egyptian porcelain". This product was manufactured in Egypt from the 4th millenium B.C. to the 15th century A.D. From the middle of the 3d millenium pearls—made similarly to Gablonz artificial pearls—and pendants can be documented which consist of glass entirely. They were without any doubt made by manual pressing, squeezing or casting. The first true glass vessels—smaller bottles, goblets or vases—that have been preserved are colored and come from Egyptian royal graves of the period around 1555—1350 B.C. They have been made by the so-called sand core technique [31] in which a sand core is stuck to a metal rod, fired or fritted, and finally coated by a thick layer of viscous glass. Presumably further forming was achieved, with frequent reheating, upon a polished marver, using simple tools such as pinchers, but certainly not by blowing. The glass was most likely not produced in one work cycle, but was first fritted and then melted again. Analyses have shown compositions resembling modern ones. The sand core technique was used in Egypt as well as in Mesopotamia in the middle of the second millenium B.C.

The 3rd and 2nd centuries before Christ are known as a period of particular propensity toward experimentation with glass. Perhaps at this time someone finally hit upon the possibility of blowing glass. This may have been suggested by the welding and blowing pipe used in metal work. This discovery initiated for glass, the first synthetic material, an entirely new period.

According to new archaeological findings glass blowing originated in the Eastern Mediterranian at the beginning of the first century B.C.; and the glass blower's pipe was introduced to the Western Mediterranian during the last decade B.C. [32]. The new process permitted much larger production of glass.

In addition to free blown glasses, pieces were also found which were produced in two- or many-part molds — in contrast to the early use of one-part molds as they were known from metal working. These were particularly glasses with ornamental or figurative reliefs. The mold material remains unknown.

There are no artisans' tools from this period. But from the glass articles preserved it may be concluded that tools were, in principle, the same as used in mouth blowing today. Only from the publications of the 16th century, such as those by Biringuccio, Agricola, Neri, Merrett, Kunckel, and, later (1772) the Encyclopedie des Arts et Metiers, etc., can one imagine, from the illustrations, what the glass maker's tools of older times must have looked like. According to these illustrations glass makers' pipes were not always of uniform length and, occasionally, were provided with wooden protective envelopes. The decisive invention of the glassmaker's stool was made in the beginning of the 17th century and spread, at times, very slowly.

For the sake of completeness it should be noted that in the middle ages window glass was blown — at least as far as the beginning of the process is concerned — by a hollow glass process, as described by THEOPHILUS. But in antiquity panes were cast, as proven by preserved glass. In the 13th century another process, a manual blow-spin process originated, the so-called moon glass process [34]. A sphere was blown, a punty was fused on at the opposite end, the sphere was cracked off the pipe, warmed, widened out, and put into fast uniform rotation to convert it into a flat disc. This pane was cooled and cut. The cut panes yielded little flat panes ("facet glass") while the center pieces with the punty spot formed the "bull's eyes". Later moon glass makers used a root for their rotating tool.

In 1688 LUCAS DE NEHOU in France invented or introduced the plate cast-roll process which was at least a modest step toards mechanization [24]. In view of beginning specialization further developments are better represented separately for each field. More detail about flat glass production is found in Sec. C 6.

2 Pressing of Hollow Glass

As shown above certain glass pressing processes were not unknown in old Egypt, even though they would be designated as squeezing processes for small ware (pearls, pendants) today. Apparently this technique remained in use, particularly for jewelry [24].

At the beginning of the 19th century—opinions are divided as to whether in England or in the U.S.A. — the so-called pressed glass was developed. Processes and tools were found which allowed the fabrication of some types of hollow ware without the pipe as a tool and without blowing as a process. The hot viscous glass was formed on hand presses. Even though these presses were primitive in the beginning, they signify an important step toward mechanization, a departure from craft work, and a relocation of human labor [9].

2.1 Technical Development of the Process

It may be assumed that in the beginning glass was pressed without a mold ring and that it was only in the course of further development that one proceeded to cover the mold with a ring to force the malleable glass in the mold to fill the entire cavity.

At the start there must have been difficulties in getting sufficiently smooth mold and glass surfaces for, in the first decades a very clear, transparent high lead glass was pressed exclusively. In other respects lead glass is particularly suitable since it is soft and malleable over a wide temperature interval. Following a suggestion by Moore (1855) in England, the cheaper soda lime glasses were often used. In the first period it was difficult to obtain really smooth surfaces. The easy way out was the application to the entire inside of the mold of a decorative engraving. Later other engravings imitating cutting were used. Starting in England and America pressed glass production spread to other countries. In 1836 the Lobmeyer glass house in Marienthal (Slavonia, in former Austria-Hungary) produced pressed ware with wreath engravings; and at an earlier time (1830) the Baccarat glass house near Luneville is mentioned as a pressed glass house. The first press mold material was brass. This material facilitated engraving (chiseling) because of its density and softness. Soon, however, many molds were made from cast iron, and, in special cases, from alloy steel. It was recognized at an early time that the lubrication of the cavity and the ring of press molds prior to filling them with glass was advisable. Bee's wax and graphite, also tallow, were used. The lubricant was to be free of ash and to have a low flash point, to avoid crusts in the mold. As a result of lubrication press ware acquired smoother, more glossy surfaces. Even in modern fully automatic production lubrication is still necessary. The low flash point resulted in the ignition of the lubricant shortly after application to the hot mold and the remaining oil char layer permitted good gliding of the glass during pressing and removal from the mold. The glass is introduced into the mold by a gathering iron or a ladle. To avoid damage to the upper rim of the mold and to the ring, the trailing glass thread must be cut by shears. Soon it was found that subsequent heating or fire-polishing in gas or oil fired glory holes leads to very smooth surfaces and rims.

When the great applicability of the press process had been recognized the simple one-piece molds were supplemented by those with bottoms that could be lifted and, later, by those made from hinged parts. Thus it became possible to press jugs with handles, bowls with feet, etc. In some glass houses pressed ware was reheated in the glory hole and reshaped considerably by flaring. By manual flaring objects were obtained that otherwise could be made only by press-blow processes. Examples are bowls with retracted edges, vases with expanded middle parts, jardinieres, and the like. To produce uniform ware certain flaring mechanisms were constructed. The press process was perfected in many directions. Bottles can, for instance, be made from two fused parts. On one press the upper part was pressed; on an adjacent press, the bottom. After taking out both parts from their molds the parts were fused using the residual heat in them and a bottle was made. The mouth opening was pressed through and the hole was opened by grinding. In another process a lamp was pressed in a multiple mold. The upper part of the mold was provided with a sort of blowhead, and the malleable glass still present in the inside of the lamp was simply blown out to the mold after removal of the plunger.

Two types of hand presses originated early:

(a) The spring press, on the lower ring of which is fastened the mold ring. The spring pressure can be regulated and causes the ring to sit tight on the mold. It is understood that the presser's muscle power must overcome the prestress of the spring.

(b) The eccentric press. If large pressures are required, as for large pressed ware, the eccentric presses are more suitable because the part of the press carrying the mold covering ring is locked immediately after contact with the mold. After pressing the mold covering ring remains locked to the mold for a short time and this time can be regulated. When the plunger is withdrawn, the ring lifts, and the product is pushed off if it sticks to the plunger. In 1881 FAHDT in Dresden patented a spring press with adjustable springs, claws for plunger locking, and air cooling.

The first fully automatic press was manufactured in 1898 by W. J. Miller in the U.S.A.

Between the old manual presses and today's fully automatic presses many semi-automatic presses were developed. One of the first improvements consisted in the insertion of compressed air cylinders which were to relieve the presser of the manual work of pressing proper. An interesting construction from the time between the two World Wars mechanically controlled the plunger from a compressed air cylinder, while the presser could exert the final pressure and remove it after pressing. This construction created an intelligent combination of craft and machine potential. The general recognition of such needs lead to the so-called "dwelling device". It was produced early by W. J. Miller's factory in Swissvale, Pennsylvania. He also originated a device which saves compressed air by a "false lid" sitting in the cylinder which could be shifted to alter its stroke. Rotating presses were next developed to accelerate pressing speed. The beginning was the semiautomatic spring press by the glass machine factory of Jean Wolf, in Cologne. The number of molds was increased, and an attempt was also made to prevent overheating of plungers by the use of several plungers arranged on a horizontal axle. At each press act a new, cooled plunger is presented. More information on contemporary presses, their construction and functions is found in Section E 1.

The necessity to cover the mold by a ring to assure the uniform filling of the cavity has been mentioned in the introduction to this section. Various attempts have been made to avoid this ring. For this purpose the gob must be placed centrally, and must have uniform weight and temperature distribution. If mold and plunger are also uniformly heated, ring-free pressing produces particularly beautiful rims: they appear like ground edges.

2.2 Development of Special Processes

Short, thickwalled glass tubes can be made successfully by providing the press mold with a yielding bottom. At the nearly lowest position of the plunger pressure increases and causes the bottom to yield and the glass is cut by the plunger at the inner mold edge. APPERT proposed another press process for making long hollow pressed objects at the end of the 19th century [18]. In his process the plunger penetrates from below, in the place of a bottom. The plunger must be pointed and the interior wall of the mold must be grooved so that the glass is not taken along. To make flat ware pressing can be combined with rolling. A flat low mold is filled with glass, and the glass is spread over the bottom by a roller, then pressed. This process produced plate shaped pressed objects. Similar to APPERT's process is the font mold "press-on" process. The glass fills the mold as usual; next the mold is covered with a lid and the glass is pressed into the mold by the lid which serves as a plunger. To produce cheap "cut-glass" the cut grooves or facets are pre-pressed and then finished by genuine cutting.

3 Pressing and Blowing of Hollow Glass

To make hollow ware by pressing only one must use tapered cavities the maximum diameter on top. The cross section may be round, oblong or cornered, but the plunger must be capable of withdrawal from the glass after pressing. To produce bulged bodies prepressed objects can be afterworked by blowing. In this case the temperature of the prepressed piece must be high enough to keep it malleable. This process permits the manufacture of jugs, vases, and container glasses.

3.1 Development of the Classical Process

The thought of molding glass by pressing and blowing in suitable press-blow machines appeared around 1860. ATTERBURY reportedly made the first attempt to mechanically afterwork pressed glass by blowing in the United States (1873). Since the American patent 139,993 (J. S. and T. B. ATTERBURY, Pittsburgh) claims a jug with a pressed-on handle, it must be assumed that press-blowing simpler hollow ware must have been known before [9, 29]. ATTERBURY's idea is remarkable because he tried to use part of the two-part blow mold as a press mold. The bottom of the press mold which after prepressing is simply drawn down, remains the bottom in both phases: pre-pressing and blowing. The compressed air required for blowing was blown through the plunger after withdrawal of the bottom. It must be assumed that the exit hole in the plunger was closed by a valve during pressing. The Atterbury process used a ring as in the pure pressing process.

Arbogast's press blow process (Fig. C 1a) originated in 1882 and was protected by U.S.A. Patent 260,819. This patent first mentions the

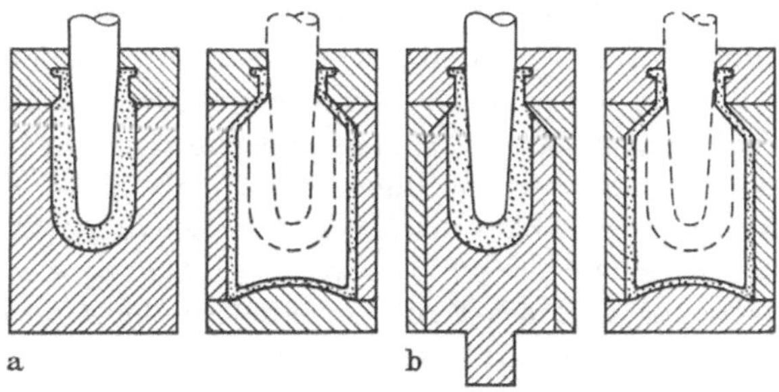

Fig. C 1a and b. The classical press and blow process.
a) Arbogast; b) Windmill.

two-part neck ring mold and the separate blow mold. This process also first made it possible to adapt the shape of the prepressed parison to the blow mold in a manner to achieve good glass distribution during the subsequent blowing operation. The parison or blank fabricated in the press mold is transferred to the blow mold by the neck ring mold and is then blown to the desired shape.

A similar process was developed by the American PYLE as shown by the Austrian Patent 4812. There the glass filled parison mold is lifted against a fixed plunger above it and the glass is pressed to shape.

The Windmill process (Fig. C 1b) also shows a special neck ring shape (British Patent 8526, 1886). It is particularly suited to objects with pronounced shoulders. In this process the press mold is inserted into the blow mold. After the pressing phase the press mold is withdrawn, the blow mold is provided with a bottom and the parison is blown. The Arbogast and Windmill processes are described schematically in Fig. C 1 [5]. Arbogast machines were widely used for some time. Once it was recognized that pressing with subsequent blowing was feasible, the process developed particularly in small glass houses in such a way that existing presses were supplemented by a juxtaposed blow table so that a combination machine was created.

Similar very simple equipment of this kind was developed by HECKER, Wevelinghoven near Cologne, according to German Patent 182,204.

Stimulated by large demand the American development of press blow machines was able to go its own way. The Miller Machine and Mold Works, Columbus, Ohio, for example, built a 5 station press-blow machine driven by compressed air according to the Windmill process. The molds are arranged around a center column. The blow molds sit on the upper part of the two floor table. In the lower floor the parison mold holders are arranged, which are withdrawn after pressing. It is worth noting that the neck ring mold was held closed during pressing by compressed air activated tongs, as they are today in automatic machines.

Development in Germany was slow. SCHILLER can be credited with the introduction of the Windmill type press-blow machine. His main development work, however, was concerned with blow-blow machines which will be discussed later. The glass machine works in Cologne-Brühl constructed Windmill type machines at an early time. The company also built Arbogast press-blow machines under the direction of the designer J. WOLF. The single air cylinders were activated by a central control valve. For manual machines WOLF invented the head mold tongs that opened by manual grip; later SCHILLER built them in similar form [4, 8, 9].

The press-on process with subsequent blowing should be mentioned as a special case. The parison mold is filled with glass and closed with a lid. The plunger is pressed up from below: the parison so pressed is brought to the blow mold for blowing.

Further development in America led to rotating presses with several press molds grouped around a center column and with coordinated blow tables having several molds. Two table machines which are still used were constructed. These machines do not have the two-part head mold rings. It is true that there is a grip rim on the prepressed parison, but this can only be seized after a small lift of the parison from the bottom, after which move it is placed into the adjacent blow mold. This type of machine can work flawlessly only if the parison has assumed enough rigidity to resist the pressure exerted in this lift, but it saves many mechanical aids which were absolutely necessary when two part neck ring molds were used.

A more recent machine is the Hartford Empire Press-Blow Machine, from about 1926 [8, 9]. It must be considered a twin rather than a two-part machine since it consists of two independent units in combination. This becomes evident also from the fact that each table carries both parison and blow molds. The press molds are solid. This is why the machine is advantageous only for slender containers such as milk bottles. At one time the output of the two units was 35 quart bottles per minute.

For this production 16 mold parts were necessary. If one assumes that the bottle then weighed 700 g a contemporary IS four station press blow machine would make about 20 per minute. It should be noted, too, that the press blow process was also used for thin walled tumblers, the heavy glass part at the head end being cracked off. Similarly tumblers are produced by the Hartford M 28 machine.

In the press-blow process one has to take care that the parison expands uniformly during blowing so that thin spots in the finished ware are avoided. This limits applicability. Bottles with conventional neck diameter can not be made, at least not without great difficulty. The process thus is restricted to wide neck containers, and may tolerate milk bottles or ointment jars. On the other hand the process is distinguished by the exactness with which mouths of large diameters and complicated profiles can be formed in the pre-stress stage. This requires that the glass must no longer be malleable after pre-pressing and, particularly, at the onset of blowing. In spite of these inherent limitations an attempt was made to apply the process to narrow-necked bottles. According to the U.S.A. Patent 178,819 (Weber 1876) the press mold was divided in height to facilitate filling. The long, needle-like plunger, because of excessive heating, prevented a very large output.

3.2 Development of Special Processes

The Arnall-Ashley process (British Patent 8677, 1886) has to be regarded as a press-blow process. An upside down part mold, with a neck ring and a plunger inserted from below, was filled with glass. To fill all parts of the mold the glass was pressed down by a punch which covered the whole cross section of the mold.

A proposal made by Weber in 1876 is also of interest. Wendler [8] calls this machine an injection machine. After removing the upper part of the mold and filling it with glass, the counterweighted bottom pressed the glass upwards towards the neck. Next a plunger-like staff is withdrawn, compressed air introduced and, after lowering the bottom, the bottle is blown.

4 Blowing of Hollow Glass

The glass blowers pipe which had been introduced as described above, at the beginning of our era, and which must be recognized as a tool, has not changed in centuries. The invention of the "piston robinet", a manual air pump, by a Baccarat glassmaker in 1821 was a start toward easing the glassmaker's labor. Historically, this tool must be considered a first scouting experiment in the direction of blowing machines [17].

A further development of old tools and techniques towards machines is the use of multi-part molds connected by levers and hinges to permit opening and closure by foot. An example is the old iron bottle mold illustrated in Fig. C 2. The lower section molding the bottle shank is in one part; neck and mouth sections are in two parts. The glass suspended from the pipe is inserted into the open mold from above. The upper part is closed and the gob is blown.

The first attempts to blow bottles and other hollow glass ware with simple machines started around 1850. In England MEIN obtained British Patent 1680 (1859) covering a bottle blowing machine. The

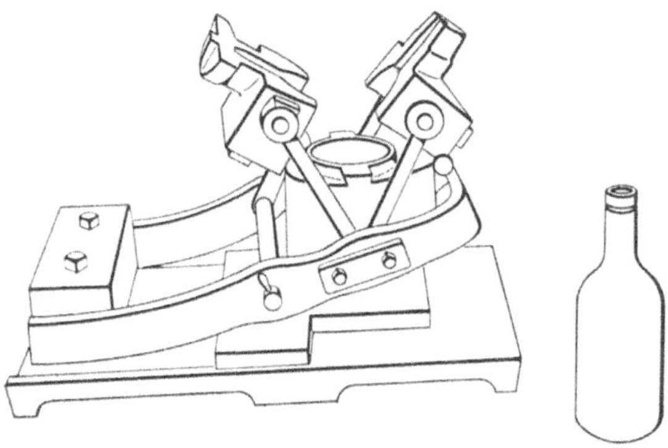

Fig. C 2. An old iron glass bottle mold (1827).

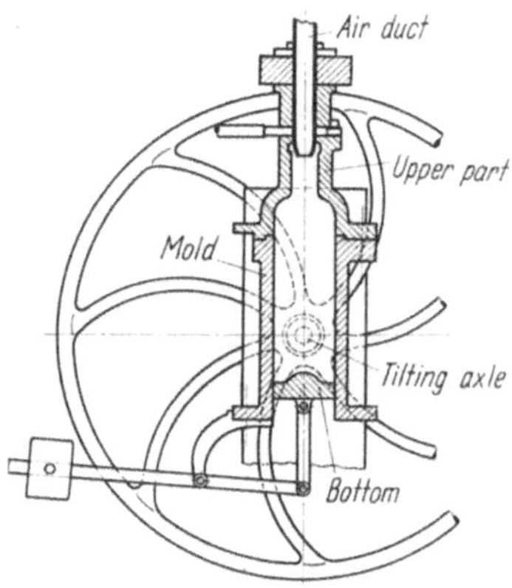

Fig. C 3. Blowing machine after the tipping system following a proposal by BOWRON.

fluid glass was filled into a mold corresponding to the bottle shape. The mold was upside down: the compressed air pipe required for blowing stuck into the mouth like a tip. The liquid glass thus was to fill the bottle head without pressure, then was to cool a little, and finally was to be blown with compressed air. The temperature conditions of the gob in this mold did not permit the manufacture of useable bottles. Equally imperfect was a proposal by KILNER (British Patent 2079, 1860), to fill the fluid glass from above. Further thoughts were developed by BOWRON [British Patent 354 (1861) (Fig. C 3)]. He wanted to fill the glass into an upside-down blow mold, next to reverse the mold and to blow. The conterweighted bottom, however, was not used in the filling of the upside-down mold; it was used to press the glass towards the neck. WEBER, in a British Patent granted in 1876, placed the upright mold into a one part undermold and a two-part upper section, the bottom of the under part being turnable (Fig. C 4). The glass is inserted into the lower part, pressed, by means of the bottom, toward the neck, and finally blown by compressed air while the bottom is withdrawn. The process of ARNALL and ASHLEY (British Patent 8677, 1886) brought a very remarkable improvement. The older Weber process (Fig. C 4) was so-to-say rotated by 180° and the pressing of the glass into the mouth was done by the bottom, which could be moved up and down above it (Fig. 5). During the next two years ASHLEY added those characteristics of a filling machine which are still valid in the most up-to-date equipment. The Ashley process is the transition from the pure press-blow process to the bottle machine proper. ASHLEY first used the parison and blow molds in bottlemaking. The parison mold was open on top, the mouth was below which greatly facilitated (Fig. 6) the insertion of the glass (Brit. Pat. 3434, March 7, 1887) [13—15]. This was ASHLEY's decisive breakthrough to mechanical hollow glass manufacture as it has remained to this day. It is remarkable that ASHLEY was manager of a foundry and ARNALL, who gave him the suggestion for building a bottle machine, a postmaster. The characteristics of the last Ashley machine (British

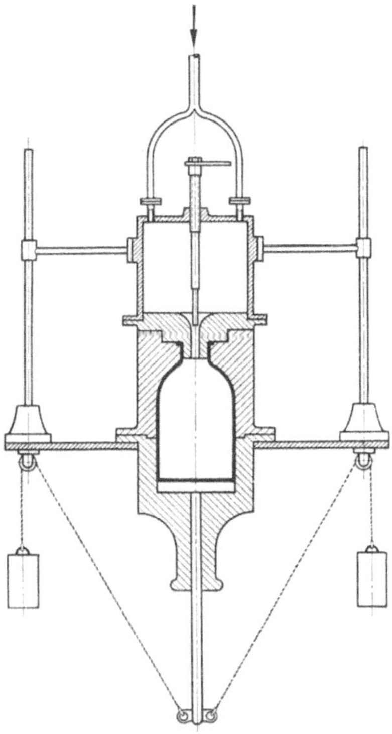

Fig. C 4. Injection mold machine after a proposal by WEBER (1876).

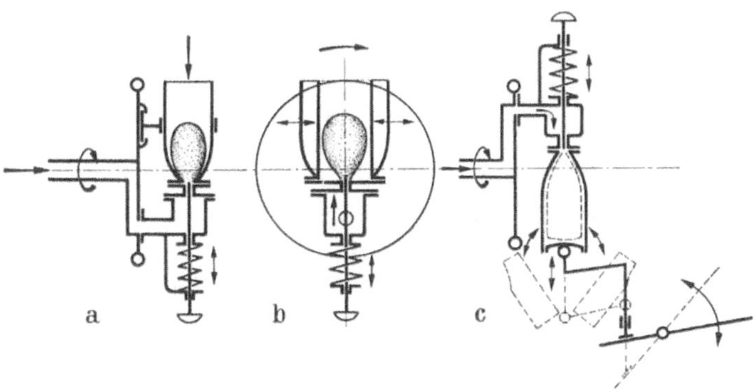

Fig. C 5a—c. Working principle of Ashley machine (1886/87).

Patent 3434, March 3, 1887 and 3673 as well as 11,288, 1889) are (1) separate molds for neck, parison and blow-out, (2) insertion of the glass into the parison mold from above, and (3) blow-out in the upright position. The tip determining the inner diameter of the mouth was pushed in only after filling and was again removed to avoid excessive

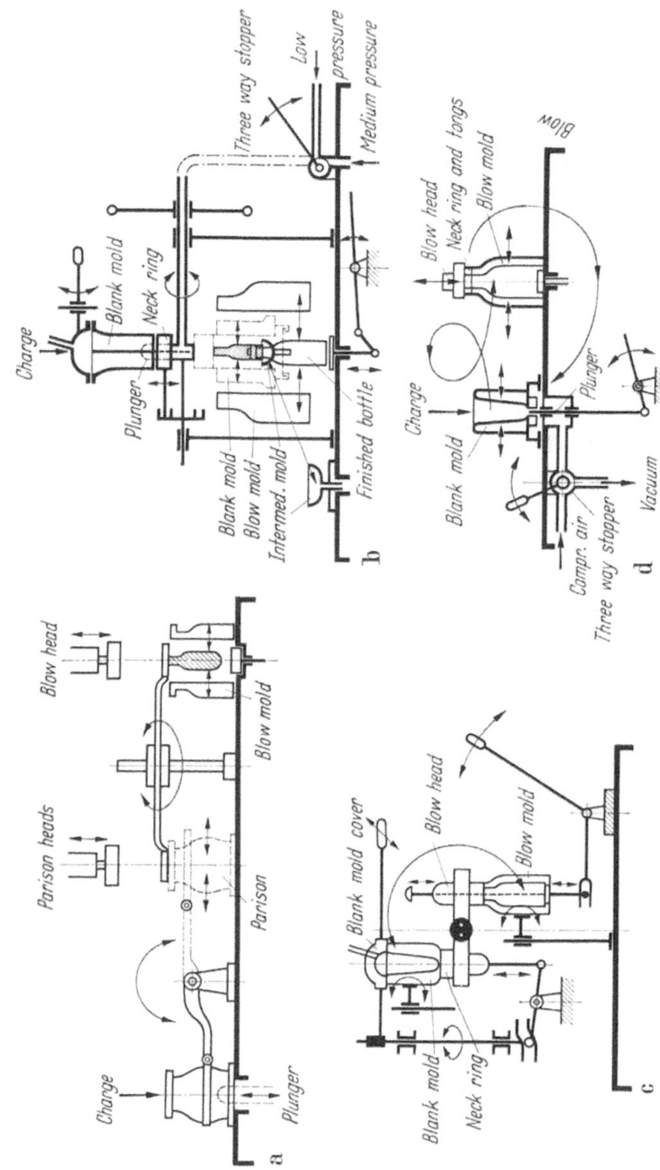

Fig. C 6a–d. Semi-automatic glass blowing machines by HILDE (1891), BOUCHER (1897), SEVERIN (1902) and SCHILLER (1904).

cooling. ASHLEY used a very long plunger. A difficulty lasting through many later constructions consisted in the excessive cooling of the gob in the surface of the parison mold, causing the hot glass to push out in blowing and leading to poor glass distribution. ASHLEY had to contend with this trouble, but for small bottles he succeeded in cooling the bottom part of the gob by impact on a bottom plate. In this he did exactly what a glass maker would do today in the same situation. WENDLER mentioned [20] that it was not necessary to mechanically press the glass into the parison mold, and that the Ashley machine is no longer to be classified as a press-blow machine.

DRALLE [9] reports that in 1892 he saw 22 Ashley machines during a visit to a Siemens tank in Castleford, England.

In 1889 BAUER applied for German Patent 45,912, describing a machine resembling that of WEBER. Nothing is known about its merits in operation.

HEERDT patented a hollow glass machine (German Patent 105,842-22-5-1898). From 1891 HILDE [19] of London, was occupied with the construction of a bottle blowing machine (Fig. 6a) applying ASHLEY's basic principle of parison and finish molds. He cut, with shears, the soft glass coming from the inside of the parison during pre-blowing. (Patents 1891—1906, starting with German Patent 58,961-1-1-1891). According to WENDLER [20] a machine of this construction yielded good bottles in the Dresden Siemens Works, but too much glass was used because of the shearing of the gob. The year 1896 is important because construction of other modern bottle blowing machines started at this time. GROTE in London was granted British Patent 12,392, 1897; and BOUCHER in Cognac was granted French Patent 262,149-7-8-1896, with an addition of 12-9-1897.

GROTE's constructions are all based on the Ashley principle and differ only in the activation mechanisms facilitating manipulation. The last construction by GROTE is noteworthy because of the pneumatic opening and closure of molds. BOUCHER's first construction shows a thick, pot-like blocking plate below the freely suspended gob (Fig. C 6b). In a later construction BOUCHER used a double headed mold. This machine did not succeed in Germany, but it was much used in France and other romance countries. Its output is given as 120 bottls of 700 g weight per hour. SEVERIN of Achern, developed a blowing machine in 1902 (German Patents 136,895, 127,298, 156,964, 161,557-5-6-1903). The main difference was that he opened the parison mold as quickly as possible, left it in place and simply reversed the blank. This avoided the large temperature differences in the gob, and the cooling devices developed by his predecessors could be eliminated. The glass could also be worked cooler and the output was higher. The yield is reported as 900—1150 bottles of 450 g weight in an 8 hour shift. The Severin machine has two molds (Fig. C 6c). One is in filling position, while the other has already turned the blank to the blow mold. DRALLE [9] reports that in 1909 the Severin machine worked in Nienburg, Achern, and Hostomitz, and in France, South America and Russia. For the sake of completeness it should be noted that the double head mold was used before SEVERIN by

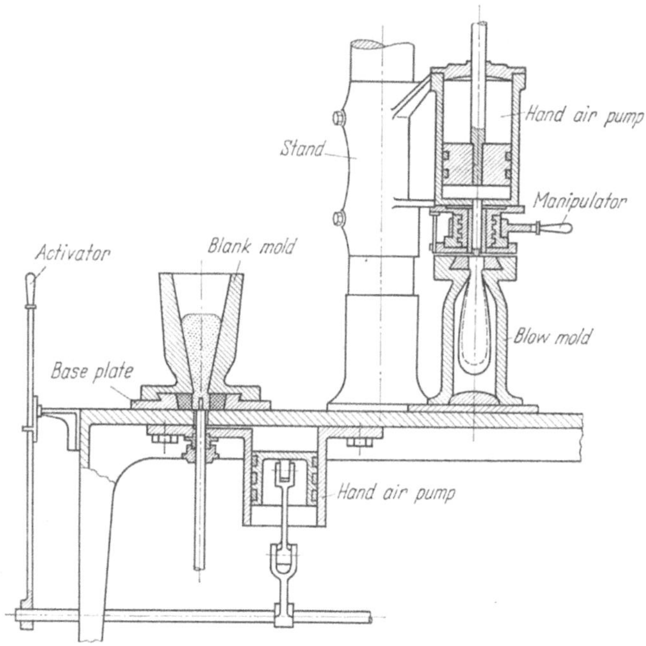

Fig. C 7. Original Schiller machine.

LEISTNER (German Patent 151 628-26-10-1901), but it was improved by SEVERIN. Finally a very simple machine by the firm A. Tourres in Graville-St. Honorine, France should be mentioned. Its operation is similar to the Boucher and Severin machine. It turns the blank, like the Severin machine, but has only one neck mold. It was claimed that it could produce about 180 bottles of 450 g weight per hour.

All machines listed so far invert the blank. SCHILLER, again a pioneer in semiautomatic bottlemaking, placed parison and blow molds next to each other on a table (German Patent 157,520-1-10-1903) (Fig. C 6d). Before SCHILLER, the American PROEGER had patented a similar arrangement (U.S. Patent 744,009) (1902). This construction was simplified and improved by SCHILLER and became the prototype of the Schiller machine (Fig. C 7). These arrangements have the advantage that "presser" and "blower" stand next to each other, at the machine. While the presser blows the parison in the parison mold the blower can blow the bottle in the blow mold. The transfer from parison to blow mold is manual in the Schiller machine: no special mechanization is provided.

All earlier constructions had simply permitted viscous glass to flow into the parison mold or pushed it in through a lid through which compressed air is introduced. SCHILLER, however, vacuum sucked fluid glass into the neck mold. This is probably the first use of air suction in a bottle machine.

In the Schiller machine, after filling, the piston is drawn down in the cylinder; the resulting negative pressure sucks the glass into the mouth section. Then the plunger is pushed downwards from the mouth, the parison is preblown, the parison mold opens, and the parison continues to be pre-blown. Next it is taken out with the head mold tongs, and is finally inserted into the slide above the blow mold. Above the latter a cylinder is mounted, by means of which the blower can blow air into the blow mold by activating a lever. To obtain airtight connections SCHILLER interposed a simple threaded piece which is impressed on parison or finish mold by a small twist. SCHILLER constructed several types of these machines he equipped them with two blow molds, connected them with central compressed air or vaccum ducts, and finally created a universal machine capable of wide neck container production. He always equipped them in a manner to keep them independent of the central air or vacuum setup. In type BI he used the so-called press-on method. Glass filling the parison mold is not only vacuum sucked to the mouth parts, but is also pushed up by a press plunger corresponding with the inner diameter of the wide neck article. It is also possible to preblow in an inverted position after pressing and withdrawal of the press-on mold. This provided perfect glass distribution. A gatherer can normally serve two machines, to which are added one presser and one blower for each machine.

Outputs reported are:

Machine B II 160—180 average bottles per hour
B I 150—170 average bottles or 120—140 one liter jars per hour.

Around 1920 SCHILLER equipped the machines for small bottles with double or triple molds. A gatherer, who had to work fast and skillfully, could bring two or three gobs to the parison molds. SCHILLER gives outputs of 6000 pieces per 8 h for smaller bottles, for triple mold machines.

A construction similar to that of the Schiller machine was developed by J. WOLF in Cologne (German Patent 176,511-7-2-1905) but it was more mechanized. In contrast to the Schiller construction most motions were activated pneumatically. The Wolf machine has three plungers and two blow molds. The fluid glass is sucked into the mouth mold as in

SCHILLER's machine. The blow mold bottom can be inserted from below, prior to blowing, to support the gob. Parison and blow molds can be cooled intensively. The suppliers give outputs of 2400 420—450 g or 2000 600 g bottles per eight hours with a three man shop.

MAINZER (DRP No. 281,306; 292,167; and 211,077, dated March 13, 1901) has tried to use only one set of molds with his fully automatic tilting machine, but it still has to be fed by hand. The machine also goes under the name Koenig machine.

All the machines listed so far can not be called fully automatic because the glass was hand gathered. They also had numerous other deficiencies. It is the great accomplishment of OWENS to have brought about a truly basic change by the development of the first successful and fully automated machine [9, 20, 22, 23, 25—28]. He succeeded in building a rotating machine which supplied itself with glass by suction. The machine rotates continuously and so has all the characteristics of a modern automat. The parison molds are arranged above the blow molds on the turntable and dip several millimeters into the glass bath of the furnace or revolving pot. The Owens machine was thought out so thoroughly and developed so well from the viewpoint of mechanical construction that its appearance started a revolution in the field of hollow glass manufacture. Since the operation of this machine is connected with automatic feeding it will be described more fully in the following Sec. C 5. Further details are found in Sec. E 3.2.1.

SEVERIN developed another interesting type which WENDLER [8, 20] calls an injection mold machine. It is an automatic two table machine with three parison and blow molds. It is fed by a so-called pressure barrel made from fire clay which sits in a small forehearth and by means of compressed air pushes glass into the parison mold. Reportedly, it worked well in Achern, producing 300 bottles of 100 g weight per hour. This output, however, is small compared to the effort involved and it must have been sensitive to disturbances in tank or pressure barrel. It was not widely accepted.

SIEVERT of Deuben, near Dresden, developed a blow process which has nothing in common with any of the usual methods of molding liquid glass (German Patent 118,246). Liquid glass is cast on a flat plate; by means of an iron ring, which carries a lid and in it a connection to compressed air, the hot glass is taken up by its sticking, and then blown into any mold. The advantage of this process is that it allows blowing of containers having large and even very large diameters. Blowing equipment using wet asbestos plates is shown in German Patent 190,365.

Finally SIEVERT successfully applied the process to window glass production. As in the mouth blow process, large roller shaped hollow glass bodies are blown, cut open, and rolled flat (Fig. C 9). In contrast to the manual window glass process, horizontal glory holes were not used, but fire boxes were arranged in the floor in which the glass "roller" is heated and can stretch from its own weight.

The technical development of bottle manufacture in the last century is so enormous and so revolutionary that it may be cited as an excellent example of the mechanization of a field of production. During the centuries in which bottles were made by glass blowers pipes the output remained nearly unchanged. The appearance of the machine, developing from the gradual replacement of manual work by mechanization to a true automat, caused the output to increase enormously in comparison to earlier results (Table C 1). Hours required to properly mold from the liquid glass melt decreased within a bare century from 15 per hundred bottles to 0.1 per hundred, or to 0.7% of the original time. At the same

Table C 1. *Output and labor cost of different glass blowing machines from 1886–1956* [35]

Period	Type	Output		Effort	
		Bot./h/mold	Bot./24 h total	Number of workers	Labor for 100 bot. (h)
1886/87	Ashley	17.5	400	$2^1/_2$	15
1896	Boucher	85	1900	$2^1/_2$	3.1
1902	Severin: hand–machine	120	2700	$2^1/_2$	2.2
1904	Schiller	190	4300	3	1.7
1956	Roirant BM (one mold)	205	4700	$^1/_3$	0.17
	Roirant A 6 (6 mold)	170	23500	$1^1/_3$	0.135
	Owens AR (10 arm)	200	46000	2	0.105
	IS machine (8 station)	350	65000	2	0.075

(Net outputs refer to average drinking bottle, effort refers to labor directly assigned to machine)

time the hourly output of bottles per mold rose from 17.5 to 350, i. e., 20 fold. Today's multiple mold machines produce a multiple of these numbers operating with 2, 3, 4, 6, 7, 8, or 10 arms. The total output of a machine rose in this time from the oldest one mold machine, producing 400 bottles per 24 h, to 65000 bottles per 24 h for a modern 8–10 arm automat.

5 Feeding

About one decade before the turn of the century, that is, soon after the invention of the first bottle blowing machine, designers were occupied by the problem of supplying existing bottle blowing machines with liquid glass. Ashley as well as the Frenchman Vernay attempted to accomplish the process of gathering with the still existing gathering iron by some mechanized means. Experiments with ladle-like equipment were also made. Croskey and Locke obtained American patent 575,214 for a suction ladle in 1896. But all these experiments failed in the end and only the method of flow and gob feeding brought complete success.

The development of automatic feeding started as early as 1880. It is characterized by the following steps:

Flow feeding with shear mechanism	Proeger and Drey
Retardation by gas stream,	
Withdrawal of the cut glass stream by suction into the furnace interior	McCauley,
Feeder with plunger, which closes the flow opening after delivery of the gob,	
Glass stream conducted by a lateral joint after cutting,	Graham, Wadsworth,
Feeder in which the gob is ejected by a paddle	Homer Brooke (paddle-needle),
Feeder with continuously reciprocating motion	Peiler.

Into this period also fall the first experiments by Owens, not only to take the fluid glass from the tanks, but also to carry out the first manufacturing step at the same time. By means of a suction head which is brought under a vacuum with an affixed hand pump, fluid glass was

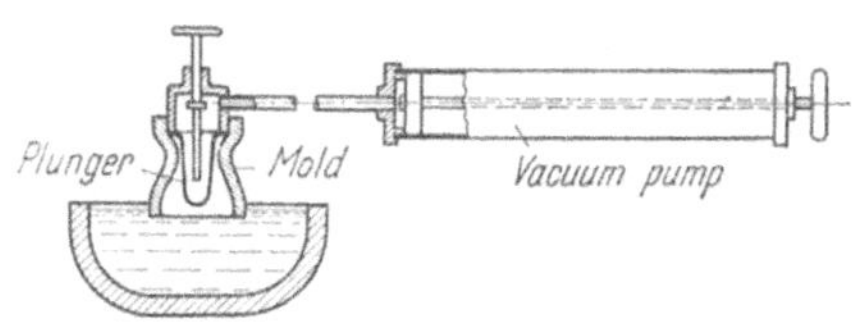

Fig. C 8
Original vacuum-blow process after Owens.

to be taken out of the tank or a large trough (Fig. C 8). Owens next imagined the blow process in such a manner that the gob formed by suciton, which already had a mouthpiece with opening, was to be blown to a bottle by blowing it into the lower part of a bottle mold. Several years later (1905) his collaborator Bock received U.S. Patent 870,664 for the basic principle of the first Owens machine with revolving pot.

Before applying the revolving pot Owens and collaborators had also tried to suck glass from the cold end of the tank. But it turned out, and was repeatedly shown in later trials, that fluid glass cools too fast at this point, which remains always the same temperature, and so the glass becomes unuseable. Through the use of the revolving pot, glass that had touched the lower rim of the suction mold, as well as the cut remainder that falls back into the pot, have sufficient time to remelt and to become "reworkable".

6 Fabrication of Sheet or Flat Glass

The rolled flat glass processes, particularly those for producing plate glass, were improved during the second half of the 19th century. In the old process traceable to Lucas de Nerou there was a rolling table which could be taken to the single box furnace. During casting and rolling the table stands still and the roller runs on rails, thus permitting adjustment of the glass thickness. The rolled plate is then shoved into the adjacent box kiln.

A further development is represented by the arrangements of the American Cruikshank who replaced the kiln by a lehr and worked with fixed tables which could be made correspondingly larger.

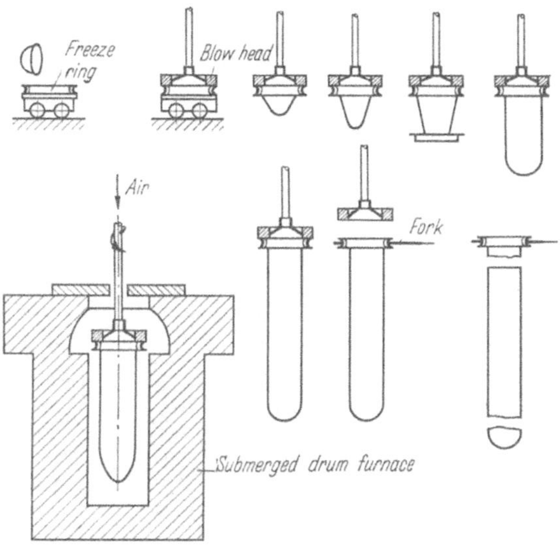

Fig. C 9. Blowing of window, glass cylinders after Sievert.

A special solution was Bicheroux' development of a rolling arrangement for glass masses up to 2000 kg. Bicheroux worked with two fixed rollers and a large table that moved below them. To control the large (up to 19 m long) glass sheet rolled from such a large mass of glass, it was cut in the rolling process by two large knives so that the single lighter sheets could be manipulated with ease.

In all rolled flat glass processes described so far the molten glass is taken out from either tank or pot with a large ladle and cast on the table or some flat surface in the rolling plant.

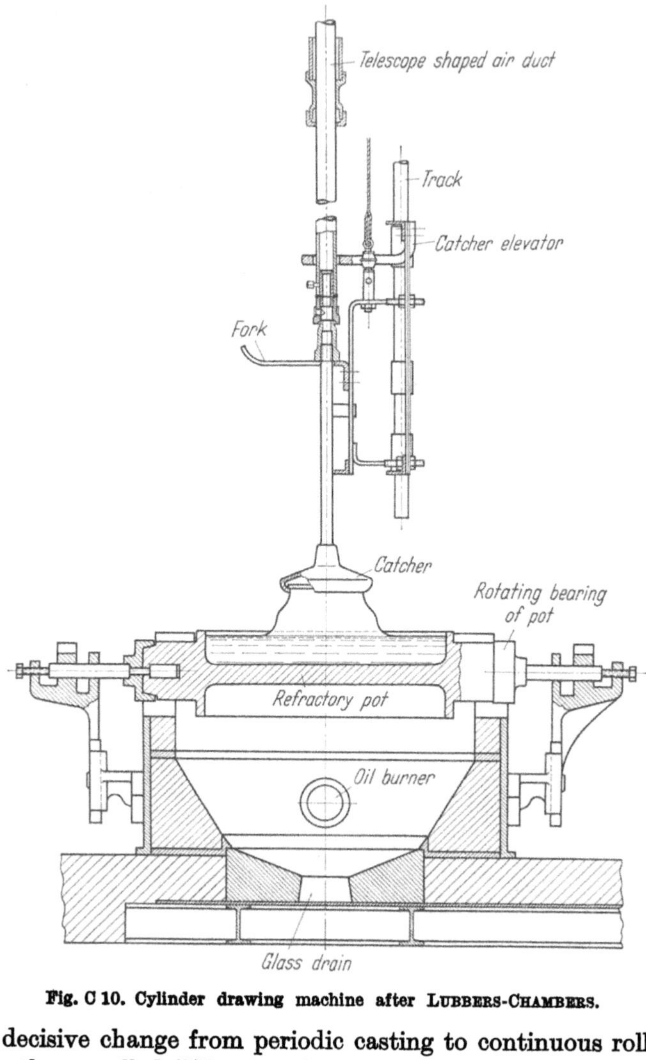

Fig. C 10. Cylinder drawing machine after LUBBERS-CHAMBERS.

The decisive change from periodic casting to continuous rolling was made by the so-called "Kontinue" process in which glass in a uniform stream flows from the tank toward fixed counter-rotating rollers. Today almost all plate glass is fabricated by this process.

Window glass developed entirely differently. Here one first attempted the mechanical imitation of blowing, then opening and stretching large glass cylinders. SIEVERT developed the process illustrated schematically in Fig. C 9. The glass is cast into a ring where it freezes in the outer zone. By means of a blow head a cylinder hanging downwards is blown in a way resembling mouth blowing.

A variation of this idea is the Lubbers-Chambers process of drawing the cylinder upwards. The glass cylinder is drawn from a container with a collar, as in SIEVERT's case, but upwards. The equipment is illustrated in Fig. C 10. A flat rotating fire clay bowl was used. Below the bowl was a heating chamber to melt glass left over after drawing. This system made cylinders 6—8 m long with a diameter of 60 cm. The process gained some importance before the First World War, particularly in the United States. After the War it was replaced by the various drawing methods which had been developed in the meantime and were more productive. They are closely associated with the names FOURCAULT and COLBURN. Details on drawing processes are found in Sec. E 5.

References

[1] LYLE, A. K.: After nine thousand years, glassmaking approaches fully automatic production. Ind. Gas 30 (1951) No. 5, pp. 12—14, 30—33; No. 6, pp. 11—13, 24/25.

[2] Schrifttumsübersicht über das Gebiet der Maschinen zur Herstellung von Hohlglas. [Abstract: Glastechn. Ber. 27 (1954) 179.]

[3] GÜNTHER, R.: Entwicklungsfragen der Hohlglasverarbeitungsmaschinen. Glastechn. Ber. 22 (1948/49) 38/39.

[4] RIEDEL, L.: Maschineneinsatz in der Glasindustrie. Z. VDI 91 (1949) 443 to 451.

[5] K. B.: Verdrängung der Handarbeit durch Maschinen in der amerikanischen Glasindustrie. Sprechsaal 60 (1927) 443—447.

[6] Anon.: Die neueste Entwicklung vollautomatischer Glasblasmaschinen in Deutschland. Sprechsaal 71 (1938) 22—24.

[7] HOWARD, G. E.: Mechanization of the glass industry. Glass Ind. 31 (1950) 75—80, 104/05, 131—134, 150, 183—190, 214, 216, 218.

[8] WENDLER, A.: Maschinelle Glasverarbeitung, Leipzig: Akad. Verlagsges. 1929.

[9] DRALLE, R.: Die Glasfabrikation, 2 vols., 1st Ed. 1911. — DRALLE, R., and G. KEPPELER: 2nd Ed., Vol. 1, 1926; Vol. 2, 1931.

[10] STEIN, G.: Glas ist schon uralt. Geschichtszahlen aus dem Glasgebiet. Glashütte (Düsseldorf-Gerresheim) 8 (1958) No. 2, pp. 3—7; No. 3, pp. 6—10.

[11] HODKIN, F. W.: Introduction to the Symposium on machinery for the fabrication of glass containers. J. Soc. Glass Technol. 37 (1953) 27—31 T.

[12] MEIGH, E.: The automatic glass bottle machine. Glass Technol. 1 (1960) 25—50.

[13] ENGLISH, S.: The Ashley bottle machine. J. Soc. Glass Technol. 7 (1923) 324—334. (A historical consideration.)

[14] SIEMENS, F.: Zu Ashleys automatischem Flaschenapparat. Sprechsaal 21 (1888) 165/66.

[15] A. S.: Ashleys automatischer Flaschenapparat. Sprechsaal 21 (1888) 203/04.

[16] BENRATH, H. E.: Mechanische Vorrichtungen zum Glasblasen. Sprechsaal 16 (1883) 354.

[17] BENRATH, H. E.: Die mechanische Glasbläserei zu Clichy. Sprechsaal 16 (1883) 464.
[18] BENRATH, H. E.: Über das Blasen der verschiedenen Glasarten mittelst comprimierter Luft auf dem Appertschen Hüttenwerk Clichy. Sprechsaal 17 (1884) 263/64, 277/78.
[19] REIBOLD, M. VON: Die Glasblasemaschinen von H. M. Ashley und H. Hilde. Sprechsaal 30 (1897) 844—846.
[20] WENDLER, A.: Die Entwicklung der Glasblasemaschine. Sprechsaal 36 (1903) 553—557, 591—593. [Reprinted from Dinglers polytechn. J. 318 (1903).]
[21] NÄBE, F.: Die Verwendung der pneumatischen Glasmacherpfeife zur Herstellung von Ballons. Reichsarbeitsbl. Part III (1943) No. 20, pp. 224—228.
[22] STEIN, G.: Die Erfindung der automatischen Flaschenblasmaschine von Owens um die Jahrhundertwende. Glastechn. Ber. 27 (1954) 15—17.
[23] SEIMETZ, K.: Aus der Zeit der ersten Anfänge im Owens-Betrieb. Sprechsaal 85 (1952) 579/80.
[24] BENRATH, H. E.: Die Glasfabrikation, Braunschweig: Vieweg 1875.
[25] SCOVILLE, W.: Revolution in glass making, Cambridge: Harvard University Press 1948.
[26] Anon.: Michael Joseph Owens. Biography. Bull. Amer. Ceram. Soc. 17 (1938) 148.
[27] MUSCHALEK, L.: Owensanlage mit umlaufender Schmelzwanne. Glastechn. Ber. 24 (1951) 16/17.
[28] WAGERS, A. F.: Die Owens-Flaschen-Maschinen. Glass Ind. 2 (1923) 104/05.
[29] WENDLER, A.: 60 Jahre maschineller Glasverarbeitung. Sprechsaal 60 (1927) 1025.
[30] Anon.: K. E. Peiler. Biography. Amer. Ceram. Soc. Bull. 19 (1940) 105/06.
[31] HABEREY, W.: Der Werkstoff Glas im Altertum. Glastechn. Ber. 30 (1957) 503—514; 31 (1958) 188—194.
[32] KISA, A.: Die Erfindung des Glasblasens. Jb. Altertumskde. (Wien) 1 (1907) 1—24.
[33] SCHULZ, H.: Geschichte der Glaserzeugung, Leipzig: Akad. Verlagsges. 1928.
[34] VÖLCKERS, O.: Glas und Fenster, Berlin: Bauwelt-Verl. 1928.
[35] GIEGERICH, W.: Ergebnisse und Probleme der maschinellen Flaschenfertigung. Glastechn. Ber. 30 (1957) 299—308.

D. Preparation of Glass for Forming

1 Apparatus for the Preparation of Glass for Forming

By W. GIEGERICH, Heilbronn

The withdrawal of a certain amount of glass from a pot or tank is aimed at supplying the exact amount corresponding to the weight of the object to be formed in the machine. The following discussion is limited to the supply for hollow ware and bottles from tanks. Analogous procedures would be valid for pots. At the present time glass is withdrawn according to three methods:

(a) by wrapping viscous glass on a gathering iron and transporting it manually to the forming machine. The gathering iron is held over the parison (blank) mold in a way to allow a gob of the desired size and shape to drop into the mold. By means of shears the gob is separated from the gathering iron.

(b) Through an opening in the bottom of a forehearth, a canal or a trough, a measured amount of glass is pushed out of the bath by a reciprocating plunger and cut. The drop (gob) falls under gravity into the mold which is located directly underneath.

(c) The parison mold of the machine dips into the fluid glass of the revolving pot in the forehearth, fixed to the working end of the tank, and the glass is sucked in by evacuation from above.

It is absolutely necessary to understand the motion of the glass in order to obtain best quality. Fig. D 1 shows these processes schematically. The first observation should cover the thermal condition of the

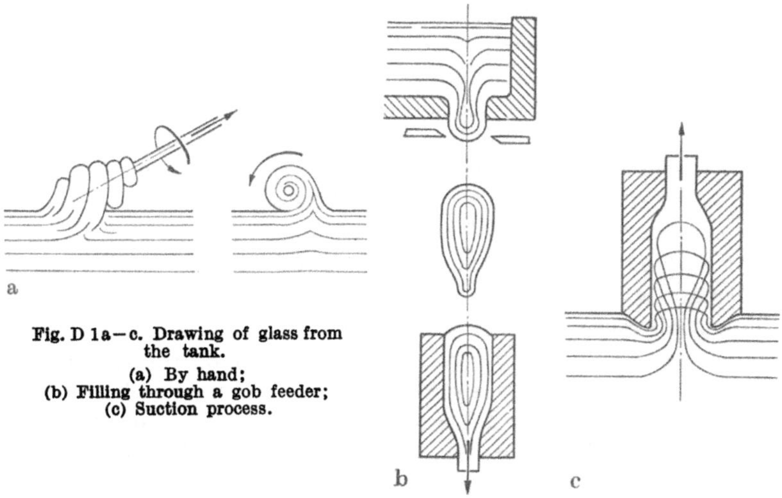

Fig. D 1a–c. Drawing of glass from the tank.
(a) By hand;
(b) Filling through a gob feeder;
(c) Suction process.

glass at the point of withdrawal from the tank, considering both time and spatial changes. Assuming constant and uniform homogeneity which is by no means assured — the picture shows that *hand gathering* (a) requires concentric isotherms around the gathering iron. If the glass weight is large, it is not sufficient to simply turn the gathering iron but it must also move longitudinally. This, in turn, causes temperature differences between the glass taken up first and last. On the way from the gathering place to the parison mold the thermal inhomogeneities equalize more or less while the iron is turned continuously. The situation is quite different in the case of the *feeder* (b). The gob is by no means thermally homogeneous; towards the older glass left from the remainder of the preceding gob, new hot glass flows rapidly, supported by the plunger. But here, too, a reasonably satisfactory equalization occurs on the way to the parison mold. While in hand gathering and gob feeding the temperature gradient from the outside to the inside is relatively small, different conditions prevail in the suction process (c). Fig. D 1 shows the complexity of the flow in this case and how the temperature gradients vary along the longitudinal axis. These introductory remarks do not pretend to be a detailed discussion of gathering and filling phenomena. They are made to stimulate further observation and to introduce the two specialized papers which follow.

2 Revolving Pots

By W. Trier, Frankfurt/Main

The vacuum machine gathers the required glass by dipping the parison mold into the fluid bath under simultaneous suction from the free glass surface. A prerequisite for perfect operation of the machine is thermal homogeneity to a high degree in the gathered glass. Unfortunately the glass is cooled considerably because of the dipping of the relatively cool parison mold into the glass and because of the cool strand of glass formed when the mold is lifted. Therefore the next mold could not be allowed to take up glass at the same place. This peculiarity of the suction machine is the reason that machines having high outputs cannot withdraw the glass directly from the working end of the tank, but demand special intermediary chambers.

For small machines with just one set of molds (Roirant B or F) a ceramic boat at the suction end suffices. For small rotating machines with several sets of molds a stirring mechanism was used successfully. It had the shape of a round disk which is pushed below the glass surface and causes fresh glass to stream to the point of withdrawal continually.

Large machines require circular tanks filled with glass which rotate slowly. These are the so-called revolving pots. They continually offer the parison molds a fresh glass surface to contact. The glass regions that have been cooled by the dipping mold are reheated by the heating of the revolving pot and the surrounding fresh glass, so that after some revolutions the same place again contains glass with a high degree of thermal homogeneity.

2.1 Revolving Pot of the Owens Machine

In the Owens machine revolving pots having a diameter of 2 to 2.5 m and a glass depth of about 25 cm are used (Fig. D 2). The glass flows in a uniform stream from the working end through a trough to the revolving pot. The open cross section of the trough can be regulated by a refractory damper and so become adjusted to the intake of the machine. Trough and damper are expendable parts which must be replaced after a certain time of operation. Usually two dampers are used in series to allow changing one at a time without too much disturbance. The pot itself is lined with refractory material and rests on a platform made of iron girders. At the edges special rim blocks are used to make sure that a ceramic container of desired depth is obtained. To facilitate skimming

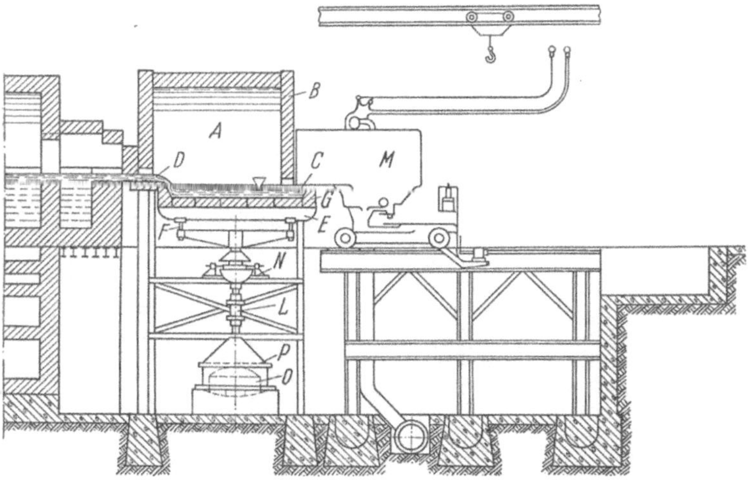

Fig. D 2. Structure of a revolving pot for an Owens machine.
A Heating chamber; *B* Head wall; *C* Bath; *D* Channel; *E* Platform; *F* Set screws; *G* Revolving pot; *L* Vertical shaft; *M* Machine; *N* Laterally adjustable bearings; *O* Pressure bearing; *P* Screw.

and draining some revolving pots possess a central drain tube. The platform rests on the several setting screws of a cross. The screws allow adjustment to a perfect horizontal position. The cross rests on a strong vertical axle which ends in a corresponding bearing. The axle is about 2.5 m long so that the bearing is out of reach of the heat of tank and pot.

The neck journal bearing immediately below the cross is constructed so that it can be shifted radially and can be adjusted with precision by horizontal setting screws. In this way the shaft can have a bump free vertical run. The shaft is driven by a variable speed motor. The drive is placed near the bearing. Because of the slow rotation of the pot worm gears are particularly advantageous. The revolutions of the pot are adjusted in a way that at the suction area of the parison mold the peripheral speed of tank and mold coincide. This usually leads to a speed of 2 to 3 rpm. The direction is opposite to that of the machine. The entire foundation of the revolving pot with bearings and drive rests in a strong steel frame.

Revolving pots of this size are always provided with heating independent of the melting tank. By suitable insulating construction the area above the revolving pot is separated from the surroundings. At the side are found a burner, and next to it, an exhaust (Fig. D 3). The flame streams over the revolving pot in U or double-U shape. There is almost always recuperation. The regenerative principle of earlier construction has been abandoned entirely. Producer gas, piped gas or oil serve as fuel. Fig. D 3 shows the heating of a revolving pot with produ-

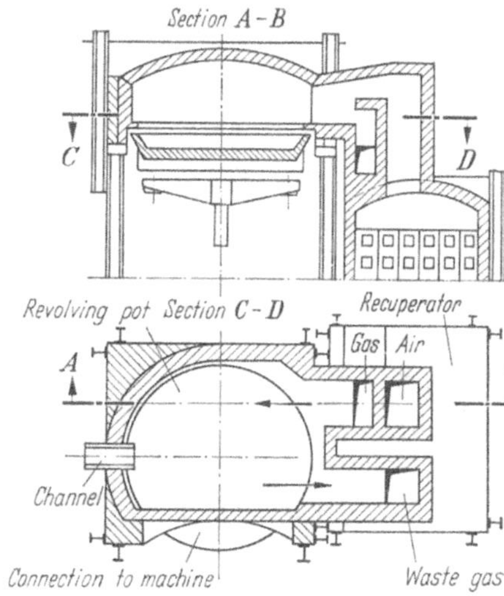

Fig. D 3. Schematic of revolving pot firing.

cer gas and refractory recuperation. Depending on local conditions recuperators are constructed in various ways. Metal recuperators are also used. While no difficulties are expected in the heating of revolving pots with piped gas, caution is advisable when oil is used. The sharp, hot oil flame is really not suited for revolving pots. One has to strive for a soft gas-like flame. This can be done quite easily with middle weight oil; but it is only during recent years that one has been able to construct practical burners for heavy oil. Proven burners split the oil stream into many small streams which facilitates the atomization of oil at the relatively low temperatures of revolving pots. One has to see that no coke particles impinge on the glass bath as they might cause bubble formation.

Various reports exist concerning the heat consumption of large revolving pots. In general one has to figure on at least $10-20 \times 10^6$ kcal/24 h. If the machine takes up much glass, heat consumption falls; if little glass is taken, heat consumption rises; this is because part of the heat comes from the hot glass itself. The relatively high heat consumption of revolving pots is conditioned by the radiation losses of the large, open glass surface outside the heating area and by the inevitable gap between pot and superstructure. For this reason attention should be given to pressure conditions above the bath.

2.2 Revolving Pot of the Roirant A 6 Machine

The revolving pot of the Roirant A 6 machine differs from that of the Owens machine and is therefore discussed in this special section. ROIRANT was quite familiar with the disadvantages of large revolving pots and tried to diminish them by developing a much smaller pot. His revolving pot has a diameter of only about 1.6 m. To warrant the reheating of the glass cooled locally at the suction place even in so small a revolving pot he abandoned the type of revolving pot that operates continuously and independently of the machine like that used by OWENS. He synchronized pot and machine. The motion of the revolving pot was

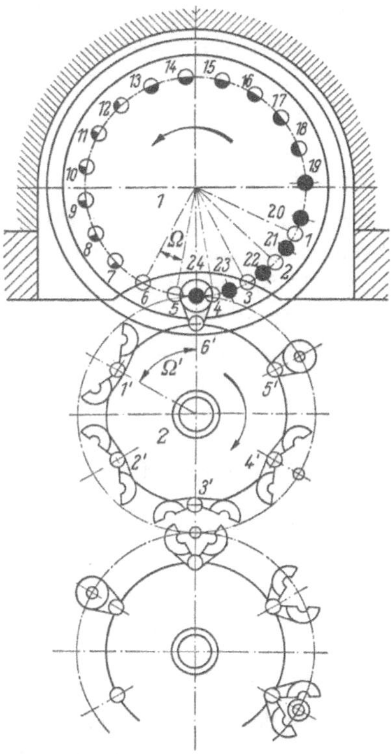

Fig. D 4. Revolving pot of the six arm Roirant-A6-machine. Schematic representation of glass delivery.

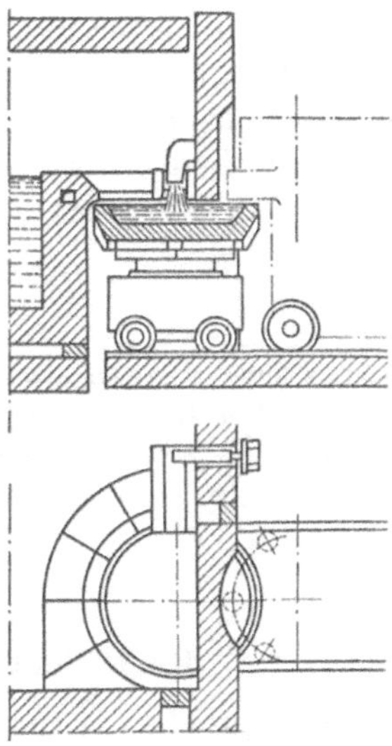

Fig. D 5. Small revolving pot built into working end of tank.

adjusted so that, in theory, the parison mold hits the same place only after three revolutions (Fig. D 4). Since the A-6 machine runs intermittently, and the parison mold stops during suction, the revolving pot was also made to move intermittently. However, one also finds revolving pots running continuously with intermittently running A 6 machines. While the machine moves over an angle of 60° from station to station (6 stations), the corresponding angle of the revolving pot is only about 15°. The next dipping place is no longer influenced by the preceding one in such an arrangement.

The basic idea of ROIRANT was to make the revolving pot so small that it could be included in the working end without requiring separate heating. In principle, this is possible, but soon several disadvantages became evident. In particular, the resistance of wall corners was insufficient (Fig. D 5) and variations of the tank temperature and pressure had unfavorable effects. This is why this construction was abandoned and a revolving pot similar to that of OWENS was placed in a separate heating chamber independent of the tank. In this form it became successful. Because it is smaller than the Owens revolving pot, the heat consumption is smaller (about $5-10 \times 10^6$ kcal/24 h). The drive has to be protected from overheating by air cooling. During shutdown the revolving pot can be shifted to an auxillary gear.

2.3 Turntable of the Pöting AM 6 Machine

To avoid the difficulties in the operation of a revolving pot PÖTING developed for his suction machine a stirring mechanism in the form of a turntable. This construction is the reverse of the conventional revolving pot but should work in principle. The stirring mechanism is discussed in Sec. E 3 which deals with the Pöting AM 6 machine.

References

[1] DRALLE-KEPPELER: Die Glasfabrikation, 2nd Ed., München and Berlin: R. Oldenbourg Vol. I, p. 653, and Vol. II, p. 1128.
[2] SEVERIN, H.: Die Entwicklung der Roirant-Maschine A 6. Glastechn. Ber. 20 (1942) 65.

3 Feeders

By E. SCHULZE and H. GANSS, Essen

A special problem in the development of hollow glass machines was presented by the delivery of glass to the machine. Manual feeding by gathering iron can be considered only for small production. It was soon recognized that the development of a productive hollow glass machine was closely associated with that of an effective feeding system. OWENS solved the problem by means of the suction principle in connection with the use of the revolving pot. The development of the glass feeder which frees the production machine proper from the task of taking the glass from the tank was a parallel development. In this way the machine became better adapted to its proper mission, the forming of glass into the desired object.

3.1 Development and Operation of Glass Feeders

Attempts to tap the glass directly from a furnace opening were tried at first but they failed, because one tried to tap it like beer from a barrel. The glass stream running into the mold started to curl in the mold and became thermally inhomogeneous, which impaired further working. A decisive advance was attained only with the gob feeder. The glass does not flow in a continuous stream, but suitable dimensions of an orifice and periodic shearing of the glass as it comes out cause the formation of drops or gobs which enter the machine in a set rhythm. As the principle of the gob feeder proved very suitable, further development centered on shaping the gob, on the control of weight and temperature, and on the gobbing rhythm.

A reciprocating plunger in the glass over the orifice was successful in controlling the flow of the viscous glass in the desired rhythm to form gobs synchronized with a shear mechanism. The operation of such a feeder is illustrated schematically in Fig. D 6.

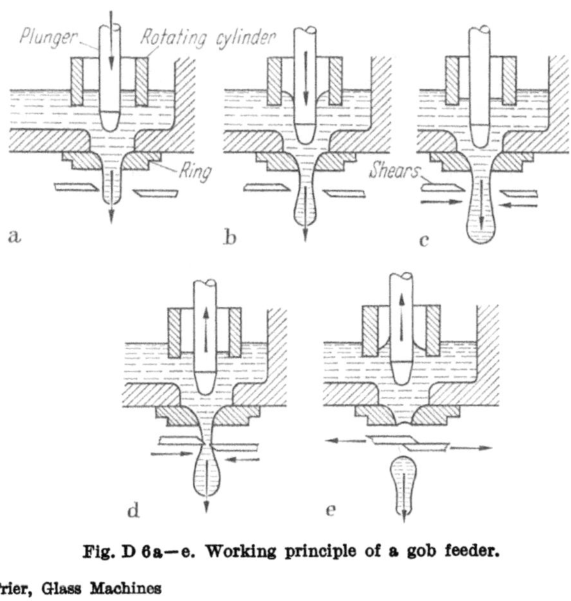

Fig. D 6a–e. Working principle of a gob feeder.

The bowl-shaped cavity in which the plunger tip reciprocates, without rising above it, hinders the lateral escape of glass during the plunger motion. To control the flow of glass even better some feeder constructions use a refractory cylinder above the opening and concentric with the plunger. Its lower rim is a certain distance from the bottom. A ring shaped opening thus is formed through which the molten glass flows to the plunger and to the cup in which the plunger tip reciprocates. The glass between the cup wall and the plunger on one side, and the cylinder and the plunger on the other side, is a little more viscous than the glass that comes fresh through the ring opening. This cooler viscous glass and the motion of the plunger exert suction and pressure effects on the fresh glass from the ring opening. These effects cancel each other so that the total flow of glass through the orifice is not much affected by the motion of the plunger. Suitable shaping of the plunger tip, suitable dimensioning of its cross section in relation to the opening, proper choice of the distance of the shears from the opening, and good timing adjustment of plunger and shear motions, permit control of the shape and weight of the gob. A feeder possesses a great many adjustments; its operation demands great experience.

The connection between the furnace and the feeder proper is the forehearth, a trough-shaped structure with separate heating equipment. The correct temperature in the feeder is extremely important. The following sections will return to this point. In general the term "feeder" includes the connecting trough (forehearth or feeder trough) between the working end of the furnace and the feeder proper, with its gob forming mechanism. Sometimes one distinguishes between the forehearth and the feeder head alone. In the following discussions feeder refers to forehearth and feeder proper, taken together.

To improve the local temperature constancy of the glass in the feeder the plunger of some feeders is rotated around its longitudinal axis while it reciprocates. It thus tends to wrap the glass in the cup around its tip, just like the gathering iron in manual working. If the glass in the spout is not uniform in temperature, inhomogeneities thus would be arranged concentrically and become relatively harmless. This rotation is an additional important feature of the feeder operation.

This same aim of homogeneization is attained by the refractory cylinder around the plunger, best of all by the rotation of this cylinder.

3.2 Construction and Function of Feeders

In the following, various feeders, particularly those of the Emhart Mfg. Co., Hartford, USA, are described. The Emhart feeders are described more extensively as they are most in use. A general survey of the construction is given in Fig. D 7.

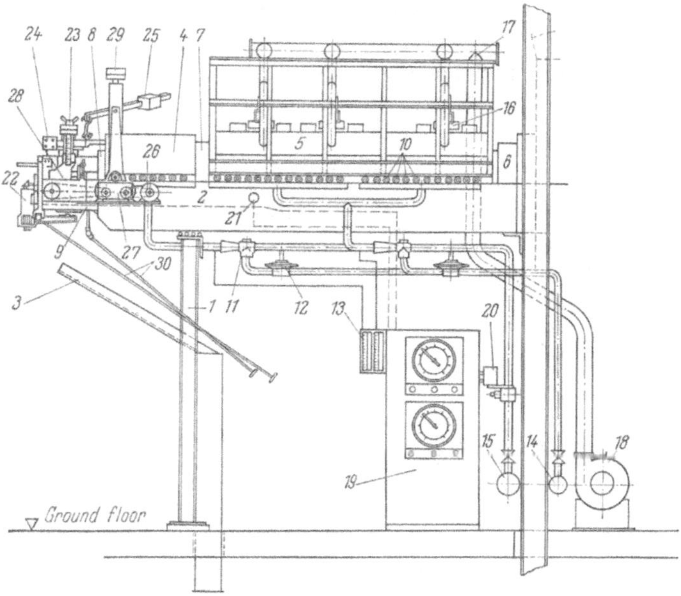

Fig. D 7. Complete sketch of a modern glass feeder [6]. (Forehearth type KW, feeder port HE 144 of the Emhart Mfg. Co.).

1 Channel support; 2 Steel mantle; 3 Drain channel; 4 Equilibrization zone; 5 Cooling zone; 6 Skimmer; 7 Back block; 8 Cover block; 9 Feeder head; 10 Burner; 11 Injector; 12 Zero control; 13 Mix pressure gauge; 14 Gas duct; 15 Air duct; 16 Throttle; 17 Piping; 18 Blower; 19 Control box; 20 Adjustment valve; 21 Thermocouple; 22 Shears; 23 Plunger holder; 24 Refractory tube holder; 25 Counter weight; 26 Motor; 27 Drive; 28 Plunger and shear mechanism; 29 Drill oil container; 30 Remote control.

3.2.1 Iron Frame. Feeder Layout

For stability the feeder channels are placed into iron frames as casings. The frame is assembled according to the tinker toy principle and can be short or long as needed. The frame is anchored on the working end of the tank and is supported by one or two girders. At its other end the cast iron spout casing is bolted on. The indented forehearth casings are held together by bolts.

The feeders are generally in a fanlike arrangement at the working end (Fig. D 8). The length of the feeders depends upon the rate of glass withdrawal. It varies between 2.5 and 8 m. The overall width of the flow channels is about 1.2 m, the depth about (0.5 m). Refractory insulating bricks are inserted between the troughs and the casing.

3.2.2 Structure of Feeder Flow Channels (Emhart Forehearth Types *K* and *Z*)

The Hartford Flow Channels or Forehearths are refractory channels heated by gas or oil, in the superstructure of which are cooling and heating systems. The forehearth consists of two zones, the cooling zone and the equalizing zone at the feeder end. The designation *K* or *Z* indicates the type of the heating (gas or oil) which also affects the design of superstructure and cooling system (Fig. D 9).

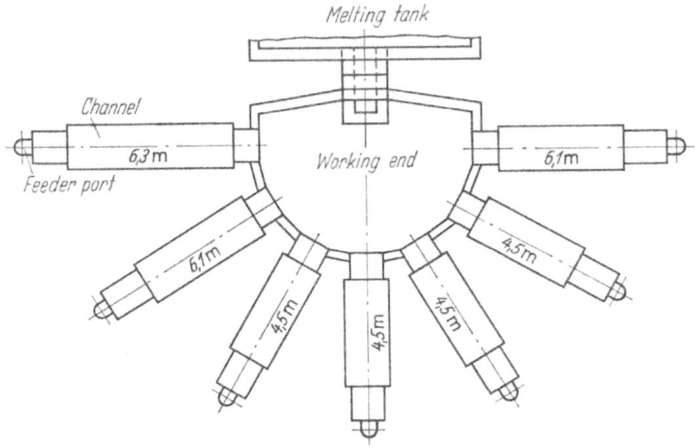

Fig. D 8. Arrangement of 7 feeders on the a working end. Lengths of feeders differ.

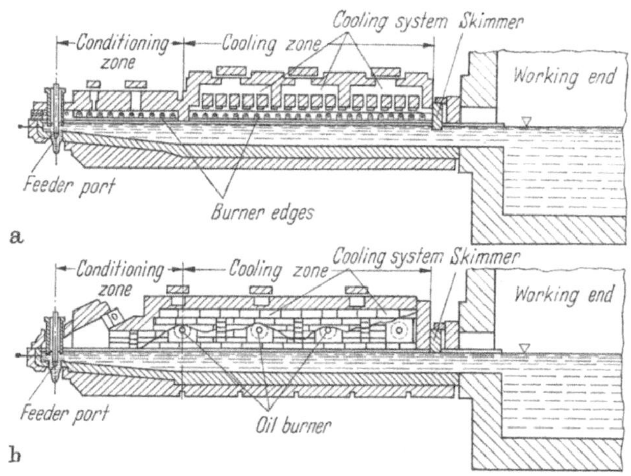

Fig. D 9a and b. Section of feeder [6].
a) Type K; b) Type Z.

3.2.2.1 Cooling Zone. In this zone the glass which leaves the working end at a temperature of about 1350 °C is brought to the desired working temperature (gob temperature) by blowing cool air, if the feeder carries a heavy load (fast flow of the glass). If the load is light (slow glass flow) heat has to be added to this zone since it loses heat on the way to the feeder. This supply or withdrawal of heat is necessary since different items call for different gob temperatures while the working end always delivers glass at the same temperature. At the transition from the working end to the feeder the skimmer is placed. This refractory block dips about 2 cm into the glass bath across the width of the channel and serves to separate the feeder from the working end.

3.2.2.2 Equalization Zone. This zone has to equalize possible temperature gradients in the glass bath and supply glass of as uniform temperature as possible to the feeder. The absence of strong temperature

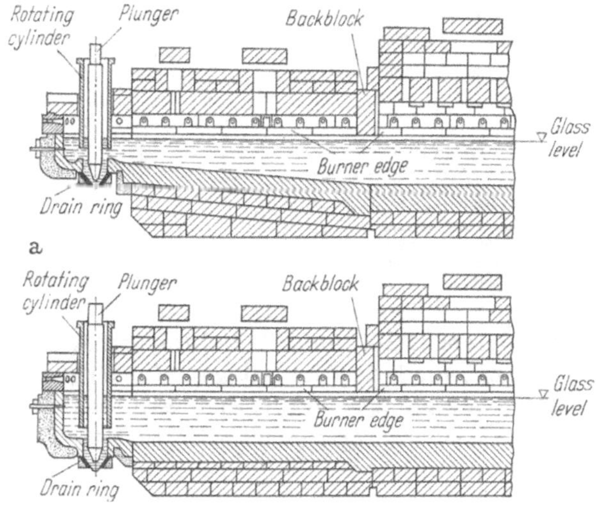

Fig. D 10a and b. The feeder port.
a) Shallow; b) Deep.

gradients is very important to avoid bubbles or foam in sensitive (colored) glasses. The baffle block separates the cooling from the equalization zone.

3.2.2.3 Feeder Port End. The orifice end of the feeder is also heated (Fig. D 10). It is made in various sizes, designated as shallow, standard, or deep spouts. The output of the shallow spout is 7 to 1420 g glass per gob, or 0—17 kg/min; that of the deep spout is 57 to 2830 g/gob or 0.113 to 34.0 kg/min. There also are special spouts.

3.2.2.4 Feeder Dimensions (Fig. D 11) [6]. The length of the feeder depends upon the desired output. It is 2.5—8 m long. The channel is made in two widths (U series and W series) [6].

3.2.3 Heating

3.2.3.1 Gas Heating, Type K (Fig. D 12). In this system low-pressure gas heating is used in a series of burners mounted around the feeder at a distance of 11.4 cm. The combustion of air (1000 mm WC) inspirates the gas which is reduced to zero pressure (12—16 inches WC). By regulating the air the burners receive the desired amount of gas. The ratio can be changed by adjusting the inspirators. This adjustment is very important and demands care. After correct adjustment the pressure of the

(mm)	Type W	Type U
Width (b)	660	406
Height (h)	179	254
Glass level		
1. in Feeder head		
flat feeder	127	127
deep feeder	229	229
2. at the working end junction	152	229

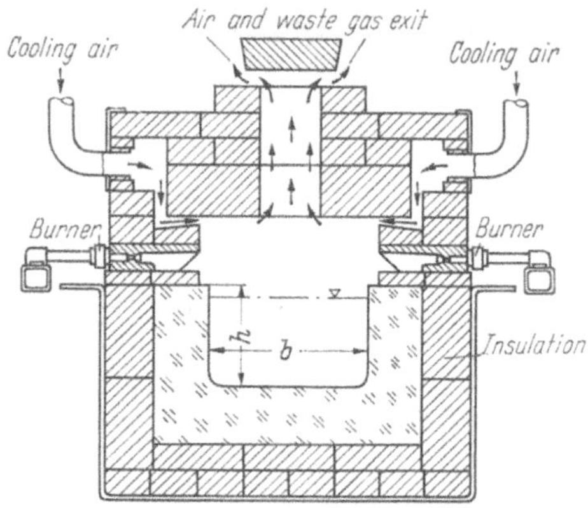

Fig. D 11. Feeder cross section.

Channel dimensions (mm)	Type W	Type U
Level	152	229
Height (h)	178	254
Width (b)	660	406

mixture should be about 100 mm WC and the flames should sting out slightly. The mixing ratio is 5 : 1 for city gas, 10:1 for natural gas. During the past few years propane and butane have been introduced. The heating system in this case corresponds with that used with city gas.

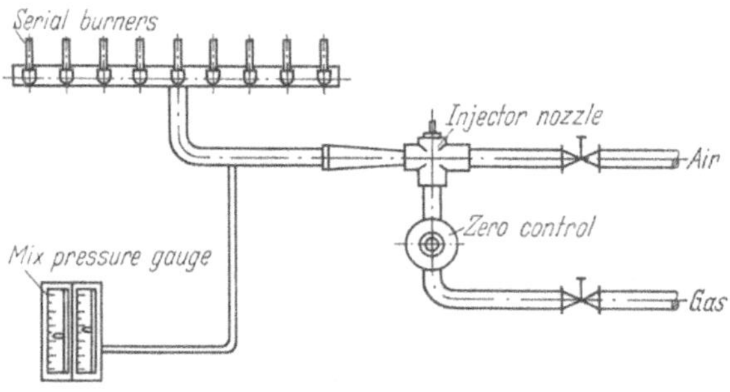

Fig. D 12. Gas firing, type K.

3.2.3.2 Oil Heating, Type Z (Fig. D 13). The Hartford feeder is oil heated at medium pressures for light oils. The combustion air is inspirated directly by the burners. The burners are located on one side of the feeder; one additional burner is located in a slanted position up in the spout end. The burners operate with compressed air for atomizing oil. The oil supply can be controlled by means of a float valve and by regulation of the atomizing air. Oil heating is not used as much as gas heating since there are some disadvantages in its use (inferior temperature constancy, higher maintenance cost, greater sensitivity to disturbances, etc.).

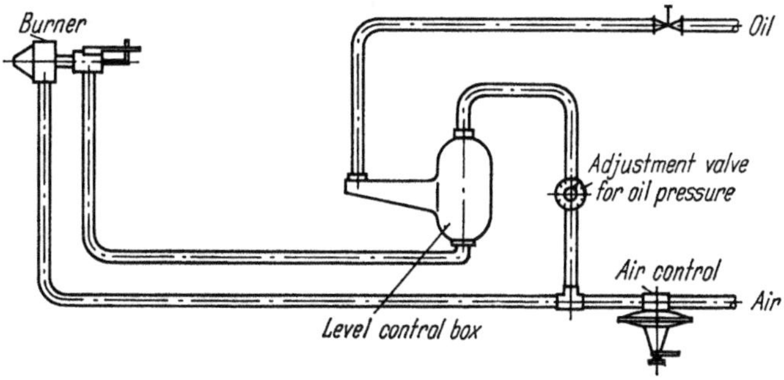

Fig. D 13. Oil firing, type Z.

3.2.3.3 Electrical Heating. Electrical heating has not yet been generally accepted for heating feeders, but experiments and development work have been carried out recently by some companies. If the electrically heated feeder shows advantages in temperature regulation and general construction it will become accepted in spite of higher cost. One American company has developed a feeder equipped with resistors. The channel is 2.44 m long, 0.35 m wide and 0.2 m deep. The spout end is the normal Hartford Type K. The resistors are arranged across the channel above the glass surface.

3.2.4 Cooling of Feeders

For high outputs the normal heat loss of glass is not sufficient to attain the desired gob temperature on the way from the working end. For this reason a cooling system is built into the cooling zone which blows cold air into the feeder to cool the glass. The air vanes are coupled with the exhaust dampers.

3.2.5 Temperature Control

The glass temperature is measured by thermocouples or radiation pyrometers. It can be read directly on a calibrated indicator. To follow the temperature changes temperature recorders are used. The control can be manual, by adjusting air, gas or oil by throttles, or it can be automatic.

In automatic controllers manual operation of valves is replaced by servomotors. The motor is connected with the thermocouple or pyrometer via an electrical or electropneumatic controller. The indicated target temperature is maintained within a certain tolerance. Depending upon the construction of the controller electrical impulses coming from the thermocouple are amplified and transferred to the motor by electrical or pneumatic impulses.

3.2.6 Stirring Unit (Fig. D 14)

To homogenize the glass in the channel (equal temperatures on the bottom, the sides, and the glass surface), stirring units are introduced; for instance, three stirrers into a W channel. These stirrers are made

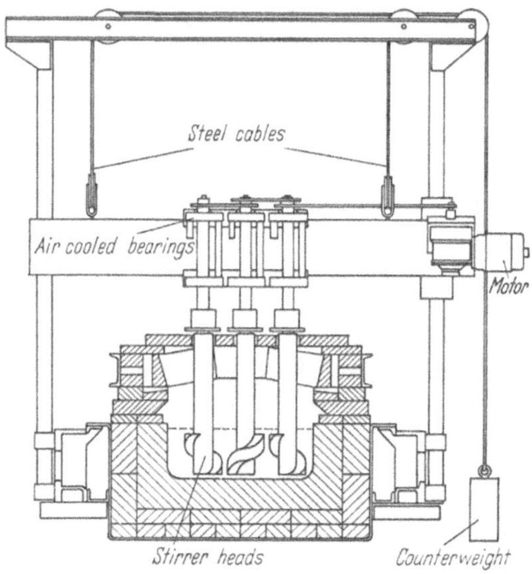

Fig. D 14. Stirring unit [6].

of refractory and have a spiralled section at the lower end. The stirrers extend vertically into the glass and are driven by a chain and sprocket drive.

3.2.7 Measurement of Glass Level

For best feeder operation it is important to have correct glass level. The level can be measured manually by a metal rod dipped into the glass. For continuous control automatic level gauges are used. Two electrodes with pyramidal cross section are dipped into the bath. The distance between the two electrodes remains constant. As the level of glass changes the electrical conduction at the surface also changes. The level can be read directly on a calibrated readout instrument. There are also instruments that work with a reciprocating electrode.

3.2.8 Operation of Feeder Mechanism

3.2.8.1. Emhart Types HE 144, HF 81 Turbex, HE 113, HE 115. The spout end has to supply a glass forming machine with uniform gobs in regular sequence. Plunger and shear motion as well as the rotation of the tube. must be timed perfectly with each other and the machine. This timing is effected by a complicated mechanism. The Hartford spout end (Fig. D 15) consists of:

1. The spout end proper with the cup for the molten glass;

2. A rotating refractory clay tube, or a rotating cylinder, the height of which can be adjusted to control flow;

3. A refractory orifice ring and plunger, to determine shape and size of gob;

4. Mechanical shears that cut the gob.

In operation the glass flows from the working end of the tank through the forehearth to the feeder spout and into the revolving tube. The clay tube causes the glass depth to be less within the tube which changes the hydrostatic pressure above the orifice. The clay tube is rotated to

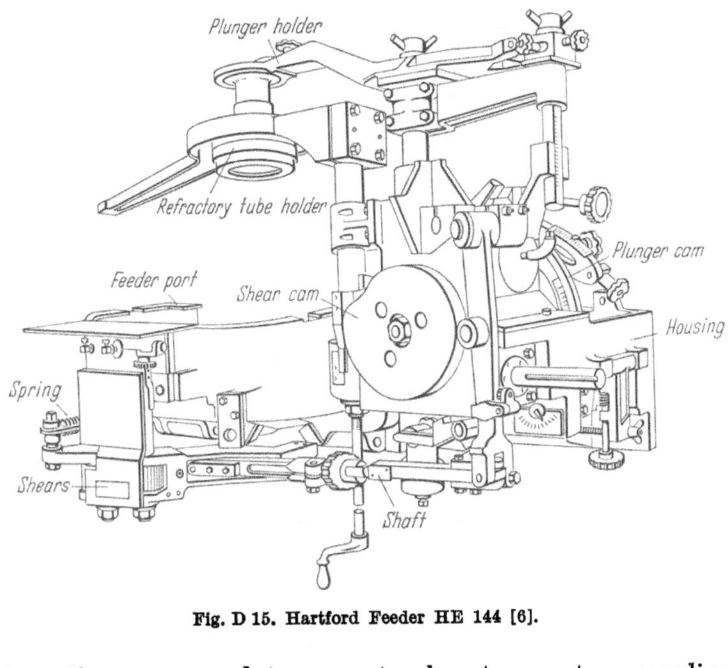

Fig. D 15. Hartford Feeder HE 144 [6].

obtain uniform wear and to promote glass temperature equalization. The refractory plunger surrounded by the the clay tube sits above the orifice and accelerates the glass flow through the orifice ring by its vertical motion.

The pulsating glass stream which is created by the rhythmical motion of the plunger is separated into single gobs by shears that are synchronized with the plunger and the machine. The motion of the plunger and the shears is controlled by rotating cams. Depending upon the desired size and shape of the gob, and the production speed of the forming machine, variously shaped cams are used. The relation between plunger and shear motion can be adjusted so that gob shape can be changed independently of weight and temperature of the glass.

The feeder mechanism is offered with standard aspect for outputs from a few grams to maximum of 17 kg/min or with a deep spout for outputs up to 34 kg/min. The feeder can be used for single or double gob operation. In this way type HE 144 satisfies most requirements of hollow glass machines. The feeder mechanism is divided into 5 major groups:

Plunger Mechanism,
Revolving Tube Mechanism,
Shear Mechanism,
Timing Valve Mechanism.
Feeder Drive Mechanism

Plunger Mechanism. The plunger mechanism consists of three subgroups:

1. Plunger holder system,
2. Plunger system linkage stroke,
3. Plunger cam and plunger drive linkage.

The plunger linkage transfers the vertical motion to the plunger holder mechanism. The linkage system is divided into:

(a) Mechanical linkages,
(b) Height and stroke adjustment,
(c) Lockout mechanism,
(d) Plunger-shear differential adjustment.

The mechanical linkage transfers the motion of the plunger (cam) into the reciprocating motion of the plunger holder and the plunger. Height and stroke adjustments effect changes in the starting height of the plunger and the length of its movement. The starting height is defined by the distance between the lower tip of the plunger at the low point of the stroke and the lower edge of the orifice ring. The plunger stroke is the vertical distance which the plunger covers in one motion. The linkage illustrated in Fig. D 16 is constructed so that an adjustment of stroke changes only the upper part of the stroke. The length of the stroke remains unchanged when the plunger height control is adjusted. (See also Sec. B 1, p. 43, Fig. B 12.)

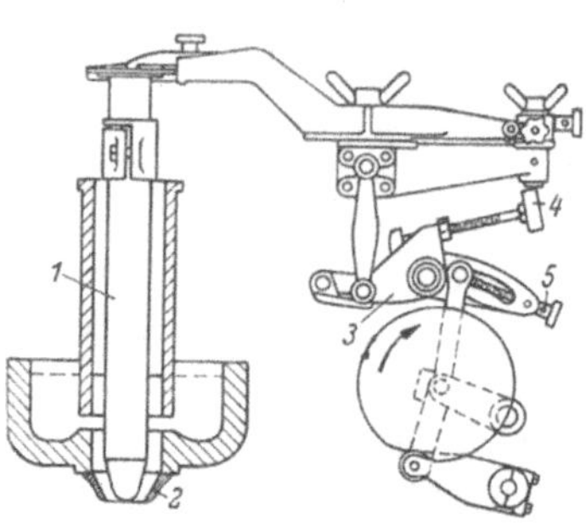

Fig. D 16. Scheme of plunger mechanism.

The plunger activation lever *3* consists of two arms which are located in the middle and can be turned. The upper adjusting screw *4* changes the angular relation of these two arms and raises or lowers the low position of the plunger. The lower adjusting screw *5* changes the lever relation of both arms and so controls the stroke length of the plunger *1*. Heavy items call for large, light ones for small stroke.

The lockout mechanism is provided to interrupt the action of the plunger without stopping the feeder mechanism.

The differential control is a very important factor in gob formation. By means of it the motion of the plunger can be advanced or retarded in relation to the shearing motion. The control encompasses 228°, with 48° being adjustable without interrupting the feeder.

Plunger cam and drive consist of:
Plunger cam,
Cam holder,
Cam holder hub,
180° cam differential control.

The plunger cam controls the plunger during its stroke. Almost all cams have the following fundamental characteristics:

(a) rapid initial rise,
(b) slow, long rise to the highest stroke position, and
(c) rapid, uniform descent to the ends of the stroke.

The selection of the correct cam is important in giving the gob the proper shape. Various cams are made to correspond with the requirements of various machines in respect to working speed and gob weight.

Mechanism of Rotating Cylinder. The rotating cylinder mechanism is subdivided into three groups: Revolving refractory tube drive, tube holder and tube adjustment. The chief task of the revolving tube is the regulation of the glass flow. It also should interrupt flow when the orifice ring is changed. The rotation of the tube helps the temperature equalization in the spout.

The revolving tube mechanism is found at the left side of the spout housing. The tube is turned via reducing gears. A friction gear is a safety device to avoid tube fracture if it meets unexpected resistance. The entire tube holding system runs on an air cooled ball bearing. The tube control mechanism permits vertical and horizontal shifts of the tube in relation to the center line of the feeder and orifice to attain uniform temperature distribution in the gob. Gob weight is generally to be effected by adjustment of the tube height. In special cases, e. g., in the production of very light fractional ounce gobs it is not advisable to control weight by the tube. In these cases it is best to bring the tube to a fixed position and to control gob weight by plunger height, and stroke.

Shear Mechanism. The shear mechanism consists of:
Shear mechanism and shear cam linkage,
Shear activators,
Shear control elements

(a) Shear adjustment,
(b) Shear overlap adjustment,
(c) Shear blade adjustment,
(d) Shear height adjustment,
Gob guiding adjustment,
Shear blade cooling.

The shear mechanism is activated by shear cam, shear cam lever, shear lockout mechanism and shear linkage. The shear cam is an important factor in gob shaping and is in close relation to working speed.

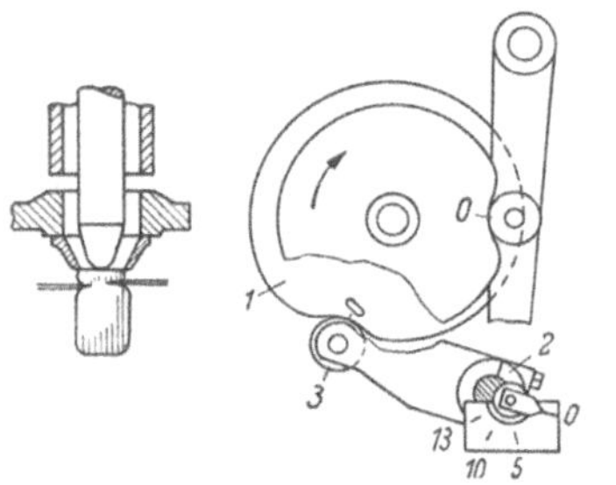

Fig. D 17. Plunger mechanism and shear mechanism without phase difference [6].

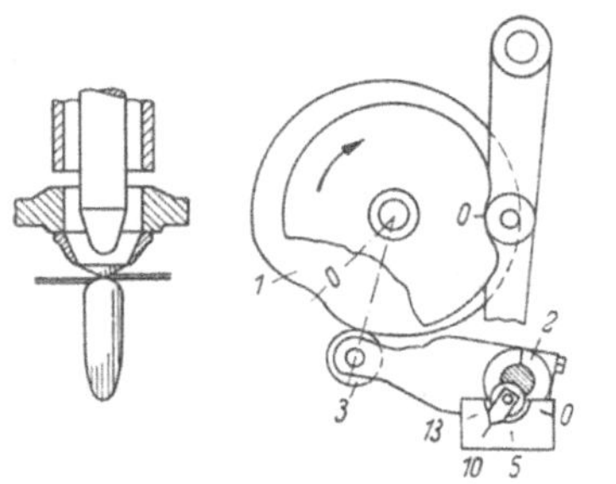

Fig. D 18. Plunger and shear mechanism with phase difference [6].

Cams of various constructions influence the stroke of the shear arms. In general shear cams are supposed to effect a fast cutting stroke, but with a smooth action so as not to cause excessive load on the shear mechanism. Shear cams are labeled with the degree number of the rotation corresponding with the cut.

The phase difference between plunger motion and shear motion has a great influence on forming. Coarse adjustment is done when the feeder rests by turning the cam shaft *1* around its axle (Fig. D 17). Fine adjustment during operation is effected by the eccentrically arranged indicator *2* which shifts the follower *3* on the periphery of the cam. If the upward start of the plunger motion coincides with the shear cut one obtains a short bulky gob. If the plunger has completed part of its upstroke before the cut a longer gob is formed (Fig. D 18). Distance, cut and speed of shears may also be adjusted. The gob shapes desired in various forming processes are illustrated in Fig. D 19.

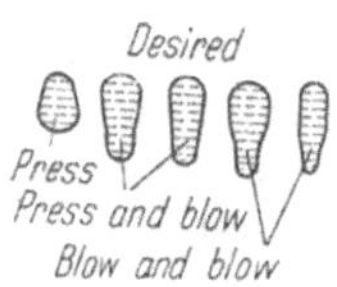

Fig. D 19. Different gob shapes.

The shear linkage consisting of cam lever connection, overlap adjustment, and shear arm connection transfers the action of the cam lever to the right shear arm. The transfer to the left shear arm is effected by gear segments.

The shear activation parts consist of the two arms which are located on vertical shafts connected by gear segments. While the right arm is fastened to the shear connecting rod, the left arm is attached to an air cylinder actuating the motion of the shear arms.

Shear Adjustment. The shear adjustment is done in two steps: shear arm adjustment, and shear blade adjustment. The shear arms are adjusted in such a way that the cutting motion of the blades ends in the vertical center line of the exit ring. The shear blades are fastened to the holders by a special gadget. Shear arm adjustment gauges serve to test their proper functioning.

Shear Overlap Adjustment. The shear blade overlap is defined by the path of the right blade over the center line of the exit ring. A maximum of 1.6 mm should not be exceeded if unnecessary wear of the blades is to be avoided.

Shear Blade Tension Adjustment. This adjustment determines the pressure between the two cutting faces of the shear blades. The right shear arm and the right shear blade are adjustable. Overlap and tension are in close relation. They should not be made larger than a satisfactorily smooth cut demands.

Shear Height Adjustment. This adjustment determines the amount of glass protruding from the orifice ring immediately after shearing.

Gob Guidance. Gob guidance prevents the lateral pushing of the gob by the left blade at the end of the cutting stroke.

Shear Blade Cooling. A shear blade spray system should prevent sticking of hot glass to the cutting faces and help maintain long blade life. Two systems can be used. In one system a coolant is continuously sprayed from nozzles. In the Emhart shear spray system the coolant is brought under pressure to a valve which is activated by the cam lever. The spraying of the blades is carried out only in the open position. Exact adjustment minimizes the consumption of coolant.

Time Valve Mechanism. Many feeders are in addition equipped with a rotating time valve which permits timed air impulses to auxiliary mechanisms, such as the lehr stacker.

Feeder Drive Mechanism. In most cases the feeder is driven by an electric motor through a P. I. V. drive. The transmission to the drive or to the feeder worm drive shaft is effected by chains and sprockets. Various sprockets permit any drive combination desired. The feeder mechanism must be timed to the glass forming machine. If the timing drum is driven mechanically, as in the IS machine, the feeder must be equipped with a bevel gear. In compressed air controlled glass machines the feeder and machine are synchronized by a timer valve mechanism. If mounting difficulties exist, synchronization may be achieved by special electrical systems. If the glass machine has its own motor like the Emhart 28, the feeder is driven by the machine.

Angular Shear Mechanism for Double Gob Operation. To permit double gobbing for all machine mountings, an angular shear mechanism covering from 30° right to 90° left can be attached to the spout casing.

Remote Operation. Almost all feeder adjustments can be effected quickly by remote control. This also permits the observation of the forming process when adjustments are made.

The HE 81 or Turbex feeder resembles the 144 in many respects. It can be operated with a standard feeder spout for a glass output of a few grams to 28.4 kg/min, or with a deep spout for outputs up to about 35.4 kg/min. The drive motor and the stepless drive are coupled directly with the feeder mechanism. Time valves for the activation of auxiliaries can be attached. The range of speeds is 6 to 160 cuts/min.

The HE 81 feeder essentially consists of a spout, a rotor system, the stationary refractory tube, the orifice ring, the plunger and strong mechanical shears. The rotor system consists of 3 rotor segments and has proven particularly good for colored glass.

As in the HE 144 feeder the plunger is lifted and lowered by a rotating cam which is interchangeable for production of different gobs. The plunger mechanism is an addition equipped with an auxiliary air cylinder which supports the down stroke and equalizes shape and size of the gob for various production rates. To assure the correct sequence of plunger and shear motion a differential covering 360° is provided. The orifice ring sizes for the to 144 and HE 81 feeders are the same.

The shear mechanism consists essentially of two cam operated shear arms made from an aluminum alloy. Cams for various production rates are provided. HE 144 cams can be used on the HE 81.

The shear stroke is activated by an air cylinder, to assure positive cutting motion for all rates. Overlap, shear blade tension and distance below orifice are adjustable. All adjustments can be effected by remote control in operation as for the HE 144. The HE 81 can also be provided with an angular shear mechanism to give more liberty in the mounting of double gob machines. The lubrication of the most important places in the HE 81 is automatic. For lubricating all individual bearings, a pump on the working floor has to be operated only once a shift.

Hartford HE 113. This feeder was developed mostly for the production of large glass gobs, e. g., for television screens and building blocks. The feeder delivers up to 11.3 kg/gob for single gobbing and $3^1/_2$ to 20 cuts/min. The HE 113 feeder has an HE 81 mechanism and uses a rotating refractory tube for stirring in the spout. Some controls and adjustments were changed to achieve a larger spout, a larger orifice ring and a larger plunger stroke. Selected parts are made stronger because of increased wear. This feeder is offered with standard spout for 42.5 kg/min. or deep spout for up to about 49.6 kg/min. The rotating refractory tube has an inner diameter of 178 mm, the orifice ring an outer diameter of 178 mm. One may also use a refractory tube with an inner diameter of 230 mm. To assure a large force in the cut, air-activated shears type 29 or 111 are recommended.

Hartford HE 115. The HE 115 feeder was developed for the double gob production of heavy containers. Gob weights to 4 kg in single, and 850 g in double gob operation are possible. The HE 115 feeder also possesses an HE 81 feeder mechanism. Some control and adjustment elements were altered to permit the use of a larger spout and a larger orifice ring. In this feeder, too, some parts are made heavier because of the greater wear to be expected. The feeder is offered with a standard spout for gobs of about 0.17 to 4.54 kg and works with a refractory tube of 178 mm inner diameter and a suitable orifice ring. The diameters of the double gob orifice rings are up to 3 inches standard. As in the HE 81 a stationary refractory tube and a motor driven rotor system keep gob weight and temperature constant. The HE 81 feeder angular shears may be used in the HE 115 to obtain greater freedom in mounting the machine for double gob operations.

3.2.8.2 **British Hartford Feeder.** The British Hartford Feeder is constructed and works like the HE 144, except for slight deviations.

3.2.8.3 **Putsch Feeder SR** 21 (Fig. D 20). The Putsch feeder SR 21 is a pneumatically driven and controlled feeder for 20 to 2000 g weights with a stepless cutting range of 5 to 50 cuts/min. The chief parts of this feeder are:

Timer,
Feeder mechanism, consisting of
  rotating refractory tube
  plunger,
  shears,
  channel,
Feeder spout,
Feeder channel and brickwork,
Heating system,
Temperature measuring or control equipment.

The timer has 6 impulse valves. All impulses can be shifted forward or backward 360° during operation. Since the timer also drives the control shaft of the machine perfect synchronization is assured. The rotating refractory tube is driven by an electric motor and simultaneously serves as a control for glass weight. Height adjustment is by hand wheel from the service area of the machine.

The plunger is driven and controlled pneumatically and may be centered with precision. All adjustments can be made from the service area of the machine as well as from the feeder platform. The shears are mounted at the front of the spout casing. The shear arm opening can be adjusted from 60 to 30°. The arms are activated by toggle levers. The shear blades are exchangeable. The cutting location can be adjusted

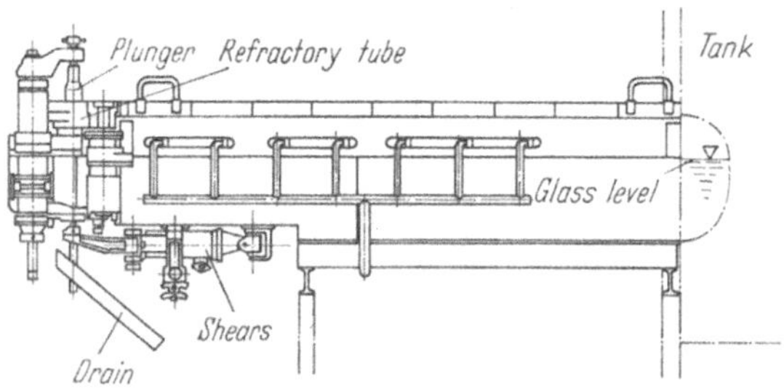

Fig. D 20 Putsch-feeder SR 21 [6].

vertically and horizontally. The trough for the temporary diversion of gobs is controlled in a way that any sudden disturbance in the air supply brings the trough automatically under the orifice. The orifice ring is exchangeable and can be adapted to gob weight. The feeder is offered for gas or oil heating.

3.2.8.4 TRB or Beatty Feeder (Fig. D 21). The Beatty feeder works pneumatically. The mechanism is constructed as simply as possible. The feeder consists of timer, plunger mechanism and shears. The timer controls the operation of feeder and machine. The plunger mechanism consists of a cylinder on the piston rod to which is fastened a cross arm to hold the plunger. The stroke height is controlled by a spindle. It is activated from the service area of the machine by chain drive. The shears consist of a cylinder to which shear arms with blades are attached. To exchange blades, the entire shears must be dismounted, since the blades can not be adjusted well while the shears are in position. The exchange of the orifice ring is time consuming since the glass flow can only be stopped by blocking the channel with a baffle.

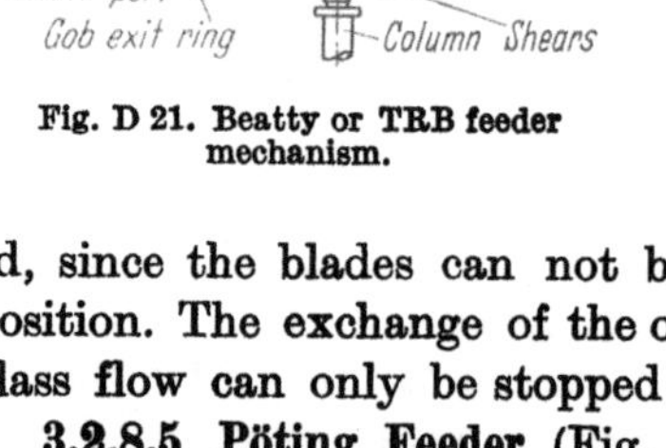

Fig. D 21. Beatty or TRB feeder mechanism.

3.2.8.5 Pöting Feeder (Fig. D 22). The Pöting feeder works pneumatically. Like the Beatty feeder it consists of timer, plunger mechanism and shears. The difference consists in the feature that the plunger turns

Table D 1. ***Working range and working process of most important feeders.*** **(According to Institut für Glastechnik in Coswig)**

Type	Manufacturer	Kind	Gobs/min	Gob wt. in g	Rotating cylinder	Plungers	Shears
„115"	Hartford	single			Fixed mechanic rotating segment	Fixed mechanical stroke	Mechanical
		double		825			
Nr. 52	VEB Glasmaschinen Freital	single	20	1500		Stroke and rotation pneumatic	Pneumatic
Nr. 37	VEB Glasmaschinen Freital	single	40	2000	Separate drive by e-motor	Mechanically driven from machine	Mechanically driven from machine
SR 21	Putsch & Co Hagen Westfalen	single	5–50	20–2000	Separate drive by e-motor	Fixed stroke pneumatic	Pneumatic
F	Pöting	single	5–70	15–3000	Normally without rotating tube, mechanically driven in special	Stroke and rotation pneumatic	Pneumatic
Nr. 191/26	T. H. Sloan Co.	single	25	2250–3650	Separate drive by e-motor	Fixed stroke pneumatic	Mechanical
„144"	Hartford	single	6–75	230–2800	Rotates mechanically	Fixed stroke pneumatic	Mechanical
		double	12–150	230–2900			
„81"	Hartford	single	6–80	355–4750	Fixed with mechanically rotating segment	Fixed stroke pneumatic	Mechanical cutting pneumatically adjustable
		double	12–160	220–2950			
„113"	Hartford	single	3.5–10	4250–11340	Rotates mechanically	Fixed stroke pneumatic	Mechanical (but pneumatic shears recommend)
		double	7–20	1240–7100			

Gobs/min scale: 0, 40, 80, 120, 160; Gob wt. scale: 0, 2000, 4000, 6000, 8000, 10000, 12000

360° after each stroke. This turn is activated by a cylinder and is supposed to effectuate temperature equalization in the port end. The stroke height can be adjusted pneumatically, the elevation mechanically. The shears consist again of cylinder and shear arms carrying the blades.

A survey of working ranges and operative processes of the most important feeders is given in Table D 1.

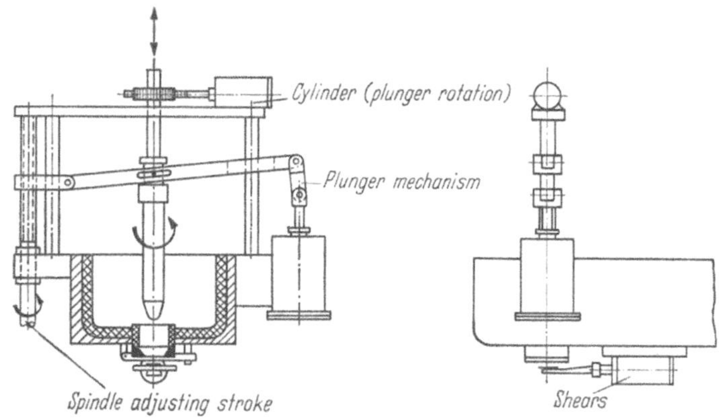

Fig. D 22. Pöting feeder mechanism [6].

3.3 Temperature Control

One of the most important prerequisites for good production is the correct adjustment of temperature in the feeder. The temperature distribution along the feeder was measured carefully. The results can be seen from Figs. D 23 and D 24. The temperature distribution for colorless and colored glass is given. The temperature distribution depends upon throughput and glass color. With high production the glass brings much heat into the feeder. To maintain optimum gob temperature the

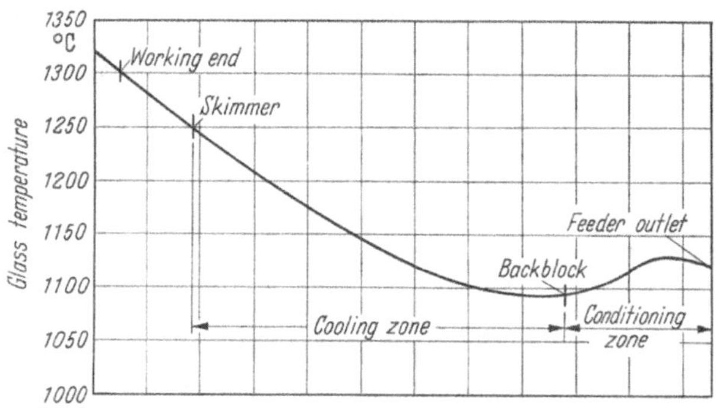

Fig. D 23. Mean glass temperature in a KW feeder (14' 7½").

glass in the feeder must be cooled a great deal in the cooling section. If the flow is diminished very much, a steep temperature gradient develops between feeder entrance and skimmer. Because of too slow flow the glass cools excessively at this place and has to be reheated which often causes seeds to form. Heating therefore must be controlled for a slow rise in temperature.

At low production (for small objects) the temperature distribution should be about linear. For every throughput a certain relation between the temperatures of the incoming and outgoing glass should be maintained which is done in part on an empirical basis.

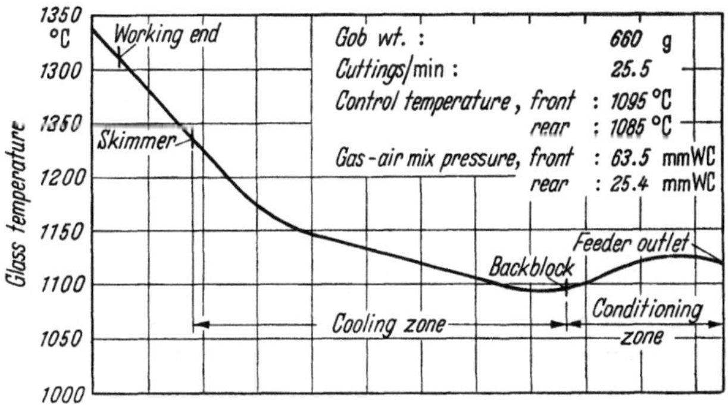

Fig. D 24. Temperature curves for a KW feeder (16′ 17 $^1/_2$″).

The temperature distribution shown in Fig. D 23 may be considered normal. It has been determined by hundreds of pyrometric measurements. The relatively strong drop right after the feeder entrance is caused by the feeder cooling system which can be throttled more or less by dampers. Recently electronic controls have also become available. The comparison of the temperature distributions in feeders under various conditions and in troublesome cases are instructive. In Figs. 24, 25, 26, 27, 28, 29 various examples are given. Fig. D 24 shows the temperature distribution in a flint glass feeder, for a gob weight of 660 g and 25.5 cuts/min. This temperature distribution corresponds practically with the ideal curve D 23. The measurements were made on a 16′ 7 $^1/_2$″ feeder. The uniform cooling between skimmer and equalizing zone is clearly visible. At the spout the glass is heated up by about 25° to reach the working range.

Fig. D 25 shows an undesirable temperature distribution. It can be seen that in spite of uniform adjustment of the cooling system and correct temperature regulation a feeder may work incorrectly. In this way pyrometric measurements assist in making correct heating adjustments. In this particular case non-uniform burner equipment had been used: the nozzle diameters had been confused in part during the installation.

After the proper changes in the burner equipment a distribution more like that of Fig. D 23 was attained which improved the gobbing.

Fig. D 26 gives a survey of various amber glass feeders none of which exhibited the desired temperature distribution since the feeders were not in the correct condition.

Fig. D 27 shows a temperature study on a flint glass feeder with excessive starting temperature. The heating up at the front end of the feeder was too sudden and caused much foaming. By a temperature in-

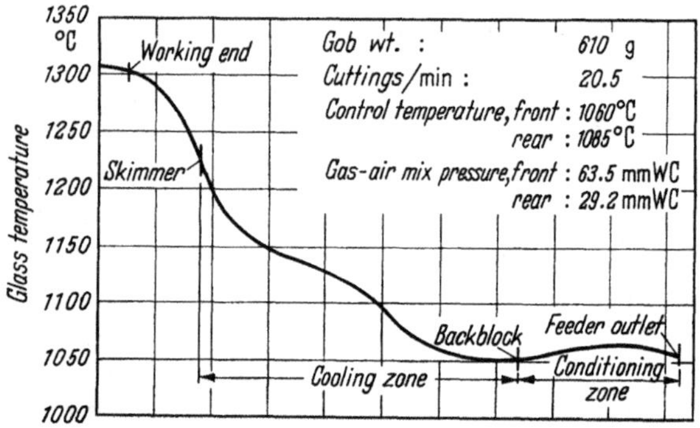

Fig. D 25. Temperature curves for a KW feeder (16′ 7$^1/_2$″).

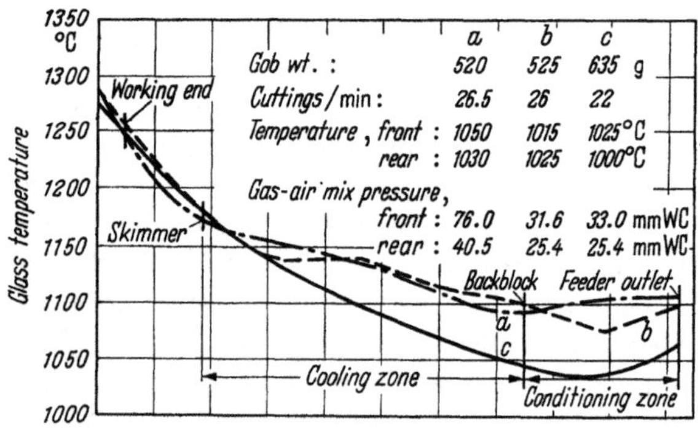

Fig. D 26. Temperature curves for a KW feeder (16′ 7$^1/_2$″).

crease in the back part and a decrease in the front part the desired working temperature was attained and foaming diminished. The check is made by stopping the rotation of the clay tube in the feeder spout. After about 5 min gob formation permits conclusions on the correctness of the temperature regulation. If the gob curls from the feeder toward the machine the back part temperature is too high. If the gob curls towards the tank the back part is too cold. The proper temperature adjustment is best made in the back where the danger of loss of control is less.

Fig. D 28 shows the temperature distribution in a flint glass feeder with too low a back part temperature. The temperature drop is too large; cooling or temperature regulation must be adjusted accordingly.

Fig. D 29 shows the temperature distribution in a colored glass feeder, particularly at a very low pull. Shortly behind the bridge block there is a relatively strong temperature drop, caused by the moderate flow in the feeder which hinders the transport of heat from the tank to the feeder. These measurements were made along the length of the feeder.

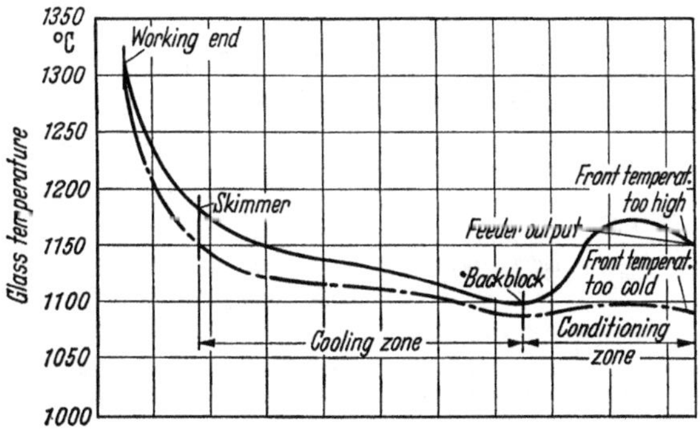

Fig. D 27. Temperature curves for a KW feeder (14′ 7 1/2″).

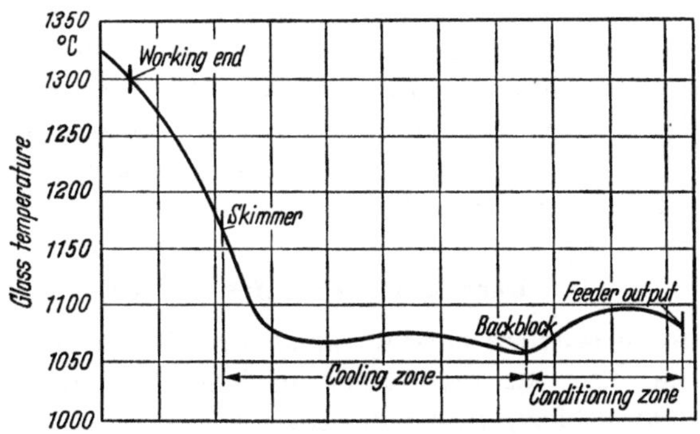

Fig. D 28. Temperature curves for a KW feeder (14′ 7 1/2″).

To find the vertical distribution in the bath thermocouples were inserted vertically. It was found that the temperature drop from the surface to the bottom of a green glass feeder was about 15°/inch; that is, over the depth of 150 mm a difference of 90° exists (Fig. D 29). To avoid larger differences three screwshaped stirrers were attached in the back part of the feeder which were to draw cold glass from the bottom. By using right-and left hand screws temperature differences could be reduced considerably.

A possible way to effect temperature equalization in green glass feeders is to use electric bottom heating, following principles used in tank heating. In flint glass feeders bottom heating and stirring is not necessary since temperature differences are small both in horizontal and vertical direction. In the case of amber glass quite different conditions exist. The most important one is the arresting of surface glass. Various experi-

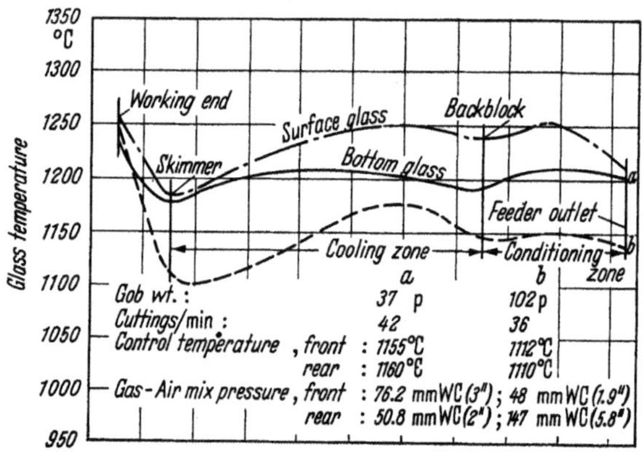

Fig. D 29. Temperature curves for a KW feeder (14′ 17 1/2″).

ments were made. First an overflow was devised which let the surface glass drain over a channel attached at a right angle. Here the regulation of glass quantities became a problem. Special heating was necessary to cause the glass to flow. A mantle block set into the arch from above and dipping about 2 cm into the bath was found advantageous. Vertical adjustments offered many possibilities.

3.4 Operational Data for a Feeder

It is hard to tabulate cutting rates since they depend on so many circumstances, e.g., the production machine, the production process (press-blow, blow-blow, pressing), glass compositions, weight-volume ratio and shape of the glass article. Usually suppliers of machines list typical cutting rates for particular items, which are then set in the feeder. Some indications are given in Table D 1.

The spout temperature depends on several factors. It is influenced by glass composition, the production machine, the production process and the type of ware to be produced. Normal working temperatures are between 1090 and 1230 °C.

Plunger cam curve and stroke determine the shape of the gob for a given cutting rate (Table D 2). For given cutting rates various shear cam curves are required (Table D 3).

Table D 2. *Plunger cams* [6]

Cuttings/min	Plunger cam
5	1, HW-30°
10	1, HW-40°, 4 Brick, M-3 Gallon 40
15	2, 3, 4 Brick, M-3
20	3, 4
25	4, 12
30	4, 5, 12
35	4, 5, 6, 12
40	4, 5, 6, 12
45	5, 6, 7, 12
50	5, 6, 7, 12
55	6, 7, 7, HS-11, HS-557
60	8, 12, HS-11, HS-557, HS-5-3/4
65	8, 12, HS-11, HS-556, HS-557, HS-5-3/4
70	8, 12, HS-11, HS-556, HS-557, HS-5-3/4
75	HS-4, HS-11, HS-556, HS-5-3/4, 23
80	HS-4, HS-11, HS-556, HS-5-3/4, HS-11-5/8, 23
85	HS-11, HS-5-3/4, HS-11-5/8, 23
90	HS-11-5/8, 23
95	HS-11-5/8
100	HS-11-5/8
105	HS-11-5/8
110	Eccentric
120	Eccentric, $^1/_2$ Eccentric

The interior diameter of the orifice ring depends upon the shape and weight of the article. This relation is covered by Table D 4[1]. But this table can only be considered an approximate guide for the selection of this diameter, since the ratio of gob diameter to orifice ring diameter varies with production rate.

Gob guides follow usually the gob diameter. For single gobs they are about $^1/_4$ to $^3/_4$'' larger than the gob diameter. Single sizes are: $^3/_4$'', 1'', $1^1/_2$'', 2'', $2^1/_2$'', 3'', $3^1/_2$'', 4''. For double gobbing usually the next larger guide is taken. Sizes are: $^3/_4$'', 1'', $1^1/_2$'', 2'', $2^1/_2$''.

[1] Table D 4 in pocket at end of the book.

Table D 3. *Shear cams* [6]

Cuttings/min	Shear cams
5— 7	14°— 50°
7— 10	18°— 53°
10— 12	20°— 60°
12— 15	31°— 61°
15— 18	41°— 68°
18— 20	49°— 73°
20— 25	58°— 90°
25— 30	61°— 91°
30— 35	65°— 96°
35— 40	67°— 67°
40— 45	70°— 70°
45— 50	80°— 90°
50— 55	91°— 91°
55— 60	106°—106°
60— 68	113°—113°
68— 75	122°—122°
75— 83	133°—133°
83— 95	149°—149°
95—102	160°—160°
102—110	171°—171°
110—	180°—180°

3.5 Conclusion

The future development of feeders seems to take the following directions:

1. Simplification of feeder mechanism,
2. Improved heating and temperature control for thermally homogeneous glass, particularly in double gobbing,
3. Automatic feeder level control,
4. Direct addition of coloring agents in the feeder,
5. Improved materials for better feeder life.

There is a tendency to operate all controls from the work floor and during continuous feeder operations.

The heating of the spout still causes much trouble. One sided heating still causes cold zones. In single feeding these difficulties can be overcome, but in double gobbing disturbing temperature differences occur repeatedly, causing gob-weight variation. One possibility is double plungers for double gobs. This however is only an emergency solution and uniform heating would be preferable. Most likely electrical feeder heating is a possible solution.

Glass level is important for perfect gob forming. For best results feeders can be lifted or lowered. But this is advisable only in extreme cases since channels or walls may crack. Automatic reciprocation without effect on the construction would be ideal.

It has become possible to a certain degree to change glass color in the feeder. This has great advantages for production since it can be

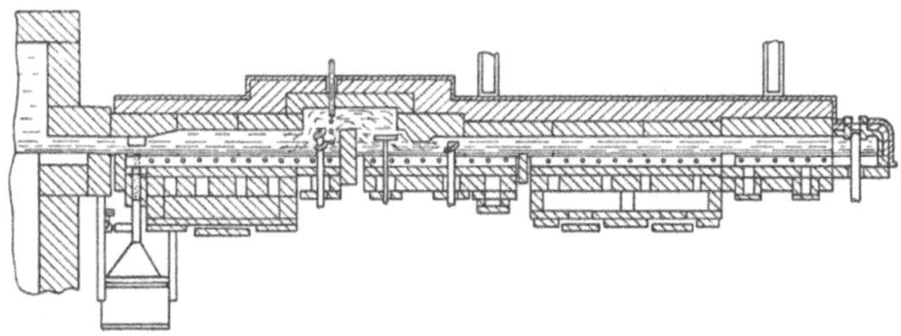

Fig. D 30. Building scheme of a mixing feeder.

adapted to the market without clumsy changes in the entire tank. The construction of such a "mix feeder" can be seen in Fig. D 30. The colored frit is sprayed on the glass where the forehearth touches the working end, and the glass flow is made to go down through the melting frit to a deeper part of the channel to accelerate mixing. The equipment consists of the channel for the flow of glass with the frit sprayer and the deep section which includes mechanical stirrers. In the deep section are bubblers or electrodes. Mixing is by up to 12 stirrers in groups of three.

The use of special high grade blocks for channel, spout, plunger, refractory tube, and orifice has led to good results. Recently, zircon containing materials proved particularly successful.

References

[1] Mejer, L.: Tropfenbildung im Speiser und die Herstellung der Flaschen im Speiserautomat. Szklo i Ceram. 5 (1954) 137—141.
[2] Griffin, T. W., F. S. Light, and A. W. Russel: Zusammenhang zwischen Glasspiegel und Tropfengewicht beim automatischen Speisérbetrieb. Bull. Amer. Ceram. Soc. 29 (1950) 183/84.
[3] Schkolnikow, I. S.: Grundlagen der Konstruktion und rationellen Auswahl von Speisern für Hohlglasautomaten. Steklo i Keram. 12 (1936) 15—19.
[4] Wardley, T. W.: Hartford feeder, Lynch machines and the Hartford IS-machine. J. Soc. Glass Technol. 37 (1953) 32—40.
[5] German Patent No. 1060109: Verfahren und Vorrichtung zur Homogenisierung von geschmolzenem Glas.
[6] Information given by the firms Emhart, Hartford (USA), Putsch, Hagen, and Pöting, Erkrath.

E. Machines for Glass Working

For the working of the hot viscous glass mass — and this is the only subject treated here — many specialized machines are available. Even though the individual processes are quite different, a common factor is that the working process proper is accompanied by the cooling of the glass mass and its concomitant solidification. Since the working properties, particularly the viscosity of the glass, change greatly with temperature, the temperature conditions in the forming portions of the machine are of extreme importance. There is hardly any material in technology that could be compared with glass in this respect. The glass working processes therefore differ fundamentally from the usual processes and a thorough discussion of glass working machines deserves attention.

Glass may be worked by pressing, blowing, drawing and rolling. Press, press-blow and blow-blow machines were developed specially for hollow ware manufacture. According to the gathering method used one distinguishes between suction and feeder machines. Tubing and flat glass are usually drawn by various methods. For certain flat glasses, particularly for so-called plate glass, rolling processes have been developed. A special position is occupied by the processes used for the production of glass fibers.

The following discussion is more or less detailed depending upon the relative importance attained to date by the various processes and machines.

1 Glass Presses

1.1 Manual and Semiautomatic Presses

By H. J. Walther, Siebenstern

Pressed glass is the designation for all glasses made on manual as well as automatic presses.

In the manual process glasses such as ordinary colorless hollow glass, lead glass, and heat resistant glass are worked. This process is still much in use, particularly if small numbers — below 20,000/month — are required, if the fabrication is particularly difficult, or if extraordinary quality is demanded.

1.1.1 Gathering

The gob is gathered by the gatherer with the so-called gathering iron (Fig. E 1). The head of the iron is coated with a ball shaped refractory material. Sometimes a ball made of non-scaling steel is used. The clay head is preferred in order to avoid iron cords introduced into the tank with the glass residue. The gatherer forces the gathering iron into the glass bath, until about half the head is submerged. By repeated turning he wraps the glass around the lower part of the head. The glass

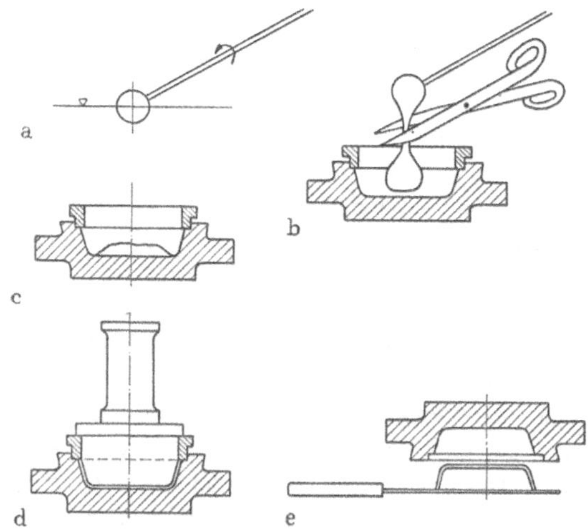

Fig. E 1a–e. Hand pressing operations.
a) Hand gathering; b) Cutting by hand; c) Moving mold to press station; d) Pressing; e) Emptying of the mold.

quantity taken out depends upon the size of the head and the viscosity of the glass. The gatherer has to put the glass into the mold at such a temperature that cords and cracks are avoided during working. Viscosity and surface tension of the glass must be sufficiently high to assure a smooth surface in the pressing process. The glass residue on the gathering iron remaining after cutting can be removed in two ways: (1) by stripping off the glass outside the furnace by turning the gathering iron on the edge of a stripping plate, or (2) by stripping into the furnace.

In the latter case the glass maker must strip off the glass as near to the furnace edge as possible to cause the residue to heat up slowly and to avoid cord formation. Clumsy introduction of residue glass also can generate seeds which may be distributed over the entire bath by convection currents.

A certain working rhythm can be maintained by the use of lights which signal at the proper intervals.

1.1.2 The Pressing Process

The press is generally served by two men, the presser, and the carry-in boy. Round table presses with Geneva drives or pneumatic drives are usually used.

The presser has to cut the correct amount of glass into the mold by means of the glassmaker's shears. It is important to see to it that the cut end falls into the center of the mold so that it reheats. The ring is placed on top, either before or after cutting. The mold is placed under the plunger manually or mechanically. A handlever exerts pressing force on the plunger by way of a crank shaft. In toggle, spring, or eccentric presses the plunger is activated by compressed air.

Table E 1. *Output of hand presses*

Job	Wt. (g)	Diam. (mm)	Av. thickness (mm)	Surface	Per-sonnel	Mold number	Pressing time (sec)	Number pro-duced/h
Plate	200	150	3— 4	Smooth	3	3	1.4	317
Dish	720	220	4— 5	Smooth	3	2	1.0	177
Large dish	870	240	4— 6	Smooth	3	2	2.3	103
Plate	225	150	4	Ornate	3	3	1.4	308
Vase	1350	90	10—15	Ornate	2	2	1.9	94
Ash tray	600	145	15	Ornate	3	3	2.0	190
Headlight gl.	180	165	3— 4	Smooth	3	3	1.2	300
Mixing tumbler	2000	100	7—10	Smooth	3	2	3.6	53

Pressing time depends on the kind of article and on heat transfer from glass to mold. Pressing times for various articles are listed in Table E 1. These are valid for a mold temperature of 400 °C. After pressing the glass goes to the take-out station and is immediately air cooled from a ventilator with a pressure of 180 mm water column. The ring must be taken off first and placed on the next mold to prevent it from cooling excessively. Otherwise the rim of the glass article may craze. Large rings with small contact areas may have to be held at the working temperature by propane, city gas or electric heating. Frequently excessive heat loss is avoided by boring the outer surfaces of molds or mold parts.

The plunger is cooled by an air blast in its upper position. To save time the press is adjusted with a cold mold so that the mold preheated to the working temperature (about 400 °C) has just to be inserted. The molds are preheated in a special mold oven which can be heated with producer gas, oil or piped gas. It is not recommended that the molds be preheated with hot glass since in the long run this operation causes defective ware (wavy glass with cracks).

The mold temperature should remain nearly constant during the entire operation as temperature variations impair glass quality. Mold temperatures during working are usually 340—450 °C. The rhythm of the shop can be disturbed considerably if glass sticks to the mold walls. If temperature variations are present glass may, for instance, enter between a too hot ring and a too cold plunger resulting in damage to both parts. If the plunger is too hot and the ring too cold a bulge forms in the interior of the glass.

The smoothness of the glass surface also depends upon the mold temperature. If the molds are too cold the glass contains cords and crazes. In spring presses the force of the spring should be adjusted to somewhat exceed the force of the glass on the ring. For thick walled objects one should press a second time. The plunger is relieved for a short while after the first pressing. The surface of the glass may freeze by excessive heat loss at the mold wall, while the interior remains soft. By the withdrawal of the plunger one succeeds in reheating the outer zones of the glass from the inside. After the outer zones have in this way warmed up, and the core consequently cooled down, the plunger is pressed into the mold once more.

1.1.3 Mold Cooling

The difficulty of each pressing process is the necessity for the fast removal of the heat introduced by the glass. Water cooling as used in automatic presses is not suited in manual operation in spite of inherent efficiency because the irregular working rhythm would be liable to introduce temperature variations in the molds. In almost all manually operated pressed glass factories blown air is used. The inertia of air cooling becomes desirable in connection with an inexact rhythm. The air quantity needed for a medium three mold press is 120 m^3/min at 180 mm water column.

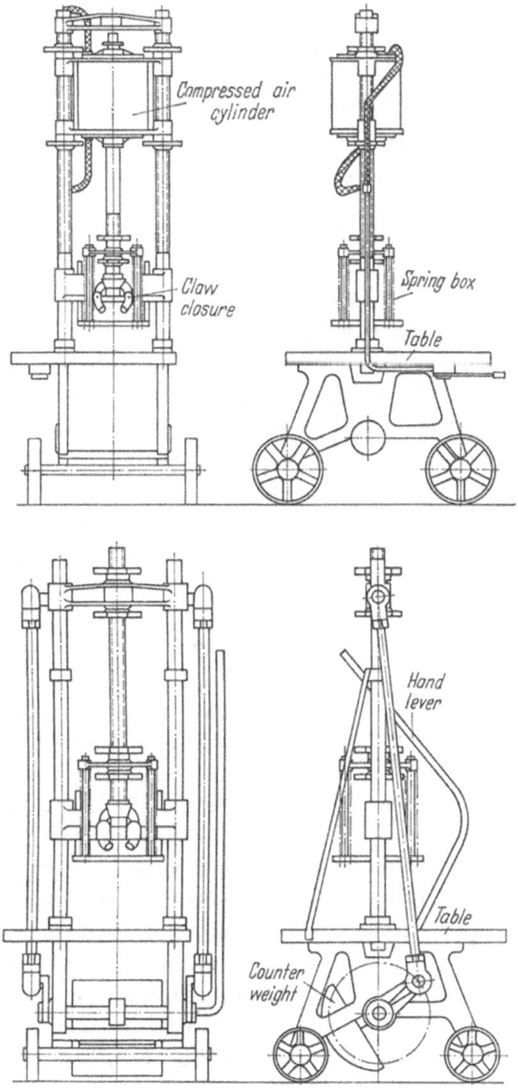

Fig. E 2. Spring box press (type Kutzscher).

1.1.4 Presses

In contemporary pressed glass manufacture the following press types are preferred:

1. Spring Cage Presses,
(a) Manual Spring Cage Press,
(b) Air Pressure Spring Cage Press,

2. Eccentric Presses,
(a) Manual Lever Eccentric Press,
(b) Air Pressure Eccentric Press,

3. Toggle Presses.

1.1.4.1 Spring Cage Presses (Fig. E 2). The manual lever spring cage press is simple. The construction sits on a heavy four wheel cast iron table which is thick enough for repeated top surface planing. This is necessary since the motion of the molds makes tracks. In one-shift operation planing is necessary every 2—3 years. The construction consists of two columns between which are actually two beams. The beams are connected with a spindle at the lower end to which the plunger is fastened by jaws. The stroke of the plunger may be adjusted by two lock nuts. To assure the proper counter pressure of the ring on the glass during pressing the spring cage is attached to the spindle. It is guided by two rods at the lower beam. Transmission goes from the hand lever to the crank shaft and from there to the beam. To keep the hand lever in the rest position the force exerted by plunger weight, spring beam, etc. on the crank shaft must be balanced by a counterweight. The weight must always be adjusted in a manner to assure that the handlever returns to its rest position slowly. The kinematics of a manual lever press are shown in Fig. E 3. Fig. E 4 shows how the plunger force varies with stroke. Stroke and plunger force can be brought into relation since they both depend on angle φ (Fig. E 3). Fig. E 5 shows plunger force as a function of stroke; it is clear that the increase in force becomes larger at small stroke. If both crank shafts of a hand press are vertical and down, theoretically the plunger force becomes infinite. The angle φ would be 90°. To avoid this

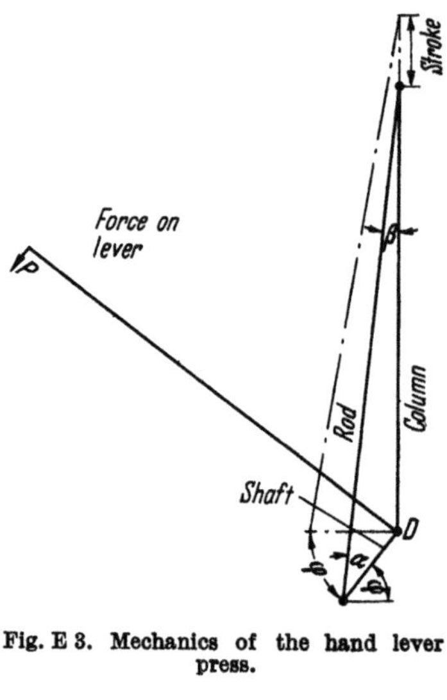

Fig. E 3. Mechanics of the hand lever press.

condition the press arm is bent so that the angle cannot exceed about 80°. The hand lever spring press is, in principle, suited for all pressed ware. Today it is often replaced by the air pressure spring press. The disadvantage that the counteraction of the finish ring under spring

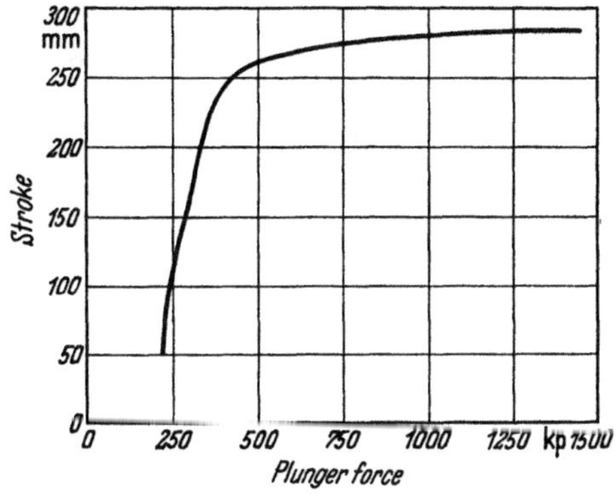

Fig. E 4. Variation of force with stroke for the hand press.

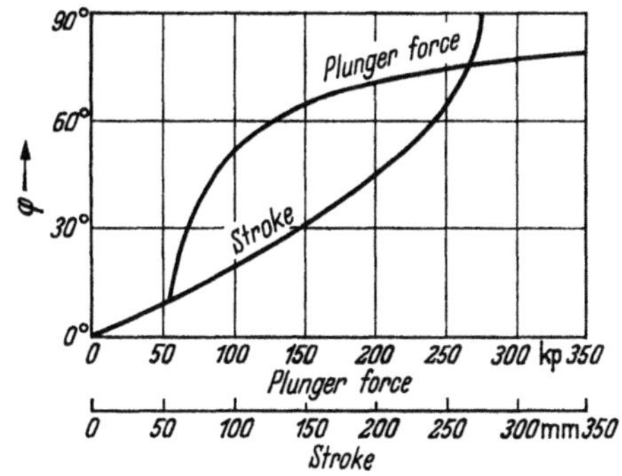

Fig. E 5. Stroke and force of hand lever press as functions of angle.

pressure wastes pressing force is of little importance. Working with hand lever spring presses requires wide experience and great physical effort.

The *air pressure spring press* is similar to the hand lever spring press. The two cross beams are replaced by a compressed air cylinder which can be adjusted by four lock nuts.

The piston rod carries the spring and the mechanism for fastening the plunger. The press is controlled by manual valves. In front of the main valve there is a reduction valve for pressure adjustment. Air requirements are tabulated in Table E 2. The pressing action of such presses is smooth as the pressure rises slowly. The air pressure spring press can also be used for the fabrication of all pressed ware. It is distinguished from the manual lever press by simpler operation. The work of the presser is limited to shearing the glass and to operating the control valve.

Table E 2. *Air requirements of air pressure-spring box presses*

Type Kutzscher	DP 1	DP 2	DP 3	DP 4	DP 5
Plunger diameter (mm)	220	250	300	375	375
Stroke (mm)	250	300	350	400	450
Pressure (kp) using 3 atü	1150	1500	2000	3250	325
Air req./stroke (Nm^3)	0.038	0.0588	0.0988	0.1344	0.1544

1.1.4.2 Eccentric Presses. The external construction of the eccentric presses resembles that of the spring presses. It is different in that counter pressure on the ring is not by spring action, but by an eccentric (Fig. 6). The eccentric is attached to the lower beam. In pressing the weight attached to a lever arm presses upon the ring via lever transmission.

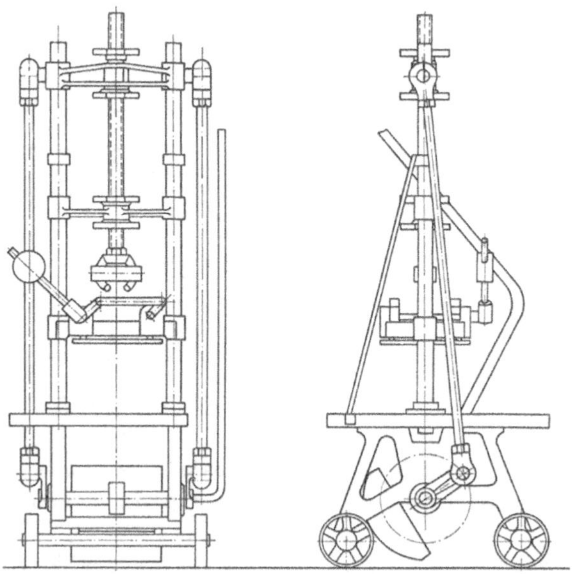

Fig. E 6. Eccentric press (Kutzscher model).

Relief is by a chain attached to the upper beam and the lever arm. When the plunger rises the lever arm and the weight are lifted. Crank shaft and lower shaft heads are provided with ball bearings in manual presses. The operational range of this press type is the same as that of spring presses.

1.1.4.3 Toggle Presses. The toggle press (Fig. E 7) may achieve a maximum force of 3500 kp at 2 atm. Between the two columns hangs the compressed air cylinder which can be adjusted in height by a hand wheel. The piston rod acts upon a lever transmission. On the lower beam the plunger spring equipment is attached as in the other press types. One has to see to it that the press is always adjusted in a way that the dead point of the toggle is not reached. Otherwise the plunger force becomes too large and joints as well as molds are damaged when the plunger reaches the lowest pressing position. The press piston is controlled by a piston baffle which is activated by a manually operated shaft. The adjustment of air pressure is by a reduction valve. The toggle press is particularly suited for the fabrication of containers of small height and large diameter, since the lever transmission permits the exertion

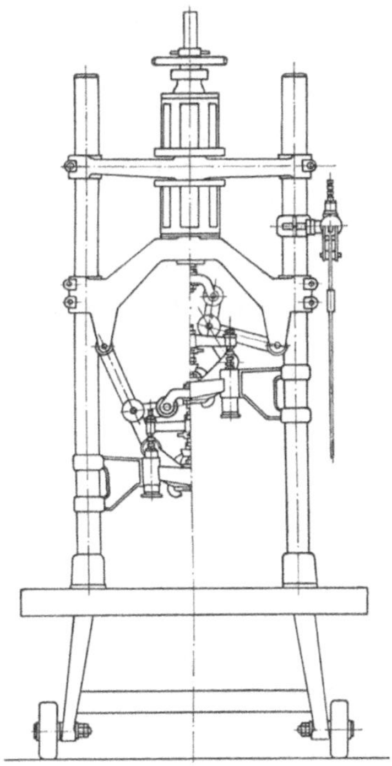

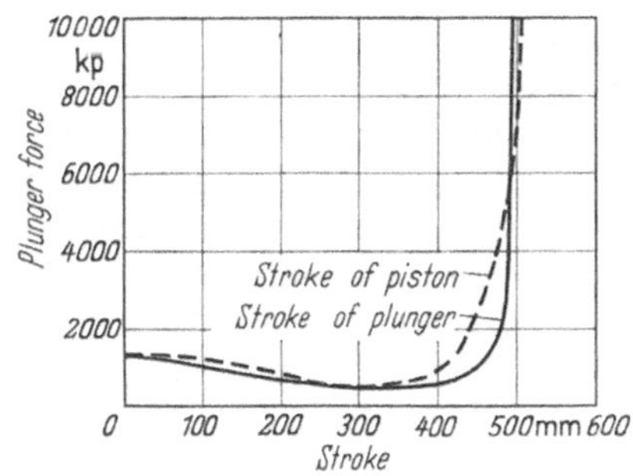

Fig. E 8. Stroke and pressure of toggle press.

Fig. E 7. Toggle press (Putsch model).

of relatively large forces upon the glass. As can be seen from Fig. E 8, the magnitude of the force varies with the stroke. For tall containers of small diameters hand filling does not permit exact control of glass weight. If the stroke is too great for too small a glass weight the article may be cracked. In all air pressure presses the stroke should not be larger than absolutely necessary, to conserve air consumption. Table E 1 contains data on number of pieces, personnel requirement, pressing times and mold number for some characteristic press ware (literature on pressing: see Sec. E. 1.2.2, p. 186).

1.2 Automatic Presses

By K. Groothoff, Mainz

1.2.1 Operation of Mechanical Presses

For pressing glass articles required in large quantities fully automated machines are used. The most current type is the rotary press with several molds on the table. The drive of the table and the control of the individual operations are automatic. The machines either are controlled by compressed air or are driven by a continuous motor via Geneva gears.

The diameter of the table varies, depending upon the number and size of molds, from 1 to 2.4 m. The machine weight depends upon the size of the articles produced. Automatic presses are used for the pressing of tumblers, plates, dishes, and similar objects of various sizes. Handles, if present, are pressed on in the pressing operation. Pressed glass is often stronger than porcelain. The weight of pressed ware that can be produced on automatic presses ranges from 30 g to several kg.

The number of molds on one table depends on weight, size and thickness of the pressed article, the cooling time required before it can be removed from the mold, and the mold cooling time. The glass type and the glass temperature also play a part. There are machines with 6, 8, 10, 12, 16 molds. The molds are always equally spaced. The table rotates

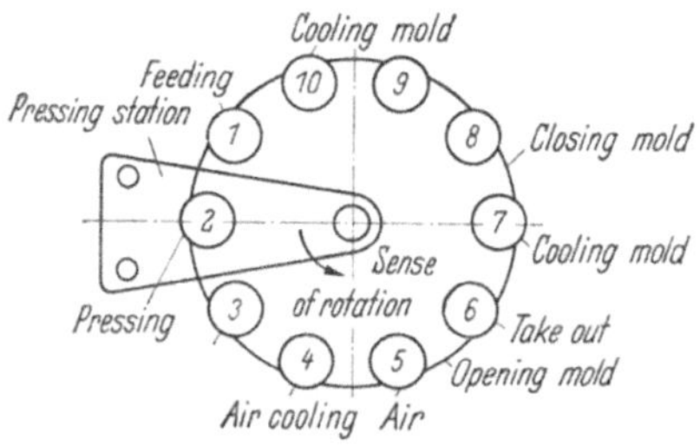

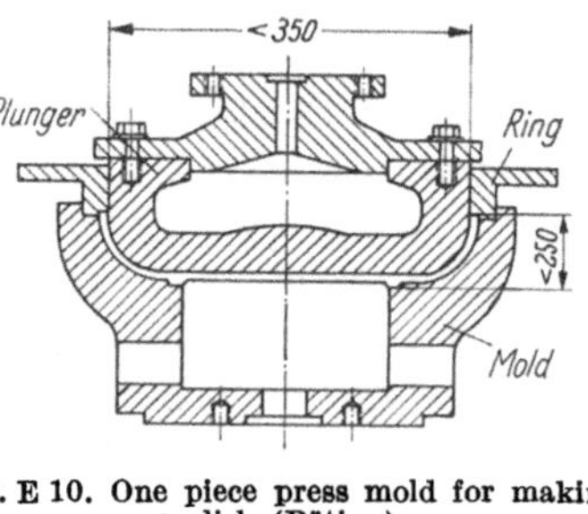

Fig. E 10. One piece press mold for making a dish (Pöting).

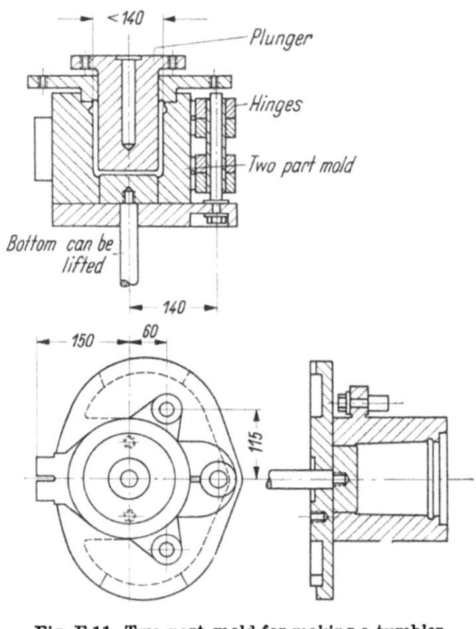

Fig. E 11. Two-part mold for making a tumbler (Pöting).

intermittently so that each mold in turn reaches a station where it stays long enough for proper operation. Drives are pneumatic or Geneva gears. Pneumatic drives prevail.

The operation of automatic presses is briefly as follows: (Fig. E 9). In *station 1* the gob is received by the mold. The cutting of the gob on the feeder is automatic in the rhythm of the table rotation and is usually controlled by it or by a special timer. Since the flow of the glass to the automatic press must be in the rhythm of the machine the flow must come from a tank of sufficient capacity. When the mold has received the gob the table proceeds one station and the mold reaches *station 2.* Here pressing proper occurs. The plunger with the ring which is used in all molds is pressed into the mold by compressed air or toggle lever. In this forming operation the mold forms the outer contour of the article, the plunger the inner contour, and the ring the upper edges. In general massive steel molds are used (Fig. E 10). If undercutting is present in the shape, two or more part molds are necessary (Fig. E 11). The kind and

Fig. E 12. Mold mounting.

Fig. E 13. Automatic take out on a press for jar lids.

shape of the article determines the shape of the mold. Press molds are fastened to the press table via suitable intermediate plates. The exact adjustment of press molds is carried out by means of nuts having gliding parts (Fig. E 12).

It is of decisive importance that the pressure during the pressing process is scaled correctly as to magnitude and duration. The plunger must make the glass fill the mold completely and must remain in the glass until it holds its shape. The release must be gradual so that the glass s not sucked along. The automatic press is adjustable to assure optimum times and temperatures. If particularly smooth surfaces are to be attained, chrome plated molds are used. For small articles the gob is distributed to several molds by the plunger. Some articles must be pressed upside down (lemon squeezers). In general, glass is pressed at one station of a one table machine, i.e., pressing is a one step operation. Recently, hydraulic plunger machines have also been used.

Stations 3, 4 (Fig. E 9), or more, serve to cool the glass to make it rigid enough for removal from the mold and take-out. Thickness and cooling time determine the number of cooling stations and, thereby, the number of molds per press.

In *stations 6 and 7* the pressed, rigid glass article is removed from the mold. This step is strongly influenced by the mold and the glass weight. The simplest method is blowing from a lateral air nozzle and subsequent manual removal. In one part molds usually an ejector lifts the glass to make it accessable to hand or tong removal. Fully automated methods use synchronized tongs. Tilt, shove or vacuum lift equipment is used (Fig. E 13). Many variations are found in practice. Even though most articles are removed automatically, there are always items requiring manual removal.

In the last stations (in Fig. E 9, *stations 8 to 10*) the molds are cooled. In general the glass stays in the mold for 6 to 10 stations. The remainder is for cooling the molds. The productivity of the machine (cutting rate, lift (stroke) number) is principally determined by the time the plunger must remain in the glass until the article has become rigid.

To attain maximum productivity, even for difficult or particularly large objects, the number of molds is often increased to 12, and sometimes 16, molds per table. The output is generally between 10 and 60 pieces per minute. In the case of automatic machines the feeder has contributed most to the increase in pressing rate. One may even feed two presses with one feeder.

The surface quality of pressed glass depends greatly on the surface quality of the molds used. If particularly high demands are made on surface quality the surfaces of the object are reheated above the softening temperature at subsequent stations. In this manner fire polish is achieved and seams are removed. If fire polish cannot be achieved because of the danger of deformation or particularity of structure (reflectors), it is best to work with chromium plated molds.

The position of the gob in the mold, particularly the shear cut, is very important when high quality is necessary (TV screens). Special guiding plates at the shears result in giving the gob an impulse directing the shear cut to a mold location causing a minimum of disturbance. One always minimizes the time between gobbing and pressing to diminish local cooling of the gob while it rests in the mold. A remarkable solution, so far used only in press-blow machines, is exemplified in the IK machine of the Knox Corparation (Sec. E 2).

1.2.2 Descriptions of Machines

The most important data for some fully automatic revolving table presses are tabulated in Table E 3 (p. 182). In Fig. E 14 a modern heavy

Fig. E 14. Heavy table press for 6 to 12 molds (model PBM of Lynch Corp.) (Courtesy Lynch Corp.).

spring press (Lynch PBM) is reproduced. The table can hold 12 molds. Opposite the pressing station one recognizes the take-out. Glass presses are among the largest and most productive of all glass machines.

Table E 3. *Technical data for the*

Drive	Brand	Type	Max. diam. of glass (mm)	Max. number of strokes/min	Mold number	Table diam. (mm)	Wt. (t)	Drive (PS or as given)	Cooling air (PS or as given)
Pneumatic	Pöting	PI	160	29	8	1320	4.6	22	5
		PII	250	17	8—12	1800	5.7	22	5
		PIII	350	15	8—12	2000	12	62 (Incl. feeder)	3600 m^3/h 350 mm WC
	Miller (Columbus)	O	100	30	8—12	1100	1.36	1.4 m^3 air/min (2 atü)	
		SJ	80	22	8 (Block)	1100	1.5—2	30	12
		T	80	60	8 (Block)	1370	1.36	30	12
	Miller Swissvale	AGO	~250	30	8—10—12 (Block)	1500	3.6	2,1 m^3 air/min (2,5 atü)	1,5

fully automatic revolving table press

Remarks

The table is rotated by compressed air cylinders with air cushioning permitting smooth operation. The drive is activated by a centrol cam on the feeder control shaft to assure synchronization of press and feeder. Table and molds are stabilized from below by pinions which at the same time release the valve activating the plunger. Thus the plunger can not be activated unless the table is fixed in the pressing position. The raising of the plunger is activated as soon as the piston driving the table approaches the end of its return stroke.
The speed of the return stroke can be adjusted by a throttle. This permits adjustment of the dwell time of the plunger in the mold. For manual or mechanical takeout the object is lifted by a pinion. The press is suited for multipart molds. A control at the center column opens and closes the molds automatically. A compressed air cylinder activates the clamps holding the mold parts together during pressing. The press is used for the fabrication of salad dishes, jelly glasses, jar covers, boxes, tumblers and champagne glasses.

A simple change in the drive and the timing cam permits operation with 4, 6, 8 or 12 molds. Control cams and equipment are essentially identical with those on models I, II. The press is characterized by an accessory part which permits the prolonged cooling of heavier pieces. Each mold has two bottoms which are connected and can be rotated by 180°. Once a mold is filled, the hinges open, the piece is lifted and rotated by 180°. A second bottom is lowered and prepared for the next charge. The first piece pressed remains on the first bottom during one complete revolution of the table. Takeout occurs only just before the second bottom is lifted.

The structure is quite similar to that of the Pöting presses. Feeder and table are synchronized. The table can be fixed, multipart molds can be held closed, and pressed pieces lifted. The plunger is water cooled. Stroke and plunger dwell are adjustable. The machine is particularly suited for small objects up to 85 g. A larger variation serves 12 molds.

The intermittent motion of the table is activated by an air cylinder mounted in the base plate. Feeder and press are synchronized. The molds are lifted against the plunger. The plunger is held by a yoke that rests on three columns and can be adjusted vertically and horizontally. The pressed object is taken out mechanically and can be caught conveniently by tongs. Cooling air is distributed by tubes from the center column.

This press is particularly suited for small objects. The outer diameter of the molds is 305—254 mm, for 2—4 part molds 254—203 mm. Molds are opened and closed by control cams. The ring is not pressed by a spring box but by a toggle lever. Lift and takeout are adjustable alongside the table.

Table E 3

Drive	Brand	Type	Max. diam. (mm)	Max. number of strokes/min	Mold number	Table diam. (mm)	Wt. (t)	Drive (PS)	Cooling air (PS)
Pneumatic	Miller Swiss-vale	AGT	∼350	40	8–10–12 (Block)	1930	7.2	2,8 m^3 air/min (2,5 atü)	2
	Lynch	PBS	125	40	6–12		1.35	5,3 m^3 air/min (3 atü)	
		PBM	250	40	6–12		5.19	9–12 m^3 air/min (3 atü)	1100 m^3/h 200 mm WC
		PGA	300	30	6–12	2020	7.7	10 m^3 air/min (3 atü)	
		PGC	400	25	8–10	2260	11.5	10 m^3 air/min (3 atü)	
		PGT	300	25	8–16	2180	11.5	10 m^3 air/min (3 atü)	24000 m^3/h 320 mm WC
		ADP	400/750	25	6–12	2180	11.5	10 m^3 air/min (3 atü)	
		ELP	600	10	11 (6–20)	∼2200	25	13 m^3 air/min (5 atü)	
Mechanical	Kutzscher	RP1	160	25	8	720	3.5		
	Lynch	MPL	300	50	12	1600	7.2	4	
		MPS	254	55	12	1400	4.5	4	
		MDP	300	55	12–16	1420–1830	∼7	3–5	

(continued)

Remarks
This press is designed for larger objects, otherwise it ressembles the AGO. Outer diameters: 406—330—279. Inner diameters of the air cylinder; 307—457 mm.
This press is designed for small reflectors and lenses. Takeout is by tipping molds which deliver to the conveyor. It has a conventional compressed air drive.
The press serves the production of medium sized objects such as cups, tumblers, jars, ash trays, small dishes (see Fig. E 14).
The press is driven by compressed air. About 24,000 m^3/h cooling air is required. The cylinder acts directly on the plunger. Building blocks, large dishes and plates can be produced.
These models are driven by compressed air. Cooling as in model PGA. Toggle press lever activation. Automatic takeout.
Compressed air drive. Cooling as above. Press mechanism is supplied with or without bell crank. The table is supplied solid or with air cooling. Building blocks and television parts can be produced.
This is the largest model now conventionally used. The pressure is 120 t. The table is air or water cooled. Four guiding columns improve the motion of the plunger. Large objects and television parts can be produced.
The table is activated by a Geneva drive.
The presses are driven by a motor via Geneva drive. The rest of the equipment is as usual. Medium sized objects such as tumblers, saucers, cups, dishes, ash trays, lemon presses, etc. can be made.

As an example for the outputs of a smaller press (Pöting PI) some data are collected in Table E 4.

Table E 4. *Data and output of a glass press* (Pöting PI)

Job		Wt. (g)	Ht. (mm)	Diam. of body (mm)	Diam. of finish (mm)	Pcs./h
Largest:	Dish	375	55	155	155	800
	Salad bowl	500	90	160	—	900
Medium:	Champagne	250	90	94	94	900
	Tumbler	250	90	70	—	1000
	Jelly jar	340	102	94	82	800
Smallest:	Box	50	35	50	45	1300
	Lid (jar cover)	150	—	90	—	1200

References

[1] Gastew, J. A., and E. G. Frolowa: Bewegung des Glases in den Formen während des Pressens (from the Russian). Steklo i Keram. 10 (1953) No. 6, pp. 8 to 11.

[2] Nichols, N. A.: Some mechanical problems of glass presses. J. Soc. Glass Technol. 38 (1954) 34—48.

[3] Tooley, F.: Handbook of Glass Manufacture, New York 1953. Section: Holscher, H. H.: Feeding and Forming, pp. 303—387.

2 Press-Blow Machines

By K. Becker, Obernkirchen

As implied in the designation, two phases, pressing and blowing, are combined in one machine. The press-blow process is suited for the fabrication of wide-mouth ware such as mason jars, tumblers, milk bottles, etc. The process is efficient and permits the fabrication of light weight bottles of uniform wall thickness. The beginnings of this method were in the period of development of semiautomatics. The process later was taken over by fully automatic machines. In the semiautomatic process two mechanisms are used: the press and the blowing machine. The glass taken from the furnace by the gatherer first enters the press mold; here finish and neck receive their final shapes. The preformed gob is then transferred to a second mold and blown to final shape. For this working process two methods are known (as already mentioned in Sec. C 3): the transfer process according to Arbogast, and the Windmill drop-off process. The Arbogast process permits a simpler construction. The gob is pressed by plunger, and finish as well as upper neck are fashioned in this phase. Then the plunger is withdrawn, the gob and neck mold are lifted from the parison mold and transferred to the blow mold. After closure of the blow mold the blow head is set on the mold and the article is blown to final shape. The Windmill process uses the same principle. The only difference is that the parison is not transferred after pressing but stays in place. The press mold is lowered and the blow mold with bottom plate moves to the parison. The blow head is set on the blow mold and the parison is blown out. It is an advantage that in this process the parison does not have to be moved within the production unit; this permits faster operation. Since the gob is prepressed in both cases the finish and the final object must be wide enough for the insertion of the plunger.

In the beginning of this development the Arbogast process had an advantage: the parison could be made on existing presses, and blown on the blow machine. The Windmill process required a new construction. For more detail on both processes see also C 3.

The contemporary automatics can all be traced to one of these two processes. The Windmill process is the basis for the operation of the Lynch JP machine, the Pöting PbI and 31 machine, and the Hartford M 28 machine. This latter is a special type for the production of paste-mold tumblers. It is described in Sec. E 3.3.6.1. The Arbogast process may be compared to the fabrication of press-blow ware on the Hartford IS machine, the so-called process 62. More detail on this process is given in Sec. E 3.3.3.

2.1 Fully Automatic Press-Blow Machines by the Pöting Company

2.1.1 Operation of the Machine (Fig. E 15)

The working scheme of this machine is exemplified in the following description of the Pöting PB II B press-blow machine. In station 1 the parison mold is lifted by a rod running through a cylinder along a cam controlled path as it travels to station 2. The blow mold is open, the finish or neck ring closed, while the bottom plate of the blow mold is swung to one side. Station 2 is the loading station where the feeder formed gob reaches the parison mold by free fall or from a trough. The parison mold is moved under pressing station 3 where the water cooled plunger is pressed down by an air cylinder and shapes the gob. In this phase the finish is formed. Since the neck ring is split it is necessary to clamp it closed to avoid opening of the ring and formation of seams.

Below the parison mold in the machine base is found the parison track which slopes downward from station 3 to station 4. The parison mold thus drops so that from station 4 the parison is suspended from the neck ring. Now the blow mold bottom plate is moved in radially by an air cylinder so that it lies beneath the parison, which sits on it for a few seconds as the glass stretches (Fig. E 17). On the way from station 4 to station 5 the two blow mold halves are closed by shafts with rollers

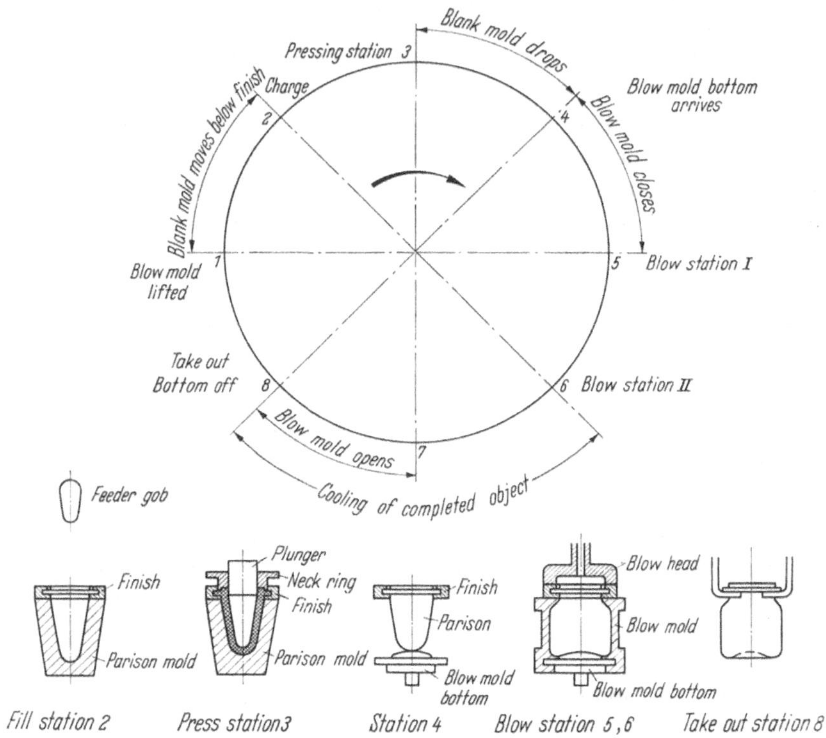

Fig. E 15. The working scheme of a press and blow (Pöting PB II B) machine.

which run on a track fastened to the center column. At station 5 the blowhead is set on the blow mold and the article is blown to final shape. At this stage the blow mold is clamped.

In station 6 the blow process is repeated and the blowing interrupted between 5 and 6 is completed. The article remains in the blow mold until station 8. On the way it is cooled and the mold opens. On the center column above station 8 is fixed a takeout mechanism which takes the article out of the mold and places it on a belt which transports it to the annealing lehr. The bottom of the blow mold in the machine is pushed outward ready for the rise of the parison mold.

2.1.2 Drive and Construction of the Machine

Fig. E 18 shows the schematic construction of the Pöting PB II B machine. Around the center column turns the mold table which is equipped with 8 molds. The machine is driven and controlled pneumatically.

Fig. E 16. Pneumatic revolving table machine with 8 molds (intermittent process) (Courtesy Pöting).

Fig. E 17. Parison just before closing of blow mold in a press and blow machine.

By a compressed air cylinder via a swinging arm and a drive cylinder the table with molds moves intermittently from station to station. During pauses the table is locked by pneumatic bolts. The scheme of this revolv-

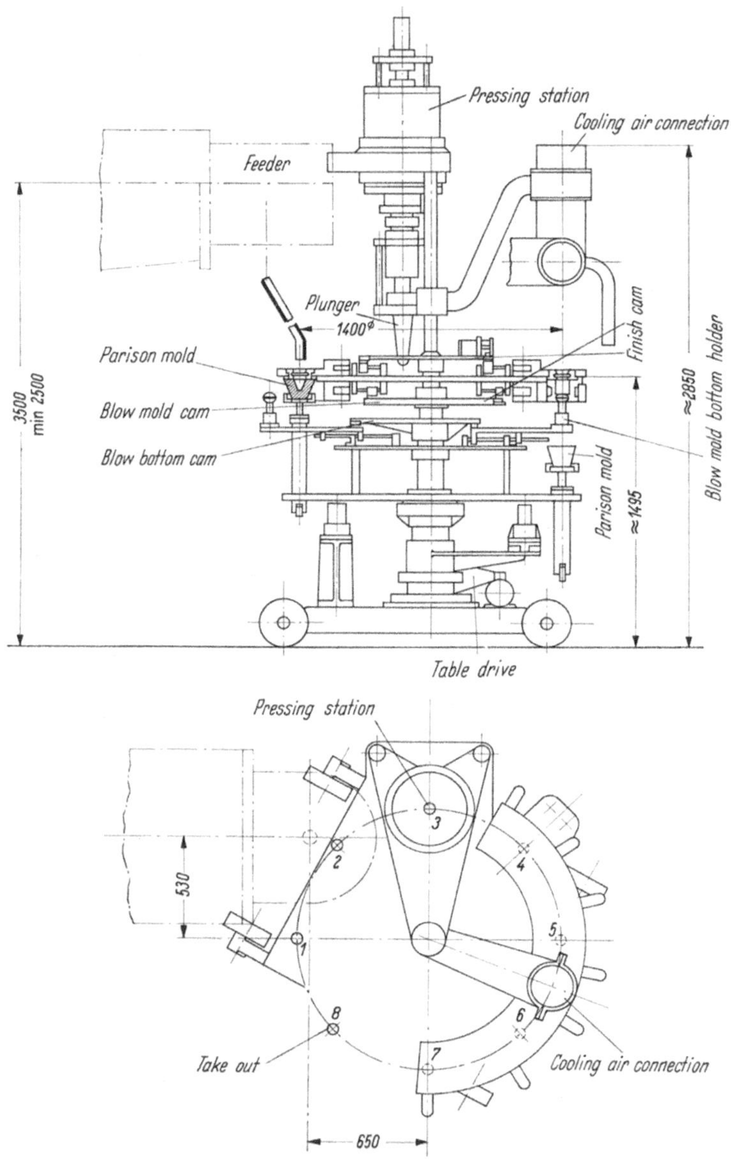

Fig. E 18. Fully automatic press and blow machine PB II B (Courtesy Pōting). Cams activating parison mold omitted to avoid confusion.

ing table drive is reproduced in Fig. B 8, Sec. B. 1.2.3. Fig. E 19 shows the air piping plan. The first impulse comes from the timer which is

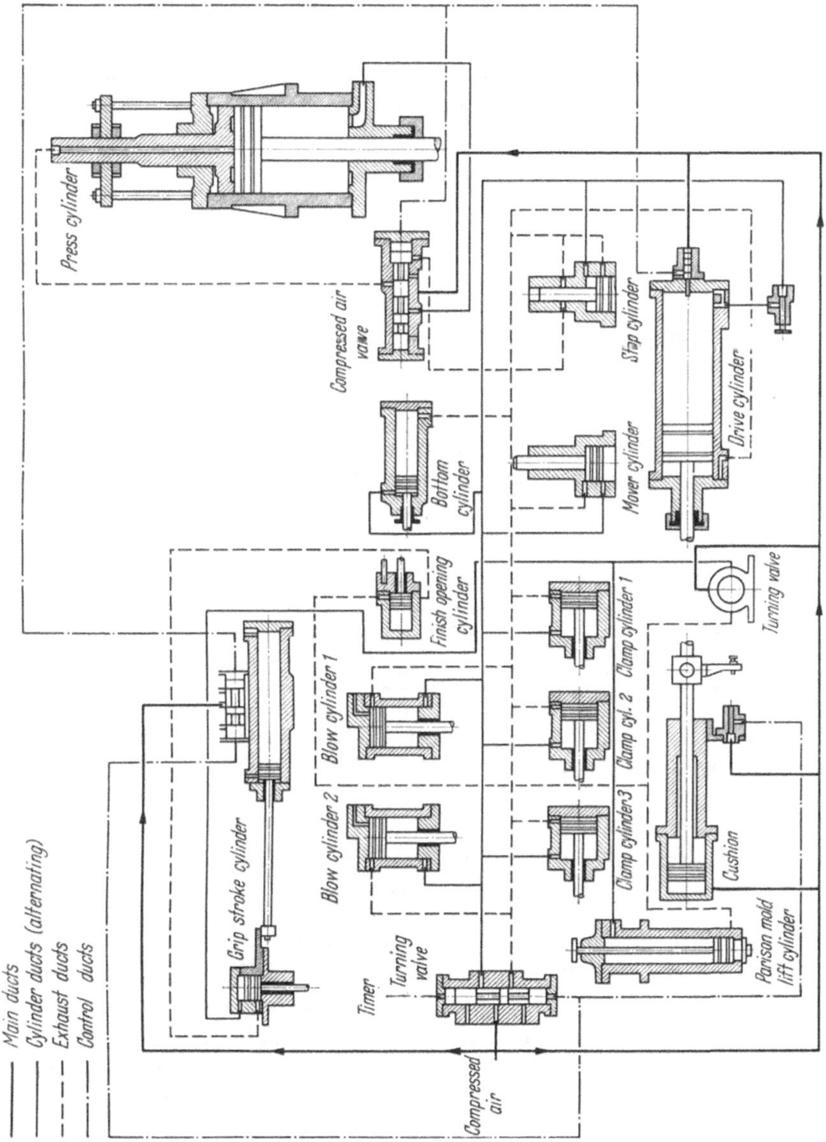

Fig. E 19. Air line plan of a press and blow machine PB II B.

synchronized with the feeder. Further control phases are induced in the machine as can be seen from the piping plan.

2.1.3 Technical Data

1. Personnel requirements: 1 machine foreman
2. Mold requirement:

8 neck rings	1 plunger
8 parison molds	1 sleeve
8 blow molds	1 pair of takeouts
8 bottom plates	

A sufficient reserve is to be provided.

3. Dimensions and energy requirements:

Base area:	2.2×2.2 m
Height:	3.0 m
Net weight:	6 t
Distance glass level-glass house floor:	2.5—3.5 m
Drive and control air:	3 kp/cm²
Cooling air:	250 mm water column

4. Maximum glass dimensions

In addition to model PB II B Pöting also makes types PB I and 31. Operation and construction are not too different. The chief differences are the dimensions: (see Table E 5).

Table E 5. *Dimensions of manufactured articles*

Type	PB I		PB II B		Mod. 31	
Dimension (mm)	max	min	max	min	max	min
Finish diam.	—	—	123	50	—	—
Body diam.	80	40	160	60	170	—
Height to below finish	110	20	240	100	300	60

2.2 Fully Automatic Press-Blow Machines

By the Lynch Co., USA

In Table E 6 technical data are given for four Lynch machine types used for a variety of wide mouth containers. The WM 10 (JP) machine is most used in practice. It is suited for round and cornered articles. The machine is driven mechanically. Stepless variable gears drive the worm gear of the Geneva drive to adjust the speed of the machine. An exterior Geneva drive is used for a movement of 36° (10 stations). The Geneva drive is located in the machine base and runs in an oil bath. Behind the variable drive is a clutch which permits stopping the machine without stopping the feeder mechanism. If the machine works with a pneumatically activated feeder the synchronization is by machine driven

Table E 6. *Technical data of different Lynch press and blow machines*

Type	WMS-10 (JPS)	WM-10 (JP)	JPM	JPL
Drive	E-Motor (3 PS, 1800 rpm)	E-Motor (3 PS, 1800 rpm)	E-Motor (5 PS, 1800 rpm)	E-Motor (5 PS, 1800 rpm)
Number of stations	10	10	12	12
Height to below finish (mm)	136,5	219.1	219.1	266.7
Body diameter max. (mm)	114,3	127.0	146.1	165.1
Finish diameter max. (mm)	88.9	120.7	120.7	127.0
Weight of machine (t)	6.8	7.6	9.5	14.0
Height of machine (m)	2.8	2.9	2.9	3.2
Base (m × m)	2.6 × 2.2	2.7 × 2.3	3.0 × 2.3	3.2 × 2.8
Press. (atü)	2.8	2.8	2.8	2.8
Air consumption (m^3/min)	8.5	5.7	8.5	8.5
Cooling air press. (mm WC)	200	200	200	250
Cooling air consumption (m^3/min)	246	283	340	340
Water pressure (atü) for cooling plunger	1.8	1.8	1.8	1.8
Stroke/min max.	55	48	58	30
Personnel (foreman)	1	1	1	1

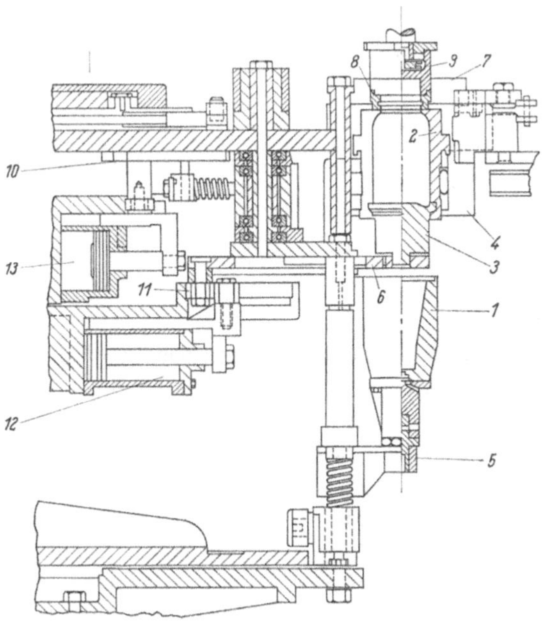

Fig. E 20. Production unit of a Lynch press and blow machine.
1 Parison mold; *2* Blow mold; *3* Blow mold bottom; *4* Mold holder; *5* Parison mold holder; *6* Blow mold slide; *7* Finish holder; *8* Finish; *9* Blow head; *10* Closing mechanism; *11* Control of bottom carrier; *12* Cylinder for bottom retraction; *13* Cylinder controlling blow and closure.

8 valve timer. In case of short disturbances or interruptions of the machine the feeder continues its working rhythm; this avoids unfavorable influences on the glass temperature that a stopped feeder would cause. For control of the air cylinder of the machine a 20 valve timer is used.

The gob formed by the feeder is guided to the parison mold by a funnel. In the press station the parison is pressed. The timing of the press process can be adjusted by the air cylinder. It is equipped with a regulator to permit synchronization of press motion and stopping time to each item of production. Otherwise, the operation coincides largely with that of the Pöting PB II B machine. After the pressing phase the parison mold drops away, the blow mold bottom plate moves below the parison, and the blow mold is closed by an air cylinder. At the next two stations the article is blown to shape. The blow molds are held closed by clamp cylinders. On further stations the article is cooled, taken from the opened mold by tongs and set on the belt that transports it to the annealing lehr. The production unit of the Lynch press-blow machine can be seen from Fig. E 20.

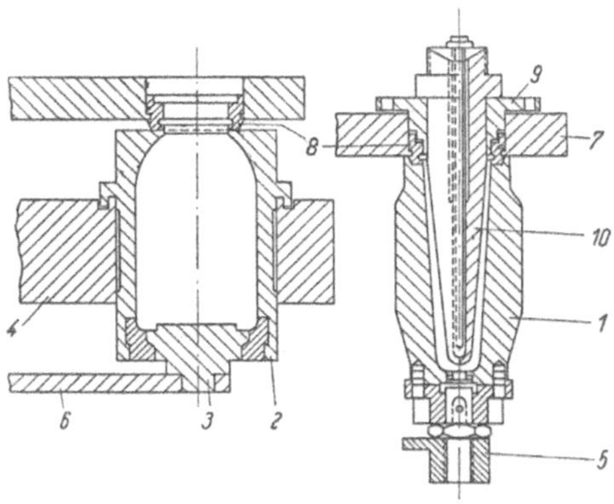

Fig. E 21
Mold assembly of a Lynch JP machine.
1 Parison mold; *2* Blow mold; *3* Blow mold bottom; *4* Mold holder; *5* Parison mold holder; *6* Blow mold slide; *7* Finish holder; *8* Neckring; *9* Sleeve; *10* Plunger.

Fig. E 21 shows the assembly of a parison mold, plunger and blow mold. The parison mold is a block mold; sometimes there is a vent in the bottom to avoid air pockets. Mold requirements are:

10 parison molds	1 plunger
10 neck rings	1 sleeve
10 blow molds	1 gob guide ring
10 bottom plates	1 pair of tongs

A sufficient reserve is required.

Output according to the manufacturer:

gram wt.	114	142	284	425	567	710	850	1140	1420	1705	1820
strokes/min	48	46	40	38	36	34	32	30	30	24	20

The description of the construction and operation of the WM 10 (JP) machine also applies to the other machines of Table E 6. They differ only in number of stations and size of articles to be produced.

2.3 Other Fully Automatic Press-Blow Machines

2.3.1 Knox IK Machine

Besides the machines already described extensively the IK machine by the Knox Co. (USA) should be mentioned. It has been developed from the Lynch JP by the Knox glass works in the USA, but deviates in its present construction from the operational scheme of the Lynch JP.

10 stations are arranged on a revolving table which can be equipped with single or double molds. The drive is mechanical, via Geneva gears. The other functions are carried out by cams and air cylinders. Parison and blow molds are arranged next to each other, in contrast to the Lynch JP. And, in contrast to all other press-blow machines, the gob of the Knox machine is dropped into the running machine a little before the press station. In this way the time between the insertion of the gob and its pressing into the parison can be reduced to a minimum; the uniformity of glass distribution is improved; and press wrinkles are avoided.

The IK machine so far has been used only in the glass houses of the Knox Co. The productivity is reported to be superior to the JP.

2.3.2 Miller MT Machine by Lynch

This is a one-purpose machine for the production of milk bottles. On the machine foundation two revolving tables are arranged to run in opposite directions. On each table 10 molds (10 parison and 10 blow molds) are attached. The transfer from parison to blow mold is by pneumatic double clamp.

Technical Data

Personnel: 1 foreman

Dimensions, weight, energy requirements (according to manufacturer):

Base area:	1.5—2.5 m
Height:	2.75 m
Wt.:	5.5 t
Control and operation air:	2.8—3.2 atm; 5.7 m^3/min
Cooling air:	200 mm water column
Cooling water for plunger:	1.8 atm
Drive 3 PS E motor:	1800 rpm.
Height of article (max.):	304.8 mm
Diameter (max.):	127.0 mm

2.3.3 Moorshead Machine

For the sake of completeness the Moorshead machine of United Glass Bottle Manufacturers Ltd. London should be mentioned. This machine is still used in England. It is a continuously rotating, mechanically driven revolving table machine with 8 sets of molds. In contrast to the machines described so far, each station is equipped with a plunger. Before the filling station the split parison mold swings out, and swings back under the plunger after the drop of the gob. The parison transfers from parison mold to blow mold by tongs. The field of application of this machine is for smaller containers.

2.4 Conclusion

In German glass houses wide mouth containers are produced mostly on the machines described here. The press-blow process of the IS machine (62 process) is increasingly used for wide mouth containers including milk bottles. This process is described in detail in Sec. E 3.3.3.

In the U.S.A. the 62 process, both single and double gob, is the one generally used for wide mouth containers. Wide use is made of another process known as the "41" process which also is applied to the IS machine. The Lynch 44 is used to some extent for double gobbing of relatively large wide-mouth ware such as quart jars for fruit juices.

References

[1] The symposium on machinery for the fabrication of glass containers. J. Soc. Glass Technol. 37 (1953) 27–96.

3 Blowing Machines

3.1 General Remarks on Bottle Production

By W. Giegerich, Heilbronn

The manufacture of a product in large quantities, in one shape, and with other identical properties, such as bottles, permits more detailed observation of individual production phases than is possible for other hollow glass ware.

3.1.1 Production Rate

The fabrication of a bottle occurs in three characteristic phases:

1. Shaping of the parison in the parison mold (parison mold time $= t_V$),
2. Reheating of the suspended parison (reheating time $= t_R$),
3. Shaping of the bottle from the parison in the blow mold (blow mold time) (t_F).

The total time (t_{ges}) is the sum of these times viz:

$$t_{ges} = (t_V + t_R + t_F) \quad \text{(sec/bottle)}. \tag{1}$$

t_V is counted from the insertion of the gob to the opening of the parison mold, t_R from the opening of the parison mold to the beginning of the final blow, and t_F from the beginning of the final blow to the opening of the blow mold.

Eq. (1) describes the time sequence of the individual operations. The cutting rate, as a measure of the productivity of the machine, cannot be calculated directly from this fabrication time, because one bottle is not formed after the other. In the same mold pair the parison shaping begins while the preceding bottle is still blowing out in the blow mold. The fabrication time per bottle t_{fab} is given by Eq. (2).

$$t_{fab} = t_{ges}/2d \quad \text{sec/bottle}, \tag{2}$$

where d is a characteristic constant of the machine defined by

$$d = t_{ges}\, 2t_{fab} = l_{eff}/l_{rel}. \tag{2a}$$

The effective output l_{eff} can be calculated from the machine constant d by means of Eq. (3a) in bottles/min per mold:

$$l_{eff} = 60/t_{fab} = d \cdot l_{rel}. \tag{3a}$$

The cutting rate L in bottles/min per machine is calculated from

$$L = z l_{eff} = 60z/t_{fab}. \tag{3b}$$

Here z is the number of machine arms, each with one parison and one finish mold.

The interrelation of these details is shown schematically in Fig. E 22. Lines *1, 2, 3* represent the functional times of three sequential operational phases: fall of gob, parison mold time, reheating and, blow mold time.

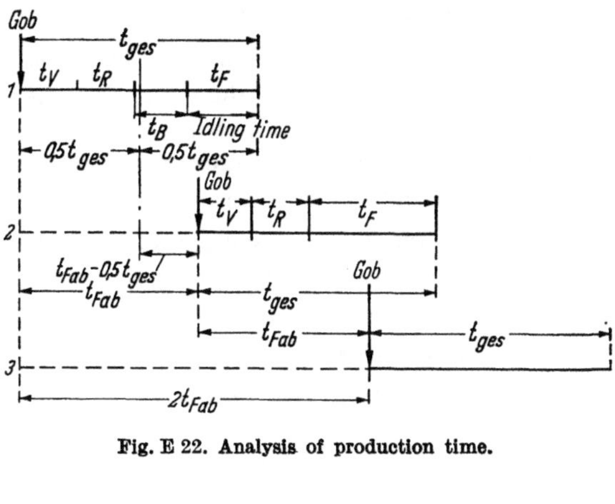

Fig. E 22. Analysis of production time.

Only part of the blow mold time is used for the blowing time proper, t_B; the little remainder $t_F - t_B$ may be termed idling time. This results from the motion required for a full turn of the machine. There is, of course, also some idling time in the parison mold phase, but it is much smaller because of the rapid sequence of suck and preblow. The start of the second gob fall is at the interval of the fabrication time t_{fab}. The second gob does not fall exactly after half the total time t_{ges}, but a little later, depending on constructional and operational detail, namely by the small time $t_{\text{fab}} - 0.5\,t_{\text{ges}}$. This time shift, which may be classified as a true loss, can be expressed by the machine constant $d = t_{\text{ges}}/2t_{\text{fab}}$ [see Eq. (2a)]. The production times thus are classified as follows:

1. Primary times: parison time, reheating time and final blow time corresponding to suction, pre-blowing and final-blowing.

2. Secondary times: opening and closure of parison, blow and finish molds; insertion and withdrawal of plunger, knife, bottom, preblow cover; idling time during blow mold time $t_F - t_B$; and loss time $t_{\text{fab}} - 0.5\,t_{\text{ges}}$.

The main portion of the loss times originates in the type of drive used. The machine constant d has about the following values for conventional drives and machines[1]:

Oscillating drive:	B, IS	0.89
Rotating drive:	R7, AR	0.66
Intermittent drive:	A6, L10	0.58.

Assuming one needs a production time $t_{\text{ges}} = 20$ sec for a certain bottle type, Eq. (2) leads to the following fabrication times t_{fab} for these three drives: 11.1; 15.1; 17.1 bottles/ minute. Future development will most likely bring a transition from intermittent to rotating and oscillating machines to attain high values of d and thus high mold outputs.

3.1.2 Time Measurements

An example of exact time studies on several feeder and suction blow machines for a bottle weight of 520 g is shown in Table E 7. Such a time study permits a comparison of the production of all machines and particularly illustrates idling and loss times. Such a comparison is illustrated in Table E 8. In detail one sees:

1. that the reheating time is on the average 11% longer for the feeder-blow process than for the suction blow process,

2. that the blow mold time is about 80% longer for the suction blow process than for the feeder process and,

3. how the times for suction and pre-blowing are distributed.

[1] For abbreviations see footnote Table E 7.

Table E 7. *Time relations in seconds in the manufacture of a 520 g bottle*

Functions		Feeder-blow[1]					Suction-blow[2]			
		IS	L 10	LA	Sev.	R 7	OS	AR	B	A 6
Start of charge and suction $t = 0$										
Settle blow (suction)	Start	0.13	0.10	0.35	1.71					
	End	0.53	0.96	1.35	2.55	0.36				
Lift parison mold							0.77	0.72	1.74	1.46
Cut							1.35	1.07	2.10	1.61
Suck ends							1.90	1.30	2.90	2.95
1. Parison blow	Start	1.20	1.64	2.70	2.70	1.90	2.40	2.00	3.00	3.40
	End	4.55	3.35	4.60	3.80	4.30	3.20	3.60	3.30	4.60
Parison mold opens		4.60	4.20	6.40	4.80	4.60	4.40	4.80	5.60	6.50
2. Parison blow	Start				5.70				4.00	
	End				7.00				4.30	
Blow mold closes		7.60	7.10	10.2	10.2	7.30	6.95	7.60	7.30	9.70
Final blow	Start	8.60	10.4	16.1	11.3	9.10	7.50	8.50	8.80	11.6
	End	13.6	14.8	17.7	18.6	14.1	10.5	13.2	14.2	17.4
2. Gob		9.00	13.9	19.4	13.2	10.7	13.9	16.2	13.1	18.2
2. Final blow	Start			19.4						
	End			20.9						
Blow mold opens	Start	14.2	15.8	22.0	18.7	14.3	18.2	19.5	18.8	21.2
	End	15.1	16.2	22.6	19.1	15.1	18.7	20.1	18.9	22.8

[1] IS = IS-Machine (4 stations)
L 10 = Lynch-10-Machine
LA = Lynch-A-Machine
Sev. = Severin-Machine (semi-automatic)
R 7 = Roirant-Machine (7 arm)

[2] OS = 6 arm Owens-Machine (type Schwartzkopff)
AR = 10 arm Owens-AR-Machine
B = 1 arm Roirant-Machine
A6 = 6 arm Roirant-Machine

Table E 8. *Time relations in seconds and mold output in the manufacture of a 520 g bottle*

Designation of times		Feeder-blow			Suction-blow		
		IS	R 7	L 10	B	AR	A 6
		Drive					
		Oscillating	Rotating	Intermittent	Oscillating	Rotating	Intermittent
Parison mold time	t_V	4.6	4.6	4.2	5.6	4.7	6.5
Reheat time	t_R	4.0	4.5	6.2	3.2	4.8	5.1
Blow mold time	t_F	5.0	5.0	4.4	5.4	4.7	5.8
Blowing time proper	t_B						
Remainder = idling time	$t_F - t_B$	1.5	1.0	1.4	4.7	6.9	5.6
Total time	t_{ges}	15.1	15.2	16.2	18.9	20.1	22.8
Fabrication time	t_{fab}	9.05	11.2	13.6	9.5	15.1	19.1
Loss time	$t_{fab} - 0.5\,t_{ges}$	1.5	3.65	5.5	0.05	5.05	7.7
Sum (idling + loss time) in % of t_{fab}		26.6	39.4	49.4	25.3	67.5	65.0
Mold output							
$t_{rel} = 120 : t_{ges}$ (bottles/min/mold)		7.94	7.49	7.4	6.34	6.0	5.28
$l_{eff} = 60 : t_{fab}$ (bottles/min/mold)		6.65	5.35	4.42	6.3	3.97	3.14

The total production time is greater for the suction blow machine than for the feeder blow machine. Striking, too, are the differences in idling time, 1—1.5 sec for the feeder machine, 4.7—6.9 sec for the suction blow machine. Time losses in the sequence are also quite different in different cases. They vary from 0.05 to 7.7 sec. Here the drive is more important than the machine type, e. g. for oscillating drive (IS, B machines) 1.5 and 0.06 sec; for rotating drive (R 7, AR) 3.65 and 5.05 sec; and for intermittent drive (L 10, A 6) 5.5 and 7.7 sec.

Another characteristic number of merit is the relation of idling plus loss time to the fabrication time. It varies from 25.3 to 67.5%. The meaning of this relation can be seen from this example: for 25.3% a production time of 14.2 sec corresponds to an effective output of 6.3 bottles/min/mold; but for 67.5% to only 3.97 bottles/min/mold.

3.1.3 Interrelation of Functional Times

It is known from experience that the first forming phases are decisive for the forming process and the quality of the product. The *parison mold time* is particularly important. Everything else being comparable, it depends essentially upon the article weight (Fig. E 23); in the figure the lines represent the average values from time measurements on various suction blow machines (Owens OS AR; Roirant B, A 6) and feeder blow machines (IS, Roirant R 7, Lynch LA and L 10 [2]. For the most commonly used machines the lines are so close to each other that crude averaging (independent of process) seems permissible, perhaps in the following form:

$$t_V = 2.88 + 0.00346\ \mathrm{g} \tag{4}$$

(g = gram weight). The dotted line of Fig. E 23 corresponds to this equation. For a given machine the values for various operations are on

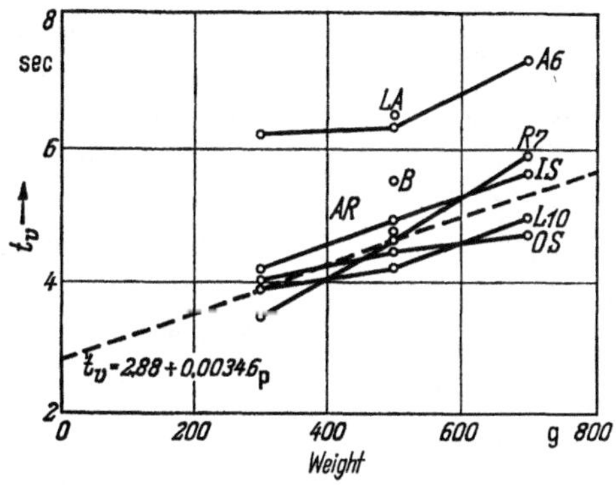

Fig. E 23. Parison mold time and glass weight.[1]

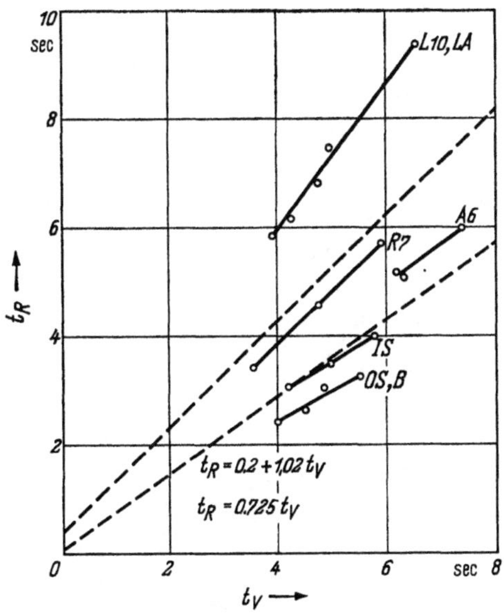

Fig. E 24. Reheat- and parison mold times.[1]

corresponding parallel lines. The intersection with the ordinate does not appear to depend much on machine construction but rather on operation.

[1] For abbreviations see Tables E 7, footnote, E 8, and text.

The *reheating time*, however, depends upon the process, the machine type, and the parison mold time. The longer the latter is extended, the more time will be consumed in equalizing inner and outer glass wall temperatures as required in the subsequent finish blow. The relation $t_R = f(t_V)$ (reheat time as a function of parison mold time) is shown in Fig. E 24. Here differences between machines are considerable. But one does see that the reheating time is longer for feeder-blow than for suction blow machines. The magnitude of the corresponding times follows from the relations:

$$t_R = 0.2 + 1.02\, t_V \quad \text{(for feeder-blow)}, \qquad (5)$$

$$t_R = 0.725\, t_V \quad \text{(for suction blow machines)}. \qquad (6)$$

The corresponding straight lines are the dotted lines of Fig. E 24.

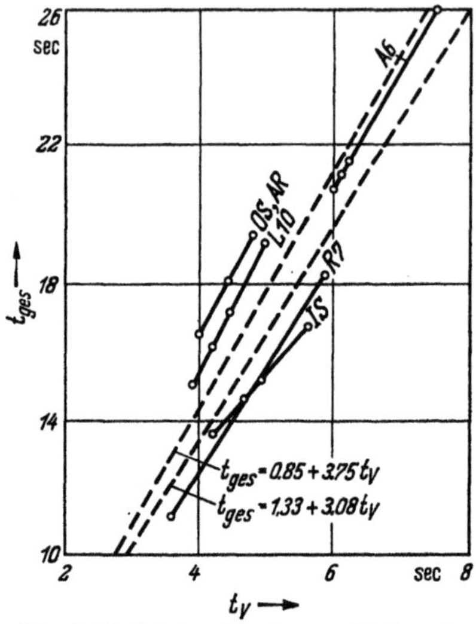

Fig. E 25. Total and parison mold times.[1]

If the total production time (t_{ges}) is plotted against the parison mold time (t_V) (Fig. E 25) a certain coincidence of processes and machine

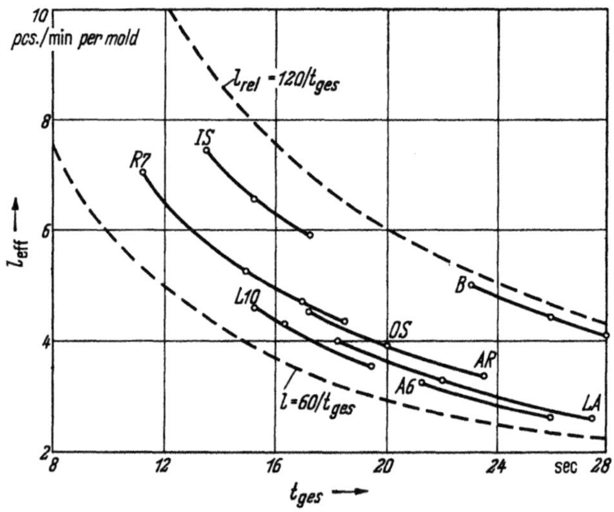

Fig. E 26. Effective mold output and total time.[1]

[1] For abbreviations see Tables E 7, footnote, E 8, and text.

times becomes apparent. It therefore, becomes possible to form averages from the dotted lines as follows:

$$t_{ges} = 1.09 + 3.32\, t_V. \tag{7}$$

The production time thus depends strongly upon parison mold time. From the production time (t_{ges}) the fabrication time (t_{fab}) can be calculated from the relations given before, and from it more explicitly *mold efficiency* l_{eff} in pieces/mold/min.

$$l_{eff} = 60/t_{fab} = \frac{60 \times 2d}{t_{ges}} = \frac{120d}{t_{ges}}. \tag{8}$$

Eq. (8) is illustrated in Fig. E 26. Between the dotted limiting curves for $d = 0.5$ and $d = 1$ lie the efficiencies of machines used today. It is striking that production in all single phases can be effected within large time ranges. The derived relations presuppose equally good operation.

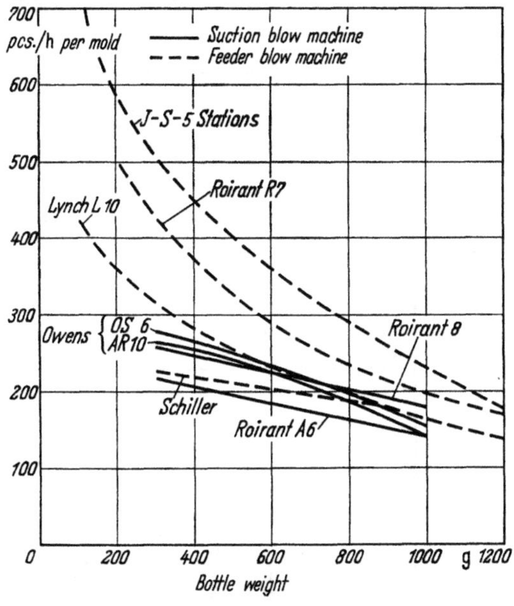

Fig. E 27. Machine output of some bottle blowing machines.

3.1.4 Comparison of Parts and Capacities of Various Machines

The obvious question as to the most efficient machine cannot be answered generally but Table E 9 shows some comparison. If efficiencies are compared (Fig. E 27) one recognizes that each machine has preferred areas of articles for which it works best. If one has to choose a new machine, he has to start from a program specifying kind and weight,

Table E 9. *Survey of most common bottle blowing machines*

Characteristics	Suction-blow						Feeder-blow				
	Owens	Roirant B and F	Roirant A 6	Hillmann	O'Neill 80	Monish Major	J-S-Machine	Lynch	Owens A Y	O'Neill	Roirant M 8 and R 7
Number of arms or stations	6, 10, 15	1	6	10	8	3	2, 4, 5, 6	6	10, 15	8	8, 7
Feeding: Suction	+	+	+	+	+	+	—	—	—	—	—
Gob	—	—	—	—	—	—	+	+	+	+	+
Working principle: Suction—blow	+	+	—	+	+	+	—	+	+	—	—
Blow—blow	—	—	—	—	—	—	+	+	—	+	—
Suction—suction	—	+	+	—	—	—	—	—	—	—	+
Transfer:											
Standing, parison and blow molds move towards neck ring	+	—	+	+	+	+	—	—	—	—	—
Neck ring moves vertically or horizontally towards blow mold	—	+	—	—	—	+	—	—	—	—	—
Parison mold or neck ring rotates	—	—	—	—	—	—	—	+	+	+	+
Motion of machine:											
Continuous rotation	+	—	—	+	+	—	—	—	+	—	+
Intermittent rotation	—	—	+	—	—	+	—	+	—	+	—
Station by station	—	+	—	—	—	—	+	—	—	—	—
Number of tables:											
One	+	+	—	+	—	+	+	—	+	—	+
Two, side by side	—	—	+	—	+	—	—	+	—	+	—

job change frequency and size of orders, and then must compare efficiencies and costs (heat, wages, power, mold cost, amortization) of various machines. Additional points to consider are condition of the glass, mobility of personnel, loss of time in mold changes, technical level of the glass house, operational difficulties of the machine, quality of mold making and maintenance, simultaneous production of different articles, controls and adjustments of the machine, wear of mold and machine parts, etc. The number of these parameters demonstrates that there is no universal machine: each glass house usually needs several machine types for its production program.

The basic development of bottle blow machines now appears completed, though some advantages may be gained from refinements through the improvement of parts such as gear and drive mechanism. But the engineering and scientific treatment of heat and flow processes is only at its beginning and progress in efficiency and quality must be expected without any doubt.

References

[1] Giegerich, W.: Ergebnisse und Probleme der maschinellen Flaschenfertigung. Glastechn. Ber. 30 (1957) 299—308.

[2] Giegerich, W.: Über technische Grundlagen der vollautomatischen Flaschenherstellung. Glastechn. Ber. 33 (1960) 441—449.

[3] Creaser, I., L. G. Creaser, and F. W. Hodkin: Suction and feeder-fed bottlemaking machines. J. Soc. Glass Technol. 37 (1953) 48—56.

[4] Densem, N. E.: A contribution to the symposium on machines for the fabrication of glass containers. J. Soc. Glass Technol. 37 (1953) 57—60.

[5] Nichols, N. A.: Improvements needed in glass-making machines. J. Soc. Glass Technol. 37 (1953) 61—70.

[6] Moorshead, W. A.: A contribution to the symposium on machines for the fabrication of glass containers. J. Soc. Glass Technol. 37 (1953) 80—87.

[7] Anon.: The symposium on machines for the fabrication of glass containers, Discussion. J. Soc. Glass Technol. 37 (1953) 88—96.

3.2 Suction-Blow Machines

The suction-blow machines were the first fully automatic hollow glass machines. This is because automatic glass supply first succeeded by using the suction principle. As the name indicates these machines suck glass from a larger reservoir by means of vacuum. The first machines were placed right at the working end and an attempt was made to gather the glass directly from the bath. For small one-mold machines this principle worked, but in the case of large machines, with several molds, experience soon taught that this method was not applicable. Next Owens created the revolving pot (cf. Sec. D 2); it was only by its use that the large suction machines became operative. For small machines with several molds special constructions such as revolving discs or paddles were used for the removal of the cut glass and the glass was cooled by the dipping mold. But the development of the feeder allowed the construction of machines which surpassed the suction machines in efficiency as well as quality. It is true that suction machines can still be found in many glass houses but their main period of use is over. Even though the center of importance thus becomes the feeder machine, the suction blow machines will be discussed in detail. They remain important and instructive.

3.2.1 Owens Machines

By I. Muschalek, Bad Godesberg, and H. Riede, Düsseldorf

3.2.1.1 Development. M. J. Owens was the first to succeed in supplying a bottle machine with glass mechanically. He created the first completely automatic glass machine. Born on January 1, 1859, the son of an Irish miner, he emigrated to West Virginia. He lost his father early and had to help as a breadwinner when hardly out of childhood. He learned glass blowing after having been carry-in boy and gatherer and worked as a glass blower in a Wheeling, West Virginia glass house. In 1888 he transferred to the Toledo Glass Co. where he found a sponsor in E. D. Libbey who allowed him to begin his basic experiments in

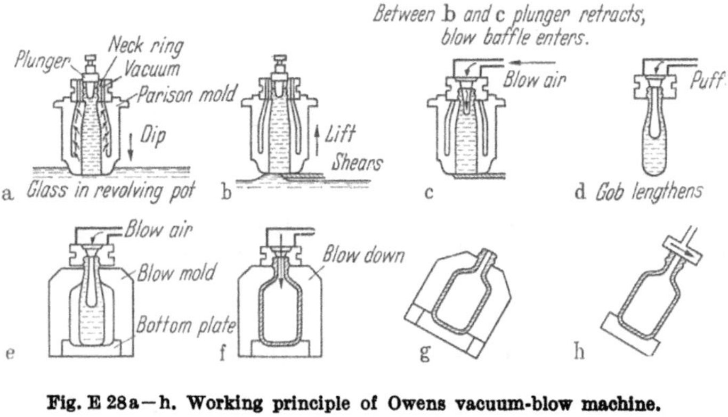

Fig. E 28a—h. Working principle of Owens vacuum-blow machine.

1898. Owens had the idea of taking a glass gob from the open glass bath by means of a special suction head, which already carried the essential forming tools such as neck mold and plunger, and then to blow the gob by placing it on a cylindric mantle (Fig. C 8). With the collaboration of W. E. Bock, C. W. Schwenzfeier and also R. La France he obtained US Patent 774 690 (1904); and in 1905 patent 870,664, claiming a six arm machine [1, 2]. This machine already possessed all the characteristics of later, mature construction.

The operation of the Owens machine is illustrated in Figs. E 28 and E 29. By vacuum the glass is taken from a forehearth, the revolving pot. The next phases are: cutting, pre-blowing (parison blowing), transfer and final blow. At first OWENS, like other designers, tried to get along without the revolving pot. But suction directly from the open glass level of the working tank always sucks along cold glass from the

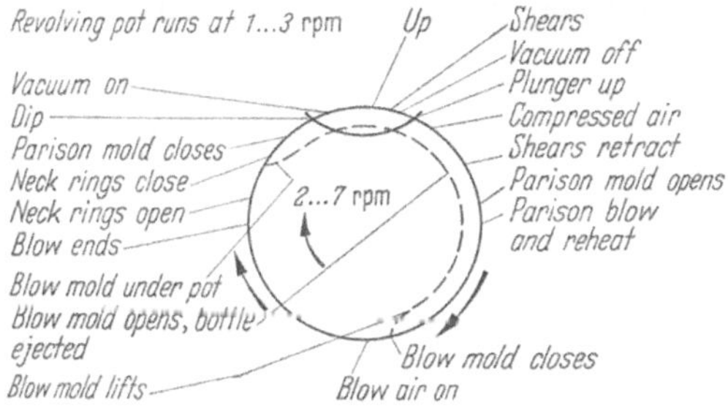

Fig. E 29. Production scheme of Owens suction-blow machine.

preceding suction at the same location. It was very difficult to maintain constant temperature and glass level. The revolving pot eliminated all these disturbances. (Details of the construction of revolving pots are given in Sec. D 2.) The flow of glass from the melting tank, with a change of level, was adjustable. By separate heating he succeeded in maintaining the optimal temperature in the revolving pot area. The location for the next suction was an untouched one because of the rotation of the revolving pot. In this manner fresh glass free of remnants from the preceding cut was available for every bottle. As can be seen from Fig. E 30, the machine is placed in front of the revolving pot in a recess in its upper construction. The machine continuously rotates around the vertical center column with adjustable speed.

The fourth model of this machine already had 6 parison and 6 blow molds. It is remarkable how OWENS solved the problems of getting the glass into the molds by lowering the entire machine at the instant of suction, the mold dipping several millimeters into the bath. This happened six times for one rotation of a six arm machine. After suction, a knife cut the trailing glass strand which fell back into the revolving pot and melted into the surrounding glass by passage through the heated inner part of the revolving pot. In the 1911 machine, OWENS dipped the arms only. In this way he obtained a much quieter operation, decreased the power requirement considerably, avoided vibrations, and was able to increase the rate of revolution. This was of particular advantage in the production of smaller bottles.

Since 1903 the Owens Bottle Machine Co. has been constructing models of the types tabulated in E 10 [2]. Even today types AE and AR run in Germany, where the first Owens installation, including a remarkable tank construction with a revolving melter, was put in operation in Sinzig, Rh. in 1908 [3].

3.2.1.2 Construction. The *construction of a machine* with 10 stations is shown schematically in Fig. E 30. The main center frame rests on rails and holds the cams. The frame can be adjusted in height to adapt

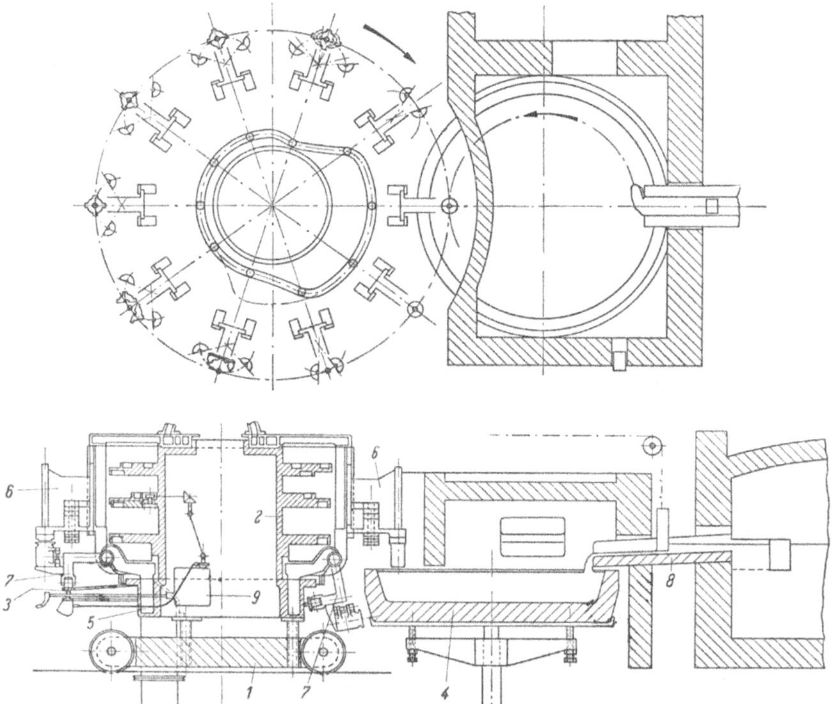

Fig. E 30. Construction of an Owens machine.

to the length of the bottle and the glass level in the revolving pot. The frame *2* also holds the mold carriers *3* which swing the blow molds down under the revolving pot *4*, and the frame channels *5* provide cooling air. Around the frame rotating arms are fastened carrying mechanism to which are attached 10 parison mold stations and 10 blow mold swinging arms *7*. The parison mold stations *6* have vertical slides which permit lowering the parison molds to the bath at the sector of the revolving pot *4* for the suction phase. The cams at the center may be adapted to operational conditions from the outside by the adjusting mechanism *9*.

The revolving pot *4* is supplied with glass from the tank by a channel which is regulated by a gate *8*. The revolving pot is heated independently of the melting tank so that the glass can be held within certain limits of temperature favorable for a particular kind of bottle (for detail see D 2).

Fig. E 31 is a schematic drawing of the *machine drive* in the base frame of a 10-station Owens machine. The drive motor drives the cog wheel shaft through the cogs of the rear shell of the frame via worm gear

Table E 10. *Working principle and specifications for various Owens machines*[1]

Principle	Design	Number of arms	Year	Production range (Wt. (g))	Output (Pcs./min)
1. Machines without automatic lifting of arms	A	6	1903/4	113—1130	10—20
	AC	6	1908	113—1130	11—22
	AE	6	1909	113—1810	12—24
	AD	10	1909	113—1810	24—26
	AL	8	1912	450—4500	8—12
2. Machines with automatic lifting in each station	AN	10	1912	to 322	20—60
	AR	10	1912	113—2260	12—45
	AQ	15	1914	113—2260	18—60
	AV	15	1917	to 322	30—90
3. Same as 2, with double molds	CA	10	1920/3	8 and smaller	80—320
	CB	10	1920/3	8—32	80—160
4. Carboy-Machines	AT	6	1915	$^3/_4$—$15^1/_4$ (gallons)	6—7

and pinion. There also exists a manual drive from the stage by link shaft to move the machine. The motor works on the large center wheel through a bevel gear. The center wheel drives three symmetrical stroke spindle gear wheels. Three large square spindles connect the upper machine, the height of which can be adjusted, with the center part. The drive motor drives the vertical pinion shaft, through a large cog. This shaft can be extended if the frame height of the machine is adjusted. A hand-wheel rotation drive with extractable pinion permits the machine to be positioned by hand, if the current fails. The activation mechanism of parison and blow molds of the Owens AR machine is shown in Figs. E 32 and E 33. The cooling air conduit is combined with the mold holders. Closure springs prevent breaking of machine parts in case of squeezed parisons or bottles.

[1] The further development of this machine by Schwartzkopff is described in E 3.2.3.2.

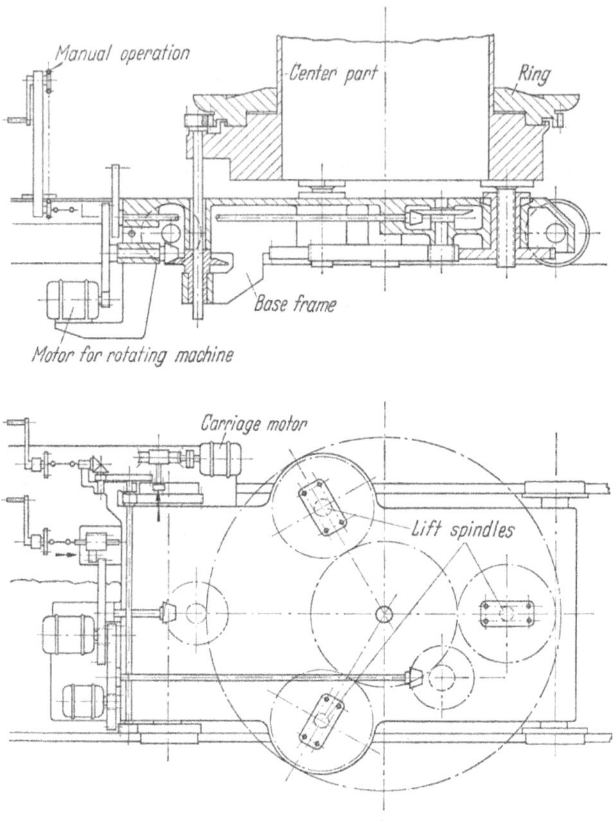

Fig. E 31. Drive mechanism of an Owens machine.

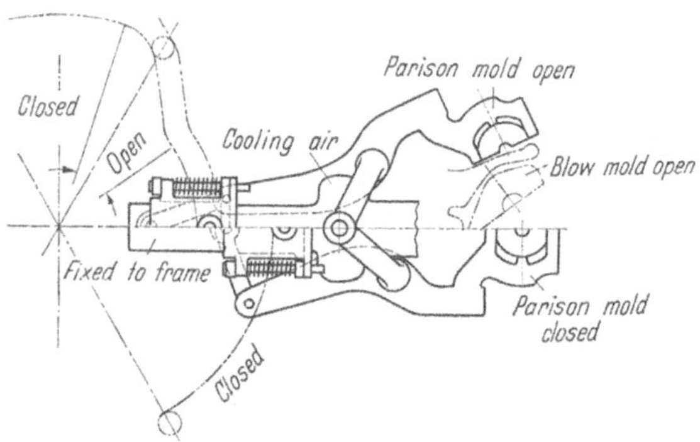

Fig. E 32. Opening of blank mold on an Owens AR machine.

3.2.1.3 Productivity. Productivities, production areas, and energy consumption of the Owens machines are collected in Tables E 11, 12, 13.

Even though the Owens machine is now dated, the work of OWENS, which measures up to the genius of all great engineers, remains unforgotten. The machine factory Schwartzkopff in Berlin built the machines for the European Union of Bottle Manufacturers Ltd.; it also supplied

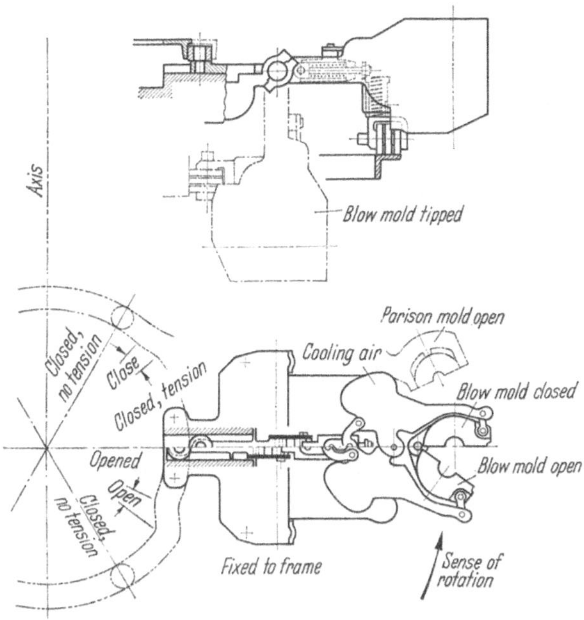

Fig. E 33. Drive of the blow mold with spring link on an Owens AR machine.

Table E 11. *Production range of type AR and AQ*[1]

	Dimensions	Single mold	Double mold
Type „AR“ with 10 arms	Min. height without finish	143 mm	143 mm
	Max. height without finish	314 mm	314 mm
	Max. body diameter	97 mm	78 mm[2]
	Max. finish diameter on narrow neck bottles	44 mm[3]	32 mm

replacements, and after 1945 developed a six arm machine designated OS which resembles Owens' original AE and AR. This machine is described in more detail in Sec. E 3.2.3.2.

[1] Type AQ has same production range as AR, but has 15 stations.
[2] Distance of double mold cavities 114.3 mm.
[3] Wide mouth containers to 82 mm finish can be produced.

Table E 12. *Average daily output of an AR machine*[1]

Bottles		Output/24 h	
Cap. (liters)	Wt. (g)	Single	Double
$^1/_3$	320	—	102000
$^1/_3$	430	54700	—
$^1/_2$	510	53200	82000
$^7/_{10}$	700	47500	72000
1	800	46000	—

Table E 13. *Energy requirement of Owens machines*

Machine type. Number of molds	Comp. air 2,1–2,3 atm (m^3/h)	Vacuum 650–700 mm Hg (m^3/h)	Air Cooling 800–900 mm WC (m^3/h)	Drive lift carry (kW)
Owens AR Machine				
10 single	900	1200	20400	max. 9
10 double	1300	2100	36000	max. 10
Owens Machine type "AE"				
6 single	600	800	18000	max. 4.5

References

[1] Stein, G.: Die Erfindung der automatischen Flaschenblasmaschine von Owens um die Jahrhundertwende. Glastechn. Ber. 27 (1954) 15–17.
[2] Meigh, E.: The development of the automatic glass bottle machine. Glass Technology 1 (1960) 25–50.
[3] Muschalek, L.: Owens-Anlage mit umlaufender Schmelzwanne. Glastechn. Ber. 24 (1951) 16/17.

3.2.2 Roirant Machines (Types B, F, A 6, BB)

By W. Giegerich, Heilbronn

3.2.2.1 Types B and F. Quite early (around 1923) the ingenious French engineer E. Roirant (1882–1955) succeeded in developing a useful automatic machine for the fabrication of smaller orders: Type B ("Bouteilles") for bottles of 250 to 1500 g wt. and type F ("Flacons") for the range 40–350 g wt. The following description is essentially one of type B, but the types resemble each other.

The machine takes the glass directly from the tank; for this purpose suction and neck ring move horizontally between their place above the blow mold and the suction place, dipping into the glass in the furnace for the filling operation. The glass is sucked from a boot while the filled parison mold withdraws, excess glass is cut off while still within the furnace. The cutting falls into the tank between boot and tank block

[1] AQ production: 50% higher.

and so is kept away from the suction position. During the transfer of the parison the parison mold stands vertically above the blow mold; the parison hanging from the neck ring drops, passing the opening parison mold, and enters the blow mold. After closure of the blow mold the

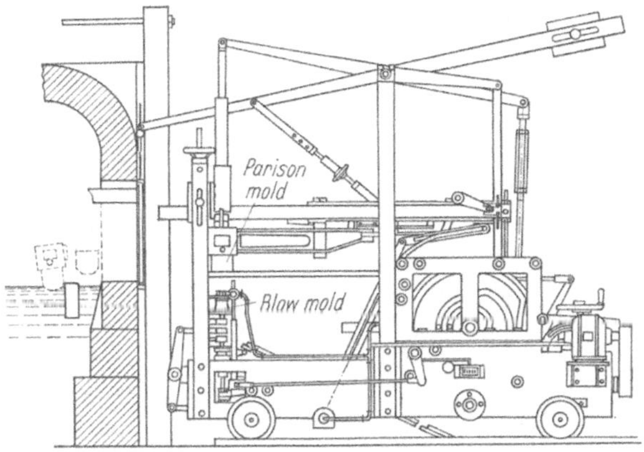

Fig. E 34. Roirant B one-arm-vacuum-machine.

neck ring opens and the parison hangs free in the blow mold. The neck ring then moves back to its initial position, above the parison mold. The finished bottle is taken out by hand or mechanically. The machine has a width of only 800 mm and so can be placed easily on the tank base (Fig.

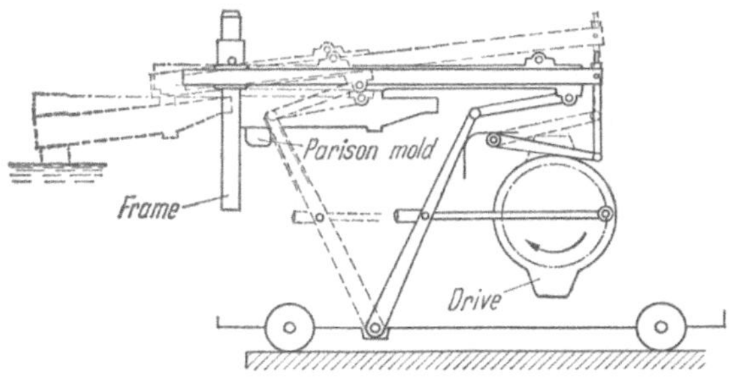

Fig. E 35. Process and dip of blank in a Roirant-B-machine.

E 34). The individual motions and functions are controlled exclusively by cams (Figs. E 35, 36a—d) [2,—6]. Even though the operation of the machine is not quiet, particularly because of gear wear, the energy requirement is low (0.7 kW). The dimensions of the B machines are 2.85 × 0.70 × × 2.70 m; its weight is 2200 kg. The cams also control opening and closure of air valves as follows:

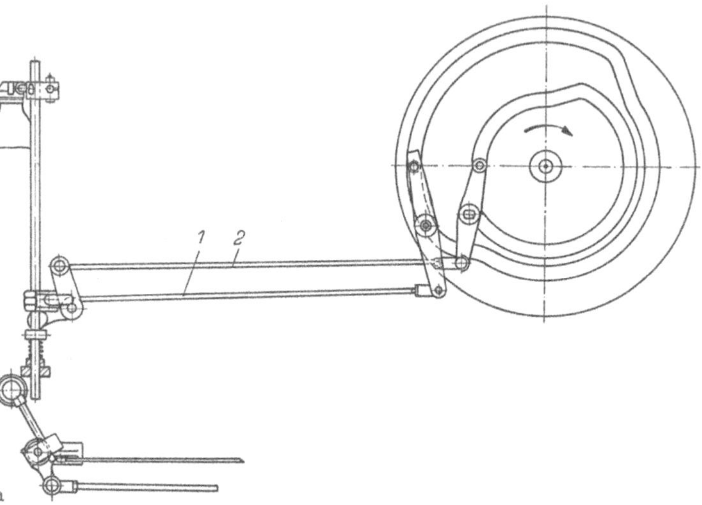

a

Fig. E 36a—d. Cam drives for the Roirant B machine.
Fig. E 36a. Drive *1* for the swing of the blow head, drive *2* for the lift and drop of blow heads.

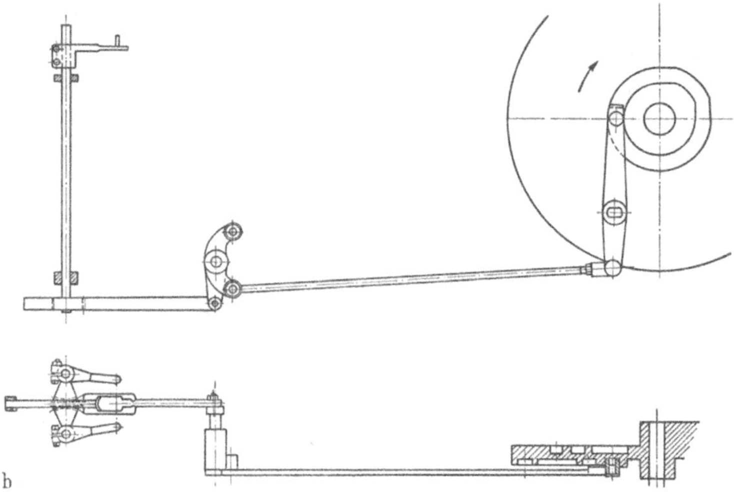

b

Fig. E 36b. Drive for opening and closing neck rings.

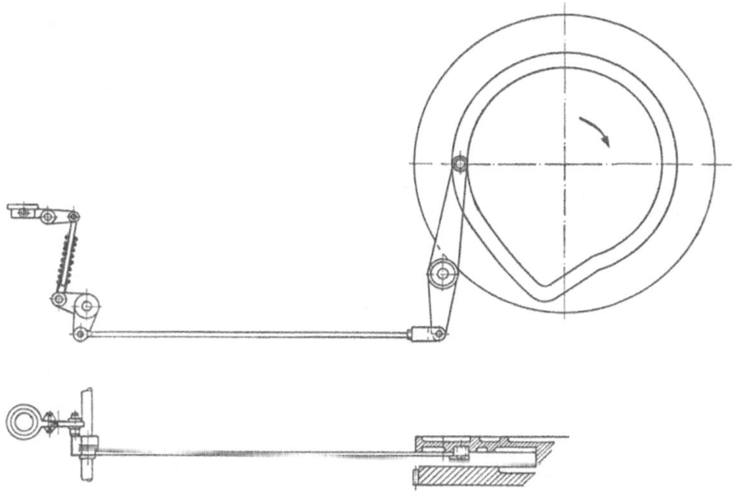

c

Fig. E 36c. Drive for the lift of the bottom take out.

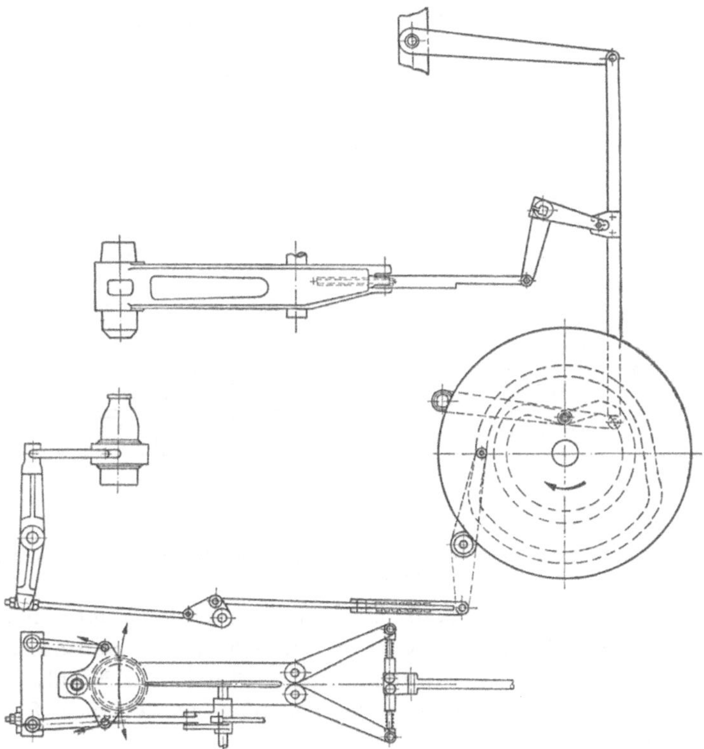

d

Fig. E 36d. Top: Drive for opening and closing vacuum mold. Bottom: Drive for opening and closing blow mold.

Valve 1 (Suction in neck ring) — Opening 10 mm before, closure 10 mm after den of dip
Valve 2 (Parison blow) (Preblow) — Opening on half upstroke of plunger, closure 10 mm ahead of stopping of lower parison mold opening by knife
Valve 3 (Gob blowing) (Parison blowing) — During dwell of filled parison mold above blow mold
Valve 4 (Suction in parison mold) — Through slots. Opening 10 mm before dip, closure as in 2
Valve 5 (Final blow) — Activation of cam in control wheel, adjustable
Valve 6 (Plunger cooling) — Opening 20 mm before down motion of head mold. Closure immediately before dwell of left cam group

With the exception of times of duration of suction and blow, times are not adjustable, but air pressure can be changed. This still gives many operating combinations. Eig. E 37 shows the limits of adjustability for a total time of 18 sec and how the adjustable times are coupled.

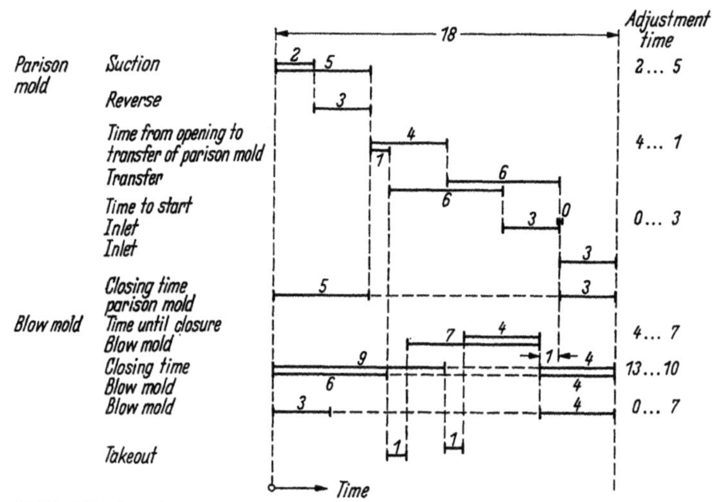

Fig. E 37. Adjustment ranges of the Roirant B machine. Example: Total time of 18 sec. (Times in seconds.)

The design of the suction head is remarkable in its construction (Fig. E 38). Its changeable functions in suction and parison blow are controlled by a vacuum controlled piston. Final blow is only by compressed air in older models, but newer machines have a choice of compressed air or vacuum.

Type F differs from B by the steady rotation of the cams which operate only intermittently in type B. Therefore F is less adjustable.

The dimensions of the F machine are 2.60 × 0.75 × 2.40 m; its weight about 2500 kg. This machine can also be equipped for double molds, with 1.8 times the productivity of the single mold machine.

In the installation of both types each machine requires a vacuum air tank of 1200 liter content and 3 compressed air tanks of 1200, 600 and 600 liter content, with proper cleaning openings, oil drain valves, measuring gauges, etc.

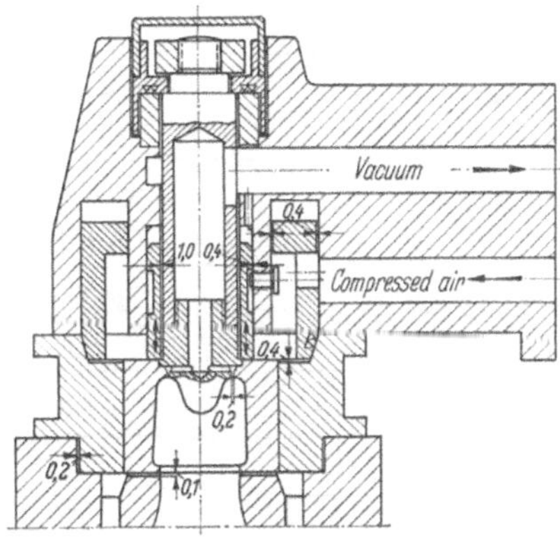

Fig. E 38. Vacuum neck ring of Roirant B machine.

The machine was widely used in Western Europe between 1925 and 1950 (more than 500 type B and 300 type F machines were sold). In many glass houses it represented an important transition to automation. Even so the machine was not used in USA. Today its importance has been diminished by the modern feeder fed blowing machines.

Operation requires 1 man for manual take-out; for mechanical take-out one operator can handle 2—3 machines. Several machines need in addition one operator and one mold maker, who also does the fine adjustment and supervises production. The complete mold changeover (parison mold, neck mold, blow mold, knife) can be done in about 45 min. The mold must be changed in the following sequence and must be then taken care of in the shop (cleaning, polishing, retouching):

Parison mold:	every 2nd or 3rd shift
Head mold:	every 4th to 6th shift
Plunger, blow mold:	after about one week of production
Knife:	every 24 h

The mold life is 80,000 to 100,000 bottles for parison, 500,000—800,000 bottles for blow molds (500 g wt. bottles). Energy consumption, productivity, and dimensions for both machines, are tabulated in Tables E 14 and 15. The data are guideposts for normal operation; for higher productivity energy consumption rises sharply, particularly for mold cooling.

Conditions for good work, satisfactory productivity, and low rejections, are careful mold making and maintenance, including proper consideration of the characteristics of the mold material and its heat wear, for the establishment of tolerances and temperature relations (furnace-best suction place-machine). The latter task is often quite difficult and may require years of development work, but it pays to carefully study the many requirements.

3.2.2.2 Type A 6. The development of the smaller Roirant machines led to the construction in 1936 of the larger type A 6, a six arm two table machine which is still widely used in Europe (about 60 machines) [7].

Table E 14. *Energy requirement and output of Roirant B machine*

a) *Energy requirements*

for	Pressure (atü)	Quantity (m^3/h)	Energy requirement (kW)
Compressed air	Parison blow 1.3–1.0 Blow 0.4	30	1.7
Vacuum	82% on machine 85% on pump	120	2.3
Cooling air	420 (mm WC)	6700 (Nm^3/h)	15.25
Machine drive	—	—	0.75
		Total	20.00

b) *Output*

Wt. bottle	Total (Pcs./min)	
	Theoretical	Attained
300	4.6	
400	4.3	at 350 g 4.7
500	4.0	at 500 g 4.2
600	3.7	
700	3.5	at 700 g 4.0
800	3.3	at 850 g 4.0
900	3.1	

(Max. working rate 4.7pcs./min)

Limiting dimensions of bottles:

Height with finish	140– 360 mm
Outside diameter	to 120 mm
Finish diameter	to 60 mm
Weight	250–1500 g
Bottom push-up max. height	50 mm
Plunger max. diameter	49 mm

Here again the machine was not used in U.S.A. It produces bottles weighing 300–1500 g. This machine takes the glass from an adjacent revolving pot (see D 2). Its special characteristics are:

(a) A small revolving pot 1650 mm in diameter usually driven intermittently by the machine; one machine table revolution corresponding with 0.25 pot revolutions; suction occurs while machine and pot are stationary; the cutting drops back into the pot. Reheating time for the cutting is remarkably long (about 90 sec).

(b) The use of the known 2-table arrangement with intermittent Geneva drive using 6 (6 parison and 6 blow mold) stations. The parison mold in addition is moved up and down to dip it into the glass bath of the revolving pot.

Table E 15. *Energy requirement and output of Roirant F machine*

a) *Energy requirement*

for	Pressure (atü)	Quantity (m^3/h)	Energy requirement (kW)
Compresssed air	1.3	30	1.7
Vacuum	82% on machine 85% on pump	120	2.3
Cooling air	285 (mm WC)	5400 (Nm^3/h)	8.5
Drive	—	—	0.75
		Total	13.25

b) *Output*

Wt. (g)	Total (Pcs./min)		
	Theoretical		Attained
	Single molds (EF)	Double molds (DF)	
50	8.0	14	at 60 g 13 DF
100	7.1	12	at 110 g 9.0 EF
150	6.3	11	
200	5.7		at 200 g 6.5 EF
300	5.8		
400	4.2	4	at 400 g 5.0 EF

(max. rate 9.0 pcs./min for single molds (EF), 16 pcs./min for double molds (DF)

	Single mold	*Double mold*
Height with finish	50—220 mm	50—220 mm
Outside diameter	to 100 mm	45 mm
Finish diameter	to 55 mm	25 mm
Plunger max. diameter	46 mm	21 mm
Weight	40—350 g	40—150 g

(c) The final blow in the blow mold is accomplished by vacuum from the outside through bottom and seams as well as by compressed air through the plunger.

(d) The possibility of simultaneous manufacture of different bottles of varying shape, size and weight within certain limits.

(e) Opening and closure of molds during dwell, cutting at the parison mold by a single knife.

The operation of the machine is shown schematically in Fig. E 39. The single phases at each station are:

Station 1.	Dipping of parison mold into glass mass of revolving pot, suction, lifting and cutting, and, on the way to the next station, plunger withdrawal;
Station 2.	Dropping of the filled parison mold onto a hollowed shaping block, first parison blow, lift while turning, opening of parison mold;
Station 3.	Dropping and ramming of parison, second parison blow by a small blast, lifting;
Station 4.	Dropping, closure of blow mold, opening of head mold, lifting;
Stations 5.—7.	Blowing of parison in blow mold, by vacuum;
Station 8.	Opening of blow mold and bottle ejection.

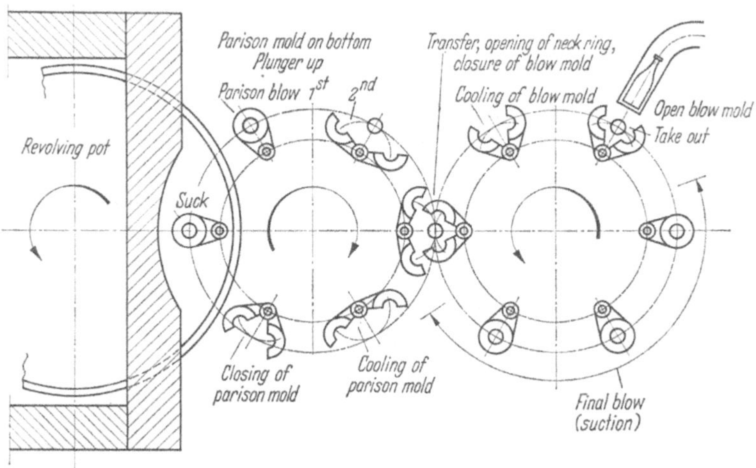

Fig. E 39. Working principle of the Roirant A 6 machine.

The parison molds are cooled by air blast on the inner mold walls during dwell times in stations (5) (6), blow molds in station (9).

The construction and the gear mechanism of the machine are shown in Fig. E 40, 41. An electric motor drives a bevel gear and gear wheel pair in serial arrangement by a stepless gear (not drawn) which is adjustable within 1:2.5. The Geneva sits on the axle *3* of the blow mold table *2*. The parison mold table *1*, which is provided with a shaft that moves up and down vertically, is driven by the gear pair *7*, *8* which is placed above the machine. The up and down motion is effected by the end of the parison mold axle which is terminated by a roller and is guided by cam *6* which rotates in synchronization with the operation. Opening and closing of parison molds is effected by the bottom side of a cam *18* which does not rotate but oscillates. The upper side of this cam activates plunger and head mold. The blow mold table has only one cam *13* for mold closure. These cams which are not fixed to the frame permit the adjustment of every motion at standstill as well as during table rotation.

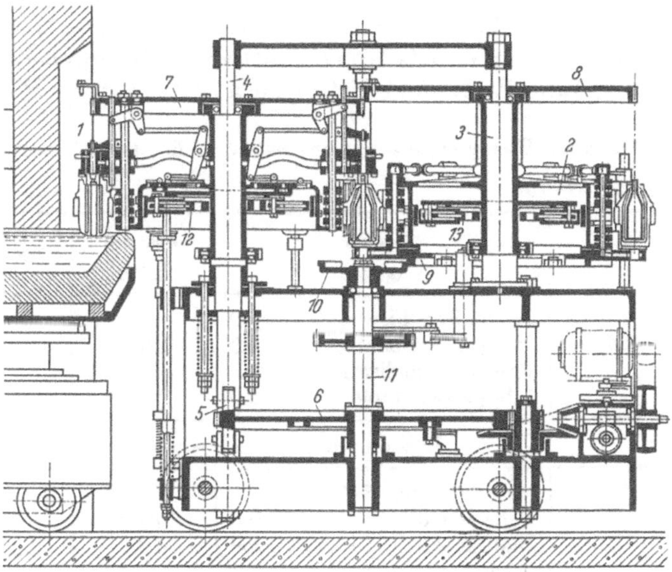

Fig. E 40. Construction of Roirant A 6 machine.

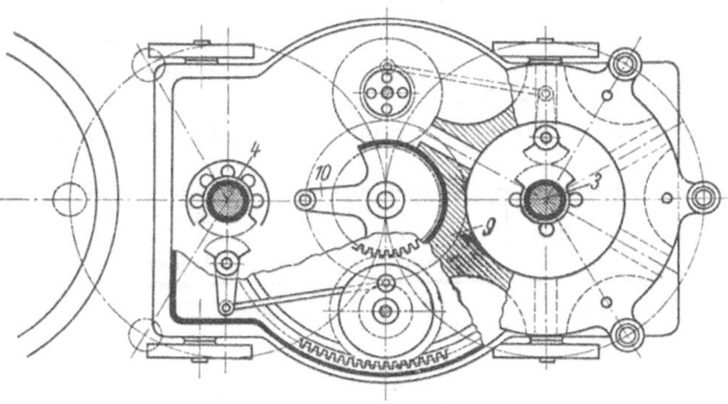

Fig. E 41. Drive of the Roirant A 6 machine.

The double drive of the revolving pot is also noteworthy; during normal operation it is effected from the machine, or by lever and ratchet wheel, by means of a worm gear that is driven by a separate motor that can be cut off. In this way the continuous run of the revolving pot is maintained during stoppage of the machine. One may also work with a continuously flowing circulating forehearth in place of a revolving pot.

The A 6 machine permits the simultaneous fabrication of various kinds of bottles on its single stations, with weights variable by as much as 150 g and also variable in shape and height. The economical limit of combining different kinds of bottles is probably limited by differences of two seconds per bottle in fabrication rate (e. g., 20—22 sec/bottle, or 3—2.7 bottles/minute. Details on productivity and energy consumption are given in Table E 16. Mold life is similar to that of the B machine,

Table E 16. *Energy requirement and output of Roirant A 6 machine*

a) *Energy requirement*

Requirement for	Pressure (atü)		Quantity (m^3/h)	Energy requirement (kW)
Compressed air	1st parison blow 2nd parison blow Pressure equalization Lift parison table	1.0 0.2 1.5 3.0	70	5.0
Vacuum	85% on pump		900	17.5
Cooling air	(mm WC) Parison mold Blow mold Shears, revolving pot	 365 250 180	(Nm^3/h) 10800 6500 2900	 21.5 9.0 3.0
Machine drive	—		—	3.5
			Total	59.5

Heat consumption of revolving pot (producer gas)
280000—410000 kcal/h
(2.0—2.9 t or soft coal briquets/24 h; $H_u = 4500$ kcal/kg)

b) *Output*

Wt. of bottle	Total (Pcs./min) Theoretical	Total (Pcs./min) Attained
300	24	at 340 g 27.0
400	22	
500	20.5	
600	18.7	at 550 g 21.0
700	17.7	at 700 g 19.4
800	16.6	at 820 g 18.5
900	15.6	at 870 g 17.5—18.0

Limiting dimensions of bottles:

Height with finish	150—360 mm
Outside diameter	to 120 mm
Finish diameter	36— 60 mm
Weight	300—1500 g
Plunger diameter	53 mm

about 80,000 bottles per parison mold, and 800,000 bottles for the blow mold; the neck mold life is about 30,000 to 40,000 bottles. The parison molds must be repaired (retouched and polished) every 24 to 48 h.

Because of the combination of the suction process with the final blow by vacuum, and the long dwell in the blow mold, the A 6 machine yields excellent bottle quality—smooth, glossy surfaces — much better than any other machine.

Table E 17. *Adjustment ranges in Roirant machines B, A 6 and R 7*

Feature	B			A 6			R 7		
	Possibility of adjustment	Time of adjustment		Possibility of adjustment	Time of adjustment		Possibility of adjustment	Time of adjustment	
		Start	End		Start	End		Start	End
	Mechanical motion								
Machine	• +			• +			• +		
Plunger in	—			—			—		
out	—			—			—		○+
Neck ring open	—			—			—		
close	—			—			—		
Parison mold									
open	—			—			—		
close	—			—			—		
Transfer neck									
ring open		• +		—			—		
Blow mold									
open	—			—			—		
close	—			—			—		
Takeout	—			—			—		
Process phase	Air pressure								
Suction	• +	○+	+	• +			• +	○+	○+
1st Parison	• +			• + ×			{ • +		
2nd Parison	• +	○+		• + ×				○+	
Blow	• +	• +		• ×	•		• +	• +	
(Suction)									
For	Cooling of mold parts Amount of air								
Plunger	• +			• +			• +		
Neck ring	• +			• +					
Parison mold	• +			• +			(+•)(○×)		
Blow mold	• +			• +			(+•)(○×)		
Shears	• +			• +			• +		

• = adjustable in operation
○ = adjustable only during standstill
× = adjustable each station
\+ = only all stations adjustable
— = not adjustable
(+•) = belongs together

The machine does not permit much adjustment. In operation, machine speed and the pressure for suction, first and second parison blow, and final blow can be adjusted. Air pressures can only be adjusted in unison for each station. The stations each receive a conduit with a valve which originates at a distribution head and alternately supplies station No. 1 vacuum, Nos. 2, 3 parison blow, Nos. 5, 6 plunger cooling air. Final blow by vacuum can be done on 1—3 stations according to choice. In operation also knife height (the same for all 6 parison molds) and the gap between head and parison mold (separately for each mold) are adjustable. Service is easy because of this restriction in controls, but area of production and productivity are limited for the same reason. Table E 17

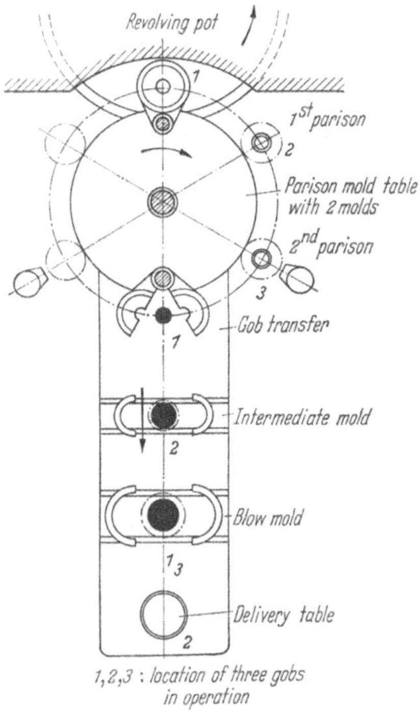

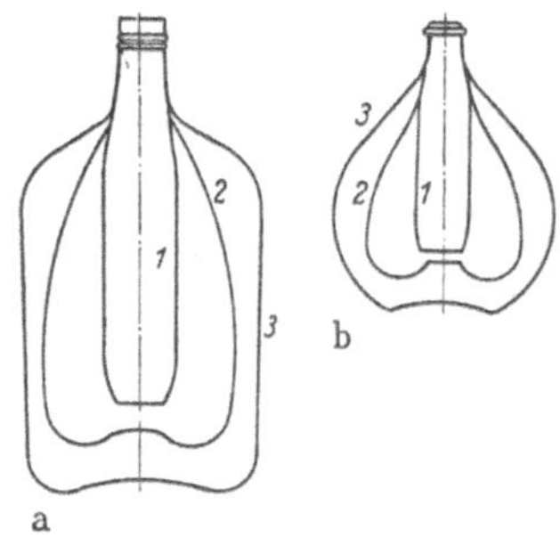

Fig. E 43a and b. Production stages of two carboys in the Roirant BB machine.
1 In parison; *2* In intermediate; *3* In blow mold.

Fig. E 42. Working principle of Roirant BB machine.
1, 2, 3 Location of three gobs in operation.

compares in detail the adjustment ranges of Roirant B, A 6 and R 7 machines.

3.2.2.3 Type BB. In 1939 Roirant accomplished the difficult task of producing large glass objects automatically by the construction of the BB machine [1] of which about 15 exist at this time. None were used in the U.S.A. It is the only machine in the world that makes carboys from 10 to 60 liter content. Its special operation is illustrated in Fig. E 42. The parison mold table has two molds and is driven intermittently.

The parison mold suctions from the revolving pot, as with an A 6 machine, at station No. 1. At stations, Nos. 2 and 3 the parison is blown. After opening of parison and neck mold at No. 4 the parison is transported by a transporter to an intermediate mold located at station No. 5 and vacuum blown to an intermediate shape. Finally it arrives at the blow mold at station No. 6 where it is finally blown into the carboy, manually extracted, and transferred to the lehr. The drive of the parison mold table and of the pot is mechanical as in the A 6. The machine demands much space; its exterior dimensions are 3.6 × 2.2 × 2.5 m; its weight is 9500 kg. The revolving pot has dimensions of 1.65 × 1.65 × 1.37 m and a weight of 2750 kg. An example of the remarkable profiling in the stages parison-intermediate-finished product is shown in Fig. E 43.

The productivity of this machine, the dimensions of the objects made in it, and the energy consumption are compiled in Table E 18. Its productivity is stunning when compared to manual fabrication: weight 4500 g; productivity: 1.95 to 1.4 pieces/min).

	Carboys Handmade	*BB Machine*
Parison mold and parison blow in parison mold	175 sec	41.5 sec
Intermediate blow in intermediate mold	44 sec	43.2 sec
Final blow	71 sec	14.9 sec
Total time	290 sec	99.6 sec
Fabrication time	310 sec	43.0 sec

The automatic production of glass articles of such large sizes is not simple and requires considerable experience in using this machine.

References

[1] Wyss, A.: Zur Entwicklungsgeschichte der Roirant-Maschinen. Gedenken zum 80. Geburtstag Emile Roirants am 27. Februar 1962. Glastechn. Ber. 35 (1962) 79–84.

[2] Wyss, A.: Die Roirant-Maschine. Sprechsaal 63 (1930) 368.

[3] Anon.: Roirant-Einformmaschinen. Sprechsaal 66 (1933) 847.

[4] Anon.: Roirant-bottle machines. Glass Ind. 15 (1934) 127.

[5] Anon.: The development of the Roirant-machines. Glass 13 (1936) 280–282, 304.

[6] Alt, H.: Getriebe bei Glasverarbeitungsmaschinen. Glastechn. Ber. 11 (1933) 170–174. Remarks concerning this paper: Glastechn. Ber. 11 (1933) 199–201.

[7] Severin, H.: Die Entwicklung der Roirant-Maschine „A 6". Glastechn. Ber. 20 (1942) 65–71.

[8] Prospectuses and techn. information of Société des Machines Roirant, Bruxelles and the Ateliers J. Hanrez, S. A., Monceau-sur-Sambre (Belgium).

Table E 18. *Power consumption, dimensions and output of carboys in the Roirant B' machine*

a) *Energy requirement*

For	Pressure (atü)		Quantity (m^3/h)	Energy requirement (kW)
Compressed air	1st Parison blow	0.8	70	6.0
	2nd Parison blow	1.5		
	Pressure equalization	2.0		
Vacuum	85% on machine		600	2.5
Cooling air	(mm WC)		(Nm^3/h)	8.0
	Parison closed	250	6500	19.0
	Parison open	365	10800	3.2
	Revolving pot	180	3600	
Machine drive	—		—	2.2
			Total	40.9

Heat consumption of revolving pot: 280000—410000 kcal/h

b) *Output*

Wt. of bottle (g)	Total (Pcs./min)	
	Theoretical	Attained production figures
1500	2.4	at 2500 g 2.4
5000	1.6	at 4800 g 1.7
		at 6500 g 1.3
7500	1.2	at 9000 g 1.0

Limit dimensions:

Max. head diam.	80 mm
Max. plunger diam.	59 mm
Max. height of body	465 mm
Max. total height	320—640 mm
Capacity	10— 60 l

Dimensions of commercial carboys (approximate values)				
Diameter (mm)	Ht. (mm)	Wt. (kg)	Capacity (liters)	
			Cylindrical	Spherical
275	400	2.5	—	10
375	500	5.0	25	30
450	600	7.5	30	50

3.2.3 Other Suction-Blow Machines

By L. Muschalek, Bad Godesberg, and
W. Giegerich, Heilbronn

3.2.3.1 Redfern Machine. The Redfern machine was constructed by Redfern, Hall and Hunt, London 1917. It works according to the same principle as the Owens machine and possesses 6, 10, 12 and 15 units. Wendler [1] supplies the following production figures in pieces/24 h:

Units (Arms)	0.756 liter bottle Single cavity mold	Double cavity mold	0.378 liter bottle Single cavity mold	Double cavity mold
6	26880	40320	35832	53760
15	67200	100800	89600	134400

The dimensions of the 15 unit machine are: height 4.5 m, diameter 5.1 m, rotational speed 1.5 to 6 rpm., wt. about 60 t.

A characteristic feature of the machine is the strong, rigid main axle, to which the units are attached in a manner facilitating changeover. All mobile parts are cushioned by strong springs to reduce jar and vibration. The blow valve of each unit can be adjusted for time and quantity in operation. A special gear assures the exactly centered position of the parison mold in relation to finish mold and blow head. The machine today is only of historical importance.

3.2.3.2 Owens-Schwartzkopff Type OS. This machine represents the latest development in the 6-arm Owens Machine, utilizing all recent experience (Fig. E 44). The guiding idea in the construction was the development of a machine adapted to the European market, by allowing economical production of large as well as small quantities. It is flexible with the ability to produce at the same time bottles of varying height, weight, and content without clumsy and time consuming changeover operations for small orders. Essentially this was achieved by an adjustable dip stroke and a compressed air control apparatus.

In contrast to the older 6-arm Owens machines of types A and AE, in the OS it is not the entire round table which dips into the glass, but only each station as in types AR and AN. But while in these latter types a constant dip stroke requires the exact setting of the lower edges of the suction molds, to assure equal dip depth, in the OS machine the dip stroke of each single station can be changed up to 50 mm. This change can be accomplished during operation by the simple manipulation of a hand wheel without stopping the machine. In older types the adjustment for varying parison mold length was possible only while the machine was stopped. This required time consuming adjustments even when worn parison molds were replaced for the same bottle, because the blow mold had to be readjusted to the neck mold. It is clear that the new adjustability greatly simplifies the service of the machine and saves time.

Fig. E 44. Vacuum blow machine design Owens-Schwartzkopff, Type OS (Courtesy Berliner Maschinenfabrik AG, vormals L. Schwartzkopff).

Experience shows that blow pressure and blow operation must be adapted to size, shape, and content of each bottle to achieve the best glass distribution and the largest rate of production. This purpose is served by the control cam apparatus. On a fixed cam disc the on and off segments for parison and final blow functions are arranged suitably. There is a unit for each station which can be varied by 30°. On the control panels are hand wheels for timing, levers for pressure control, and pressure indicating gauges. In the "puff" position forced control of the switch opens it to full pressure, while shortly before final blowing the compressed air is switched automatically by a valve, which can be adjusted to the desired pressure. For puff as well as final blow there are two levers on the panel, one each for start and stop. A right turn of the wheels advances, a left turn retards the operation. In this way a very exact regulation of the blow processes is achieved which can be undertaken during operation without any risk to the machine. This results in improved quality such as good glass distribution and perfect bottoms, as is confirmed by thermal shock and pressure tests.

Besides these improvements the smooth operation of the machine deserves mention. This is due to the elimination of the dip of the entire upper structure, the improvement of gears, and the introduction of the stepless P-I-V drive. The compensating springs at the single stations also contribute to a quieter operation. These springs replace the heavy counterweights of the older Owens machines. The spring tensions not only compensate for the weights of the stations in their up- and down motions in the dipping operation, but also maintain freedom from jolts by a remaining pre-stress. The total machine weight is also reduced considerably by the elimination of the heavy counterweights. Another advantage is the track base which permits the free motion of the machine across the floor. It is possible to easily change from single to double molds and vice versa. The machine for the alternate use of double molds is designated DS (Double Mold Schwartzkopff). The double mold equipment is based on experience with the AE machine, but has been perfected to a much lower changeover time. The productivities of the OS and DS machines are equal. Another advantage which has been stressed is that bottles of different weight can be made at the same time in double molds. The operation of the DS and the OS machine is illustrated in Fig. E 45 in which the single working cycles are indicated.

The most important data for the OS and DS machines are: weight: about 14 t; dimensions: length 4050 mm, width 2840 mm, height maximum 3400 mm; mold number: 6; power requirement: 5.5 kW efficiency and power requirement of auxiliary machines, including 30% reserve: compressed air 600 m^3/h, 2—2.5 atm, about 38 kW. Vacuum 800 m^3/h, 96% vacuum, about 20 kW; cooling air 300 m^3/min/800 mm water column, about 56 kW; personnel cooling 300 m^3/min; 70 mm water column, about 7 kW.

Production area:

(a) single molds: contents 0.2—1 liter (with special equipment 0.12—0.35 liter; height 160—320 mm; with special equipment 350 mm),

(b) double molds: contents: max. 0.33 liter, diameter max. 70 mm, smallest height 100 mm.

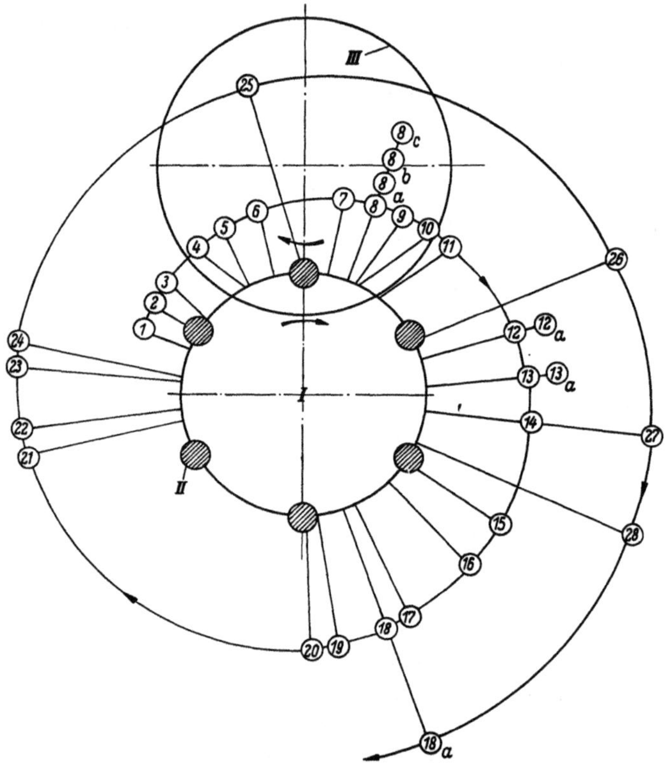

Fig. E 45. Operation of the OS machine.

1 Begin closing mold; *2* Plunger down neck ring; *3* Neck ring closed; *4* Parison mold closed; *5* Parison mold down; *6* Suction; *7* Lift and cut; *8* Lift ends; *8a* cut ends; *8b* shears under parison; *8c* suction ends; *9* Plunger out; *10* Compressed air baffle forward, neck ring closed; *11* Parison blow at full pressure begins (adjustable); *12* Blow ends; *13* Parison mold begins to open; *13a* puff releases lower gob edge; *14* Beginning of blow at reduced pressure; *15* Blow mold entirely open; *16* Blow mold starts swinging up; *17* End of blow at reduced pressure; *18* Blow mold has swung up; *18a* Blow mold takes over, starts to close; *19* Blow mold closed; *20* Beginning of blow (adjustable); *21* End of blow (adjustable); *22* Neck ring opens, releases neck; *23* Compressed air baffle returns; *24* Blow mold swings down, bottle remains; *25* Blow mold passes below revolving pot; *26* Blow mold swings up to oblique position; *27* Blow mold opens; *28* Ready bottle take out (ejection or by hand).

Production figure (by supplier):

(a) Largest rate 42 heads/min. For single molds 12000 ale bottles per 8 h shift for 640 g wt., 650 cm³ content. 12000 cognac bottles per shift for 680 g wt. and 700 cm³ content; 10500 water bottles per shift for 800 g wt. and 1000 cm³ content.

(b) For double molds 18000 to 19000 Vichy crown cork bottles per shift for 350 g wt. and 330 cm^3 content; 18000—19000 beer bottles per shift for 380 g wt. and 350 cm^3 content.

The fire-polishing machines for bottle mouths in Owens machines have been changed frequently with time. In the A machine the finished, still deformable bottle falls upside-down through a funnel on a little round table with 12 bottle holders in which the bottle solidifies while the mouth is fire polished to free it of ridges that may have formed between neck ring and plunger during suction. In later models (AE, AR and AN), which deliver quite solidified bottles, they are transferred to so-called uprighter cups. The well-known uprighters are round tables with 12 or 18 cylindrical cast iron sleeves the width of which must be exactly fitted to the bottle bottom so that no deformation through sagging can occur. Above the sleeve is a battery of burners which fire polish the mouths. The rotation of the sleeves is controlled by the machine. The sleeves are frequently equipped with vacuum and cone shaped bottoms to pinch hollow bottoms. But such equipment has the disadvantage that even for smooth cylindrical bottles trouble is encountered by inexact fit of sleeve and bottle diameter. If the sleeves have become too narrow through oil scale or the cleaning of older blow molds the insertion and removal of bottles becomes quite difficult. These disadvantages are eliminated by the fire polishing apparatus of the OS machine. The machine driven rotary table and the arrangement of the burner battery have been retained. But the bottles taken from the machine are no longer placed into sleeves but on plane, perforated plates. Cool air is blown through the perforations in suitable manner. The amount and pressure of cool air can be regulated. The position of the outlets for cool air can be regulated. The position of the mouths in respect to the burners is fixed by collimation. The mouths are enclosed to prevent sag and change in position. Vacuum heads are provided for hollow bottoms. The fire polished bottles are transferred automatically to a belt on which they are transported to the lehr.

3.2.3.3 Monish Machine. The Monish machine [2] is a suction-blow machine with three sets of molds. The name is formed from the name of the producer, John Moncrieff Ltd. of Perth, Scotland, and the name of the inventor, MacNish. It works similarly to the Owens machine, at least as far as the motion of the molds to the glass and each other is concerned, but carries only three arms and stands still for a short period during suction. This standstill is necessary because the suction molds gather from a small channel which is an arch-shaped connection with the tank. As mentioned, the suction mold stops at the suction spot, dips into the glass, suctions, returns upward, and is then rotated for one third of the circumference of the table. This is why the machine gets along without a revolving pot; also, because the glass remainder cut from the mold, and the cooled glass at the suction spot, is driven back towards the working end of the tank by a refractory paddle. This pushing of the cooled glass which can be regulated also by the speed of paddling, invites a strong stream of hot glass towards the suction spot. Two types are made: Monish-Major for bottles from 200 to 1000 g, and Monish-Minor for weights from 20 to 200 g. The two machines weigh 1600 and 1120 kg respectively and have dimensions of $2.0 \times 1.55 \times 3.075$ m and $1.1 \times 1.0 \times 1.765$ m, (length $\times$ width $\times$ height) respectively. The total energy requirement is given as 13.2 and 9.6 kW respectively. The dimensions of bottles can be:

Major: Height 114 to 305 mm; largest diameter 76; mouth diameter max. 57 and 107 mm respectively.

Minor: Height 38 to 152 mm; largest diameter 76; mouth diameter max. 38 and 63 mm respectively.

Production rates for round bottles are, according to the manufacturer:

Wt. (g)	Pieces/h	
	Major	Minor
800—1000	240—300	
400—600	360—480	
200—300	600—730	
150—200		600—730
100—150		730—840
40—60		960—1080
20—40		1080—1200

One user of a Minor reports 750 pieces/h for medicine bottles weighing 95 g, of 100 cm^3 content, 80 mm height, 30 mm diameter and 23 mm mouth. There are only a few machines still running.

3.2.3.4 Lagena Machine. This machine ressembles the one-arm Roirant machine and was developed by the Mühlig Union Company of Usti (Aussig), Czechoslovakia. It has not become known beyond its domestic area but there it has been proven excellent and is said to be still in operation. The field of production is quite large, bottles of 10—1050 g weight, 42 to 362 mm height, 12—90.4 mm diameter and 8.5 to 34.4 mm mouth. The total power requirements are given as 11.4 kW, the production as follows:

Bottle Weight (g)	Pieces/h
1050	105
840	135
500—550	210
10.5	1000

The machine weighs about 2200 kg and has exterior measurements of 2.8 m length, 0.68 m width and 3.38 m height.

In its operation it seems remarkable that the parison mold does not reciprocate on a straight tiltable slide (for the suction dip), as in the Roirant B and F machine, but (see Fig. E 46) the parison mold *11* is guided

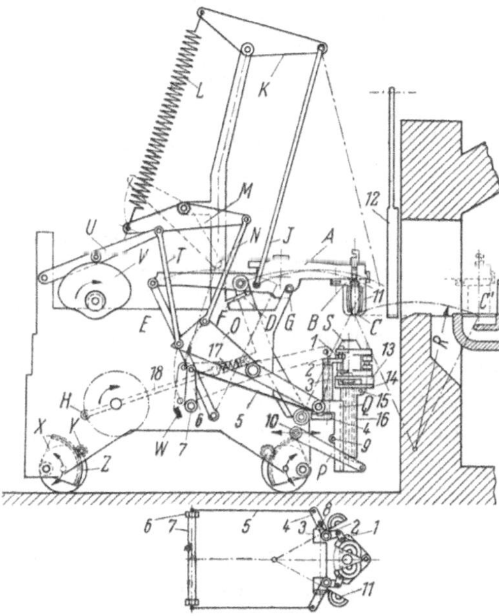

Fig. E 46. Production scheme of Lagena machine.
A Parison mold carrier; *B* Finish mold; *DE* Guide for *A*; *H* Crank; *JK* Control gears; *L* Spring; *MNOP* Gears for blow mold lift; *Q* Holders for blow mold support; *S* Blow mold; *TU* Control gears for blow mold lift; *V* Cam; *W* Stop; *XYZ* Height adjustment; *1* Connection; *2–7* Gears for blow mold opening; *11* Parison mold; *12* Work hole baffle; *13–16* Blow mold supports; *18* Push rod for parison mold support.

along a flat arch of radius *R* from the working position into the furnace. The beam shaped parison mold carrier *A* is moved in horizontal position by the joint parallelogram *EDA*; this means the parison mold is moved in vertical position and brought above the glass level. Another advantageous acceleration of the operation is brought about by the circumstance that the plunger is lifted and lowered by mechanical coupling with the machine drive and so can be lifted immediately after cutting. Thus the blowing of the parison in its mold can start earlier then in the Roirant machine in which the plunger is lifted on the return path of the parison mold by a projection on the slide. Because of this earlier blow the parison solidifies faster and is ready earlier for its transfer to the blow mold, so that the operational speed of the machine is increased.

These two circumstances at least compensate — as practice has shown — the delay caused by the fact that in the Lagena machine, in contrast to the Roirant machine, the final blow is done while the bottle head still sits in the neck mold *B*, so that the arrival of the next parison mold can only occur after completion of the final blow and the return of the neck ring to the suction head. The close similarity of the Lagena machine to the Owens machine has the additional advantage that it can work with mold sets used in Owens rotary table machines.

The cams for lifting, lowering, opening and closing of molds and baffles, as well as the crank *H* for the motion of the parison mold carrier *A*, and the control mechanism for dwell times, are all placed at the rear of the machine away from the furnace heat and are connected with the control rods only by hand power. During suction and final blow the cams are stopped and are reactivated by an automatic control mechanism. The durations of dwell can be set independently. The small centrifugal force of cams that are outwardly controlled, without backlash, together with the short path of the parison mold, the spring-supported swinging motion of the molds, and the early solidification of the parison, permit faster cam rotation and so a shortening of fixed motion periods. As a result, smallest bottles can be made in 8 seconds each in continuous operation without interfering with the smooth operation of the machine.

The machine has obtained only local importance. It is described in some detail because of its unusual and instructive construction.

3.2.3.5 Wyss AW Suction-Blow Machine. The one-mold bottle blow machine by A. Wyss [3, 4] represents a continued development of the one-armed Roirant B machine. The most important innovation is a heat protecting hood made of cast steel which constantly closes the working hole and is provided with a refractory plate closing the hole on the outside face. When the parison mold is swung into the suction position this hood comes along, forming an effective shield against the radiation of the furnace. This allows a reduction of the artificial air cooling of the suction members. Mold changeover can be effected in minimum time. Height adjustement of parison mold and adjustment of neck ring to parison mold is by rotating spindle; the blow mold is not moved. The blown bottles can be transported directly from the machine to the lehr. For this purpose each blow mold has several bottoms which move away from the blow mold with the blown bottles. Bottles not blown out or falling parisons thus are not jarred in closing mold halves but move away and can be eliminated easily. These improvements doubtless result in higher productivity and lower energy requirement.

Limits of production are: height of finish mold: 160—320 mm; diameter of bottle bottom: max. 105 mm; finish diameter: max. 40 mm. This machine has not attained much practical importance.

3.2.3.6 O'Neill Suction-Blow Machine [5]. This machine possesses 3, 4, 6 and 8 arms, was redesigned several times, and derives from the O'Neill feeder blow machine. It is a two table machine. Fig. E 47 shows such a machine equipped with 4 mold sets at the fixed forehearth of a tank. The parison mold table visible at the left is set on the base plate with the blow mold table. They both can be rotated. They are driven by an electric motor through large gear wheels. The parison mold table rotates

Fig. E 47. O'Neill vacuum blow machine.

in the opposite sense to the blow mold table. The parison molds are guided in due order over the forehearth. There is no special stirrer for the equalization of the glass mass that has been contaminated by the cutting marks, but a certain homogeneization at the suction spot results from the motion of the dipping molds. The circumference of the forehearth is so arranged that the line connecting the two table centers forms an angle of 30—40° with the tank wall which supports the forehearth. The arc between the dipping point of the parison mold and the place of transfer to the blow mold is longer than that between this transfer place and the next dip. In an earlier model the neck ring and its opening wedge as well as its guidance was attached to a slide which could be shifted radially to the axle above the parison mold, similarly to the O'Neill feeder machine. The neck ring and the wedge were shifted outward towards the parison mold path before the transfer of the parison and without crossing it.

Various adjustments can be made while the machine is running e.g. an adjustment of suction, of preblowing, of the movement of the plunger, and of the dwell time of the parison in the closed parison mold — which is very important for heavy pieces. In the simultaneous production of objects having different heights the varying height of the blow molds can be compensated by the simultaneous use of neck rings of different height.

Two users of model 80 with 8 molds report bottle production as shown in Table E 19.

The machine data are: 8 mold pairs; power requirement 5.2 kW, net wt. 11,300 kg; dimensions; 4.55×2.3×2.55 m (length × width × × height). The machine has little importance today.

Table E 19. *Production data of O'Neill Vacuum-blow machine*

Wt. (g)	Vol. (cm^3)	Ht. (mm)	Diameter (mm)	Finish diameter (mm)	Output (Pcs./h)
150	200	125	60	30	2000
190	215	165	62	28.5	1500
320	265	200	60	30	1500
650	1030	300	100	75	800
700	1000	320	84	31.5	900

3.2.3.7 Hillmann Machine. Hillmann's suction-blow machine (1920 bis 1928) improved the construction of the Owens machine by using several new ideas. However, it did not persist because of various unfavorable circumstances, particularly because its final perfection in practice was prevented by lengthy patent litigation. Hillmann's ideas have, in part, been transferred to others in the form of accepted claims, and in part have been used by later designers. Since the machine suggests useful innovations it is described here. The chief purposes when the machine was designed were simplification of both machine and furnace construction together with economy and adaptibility to simultaneous production of different bottle sizes. Fig. E 48 shows half of a vertical cross section. On the mobile machine base, and around the center column constructed on it, turns the 10 unit round table. New compared to the Owens machine is the transfer of the parison from parison mold *2* to blow mold *1* by lowering of neck ring *4* (instead of lifting of the blow mold). The neck ring is lowered between the opened parison mold halves into the opened blow mold. This becomes possible by attaching the neck ring with a cap *3* — which contains the plunger on slide *18* which is moved up and down by cam *10* at the upper column and rod *9* of the round table. The parison mold sits on slide *5* which is lifted and lowered for the dipping of the parison mold by having the stand *7*, which carries the slide, run on a correspondingly rising and descending path by means of a roller. The segment of the path which determines the dipping stroke can be adjusted by hand crank while the machine is in operation and can be adapted to the glass level. The zero point in the height adjustment of the mold is knife *8* which is swung outward by the opening parison mold,

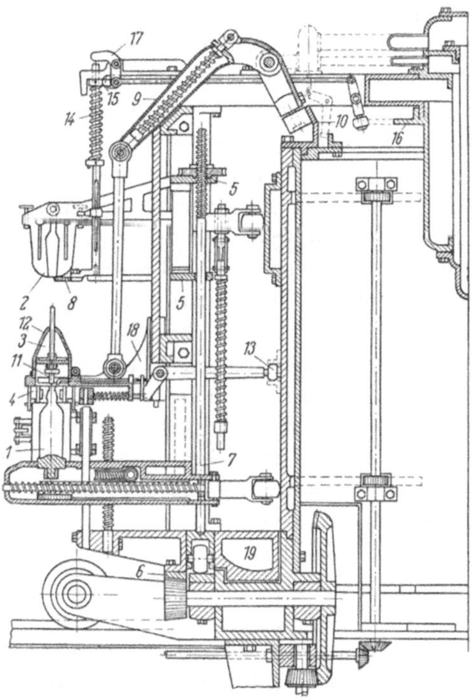

Fig. E 48. Section through a unit mold for Hillman machine.

fixed in the outward position by bolt *15* which is released for cutting by cam *16* at the center column and by a rod leading to this bolt. The bolt moves below the parison mold by means of spring *14*. Before the knife moves back finger *17* of the lever activated by the above mentioned rod pushes the knife down from the lower parison mold face to decrease wear. In order to correctly adjust parisons of different heights to the knife the zero point slide *5* can be adjusted in height on stand *7* by hand wheel and worm gear. The blow mold and its base plate can be adjusted in relative height to the base plate of the round table by worm gear and hand wheel. This adjustment is necessary to adapt blow molds of different height to the lower stroke limit of the head mold. Neck ring, parison and blow mold halves all close under spring pressure. This led to trouble in practice because the released springs yield a little during preblow. The parison and blow molds are opened by thumb cams at the center column and the neck ring is drawn open by a rod controlled by cam *13* (Fig. E 48). In the practical realization of the machine all cams are in the base, so that the center column is eliminated and the machine can be entered while in operation even though the diameter of the mold circle is only 2.5 m.

The described arrangement permits the simultaneous production of bottles of different dimensions. This becomes possible particularly because the vacuum and air can be controlled automatically for each mold unit in operation, in this case by electrical control. For this purpose the upper end of the machine has a control drum with a switchboard (see German Patent 552,092, inventor HILLMANN).

To eliminate the revolving pot and all its difficulties HILLMANN was the first to provide a fixed basin shaped protrusion of the tank for the dipping of the parison molds. This construction proved itself for colorless but not for green bottles. To remove the knife marks from the falling glass there are provided either skimmers preceding the parison mold, which throw back into the tank the glass cooled by the last parison mold, or guiding walls are set into the working tank which guide the current resulting from the motion of the dipping molds in such a way as to reintroduce the cut glass into the melting area of the tank. For colorless glass the production figures were:

300 g, 250 cm³ : 1250 pieces/h
450 g, 500 cm³ : 1050 pieces/h
850 g, 1000 cm³ : 900 pieces/h

Unfortunately after a promising start further development work showed that forehearth and skimmers did not bring sufficiently homogeneous glass to the suction spot so that producible sizes were limited.

3.2.3.8 Pöting Suction-Blow Machines. For the production of medicine bottles and containers of 10—250 cm³ content PÖTING in 1940/41 developed two continuously rotating one-table machines (AM 8 and PD, later AM 6/A) with 6 and 4 arms. The AM 6/A spread considerably and has recently been brought out in an improved version, AM 6/D (Fig. E 49). This occurred because it can be operated in a simple manner and gives good results. Over a small bay it sucks glass directly from the tank. In it is a stirring mechanism for the homogeneization of the glass at the point of suction (German Patent 800,878) which possesses a refractory disc dipping into the glass and fixed to a water cooled shaft (Fig. E 50). The speed of the stirring mechanism and of the machine is adjustable. By height adjustment of the parison mold holder one may produce simultaneously bottles of varying length and weight (length of parison

Fig. E 49. Pöting small glass automatic AM 6/D machine (Courtesy Pöting).

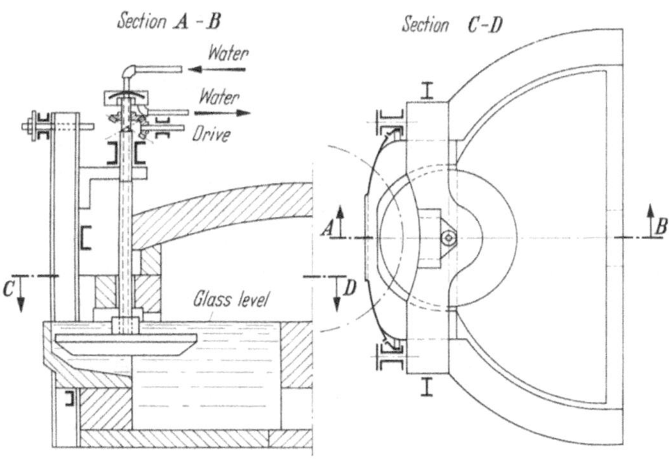

Fig. E 50. Stirring unit for the Pöting AM 6 machine.

mold ± 10 mm) as in the Owens machine. During dipping, only the parison mold holder moves, with pneumatic damping. By hand wheel the machine can be adjusted in height (dipping depth). Experiments with double molds for small bottles are underway.

The AM 6/D machine has 6 arms, weighs 3600 kg, has a mold circle diameter of about 1.75 m and is 2.3 m high. It requires 1.5 kW for the drive of the machine. Air quantities needed are:

Vacuum 95%:	150 m^3/h
Compr. air 3 atm:	60 m^3/h
Cooling air 400 mm water col.:	7200 m^3/h

Total power requirement reportedly is about 30 kW. Production limits are: height (including finish) 40—200 mm, max. diameter 75 mm (for large blow mold holders 85 mm), max. finish diameter 40 mm, wt. 5—300 g. Production figures are:

Contents (cm^3)	10	30	50	100	50
Pieces/min	44	41	38	33	27

3.2.3.9 Schwartzkopff KS 6/60 Bottle Blow Automatic. The automatic bottle blowing machine KS 6/60 is an innovation of the Berliner Maschinenbau AG, formerly L. Schwartzkopff, Berlin. It appeared on the market in 1958.

The machine produces flacons, medicine, container and beverage bottles. The production range comprises heights (including finish) of 25 to 220 mm and weights from 5 to 420 g. This is a circular table machine with 6 stations. It can be used with single or double molds.

By an ingenious construction of certain machine members, particularly of the suction-blow head DBP it became possible to avoid additional structural parts for changing from single to double molds, and vice versa. Only the usual mold change and an adjustment of the machine are required.

The new machine works according to the suction-blow process under the designation KS 6 or according to the so-called combination process (DBPa) under the designation KS 60. In the latter it is possible to work either by suction-blow or suction-suction process. The switchover between these processes can be done with a few hand manipulations during mold change. A fully automated bottle take-out device is provided.

The position of the parison molds in relation to the glass level is regulated by an electrical drive. This permits compensation for level fluctuation during operation. For horizontal motion of the machine a driving motor is available and in the case of power failure a hand crank.

The construction of the machine can be seen in Fig. E 51. The arrangement of the machine at the working end and the main dimensions of the entire production system are seen in Fig. E 52.

Fig. E 51. Small glass machine KS 6 (Schwartzkopff) (Courtesy Berliner Maschinenbau AG, vormals L. Schwartzkopff).

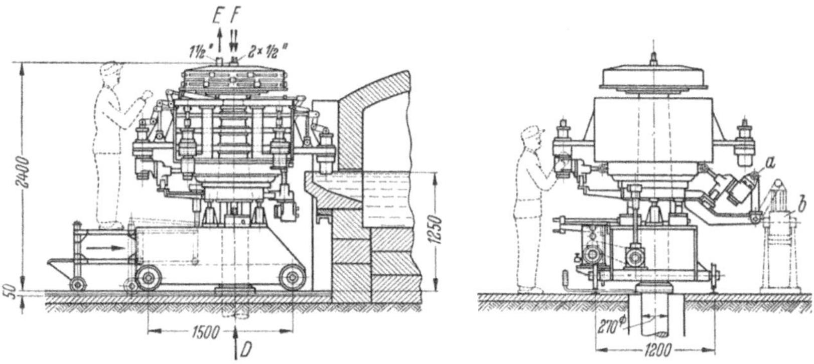

Fig. E 52. Outer dimensions of machine KS 6/60.
a Takeout; *b* Conveyor.

The drive of the machine is by motor via stepless regulatable gear and a further mechanical gear acting on the main driving gear of the machine. For optimum operational safety the control of the individual phases at the perfectly equal stations is effected by an open cam.

As mentioned before the adjustment of machine (or parison mold) to glass level is electrical. But each station also possesses an arrangement for individual adaptation to the glass level. With this arrangement — the regulatable dipping stroke-and the blow control drum on top of the machine, it becomes possible to produce simultaneously bottles that differ to a certain degree in shape, weight and length. If various kinds of bottles are made simultaneously it is necessary to adjust only the dipping stroke of the single stations according to the differences in parison mold length. Depending on absolute size and weight permissible weight differences may be between 15 and 30% ; length differences may be about 25 mm.

All working functions of the machine can be adjusted within wide limits. While blow phases and mold cooling can be changed in operation, the remaining functions must be adjusted before starting the machine, using available tables.

The suction-blow head of the machine is constructed so that a single, cam-controlled rod arrangment starts all work phases: suction (beginning and end), plunger lifting and lowering, blow air, closure and opening of mouth carriers, as well as activation of the suction valve for the blow mold. The timing can also be adapted to operational conditions.

The knives whose brackets are in fixed connection with the slide guide may be adapted to each parison mold in vertical direction as usual. A special fine adjustment of the knife to the suction mold can be made

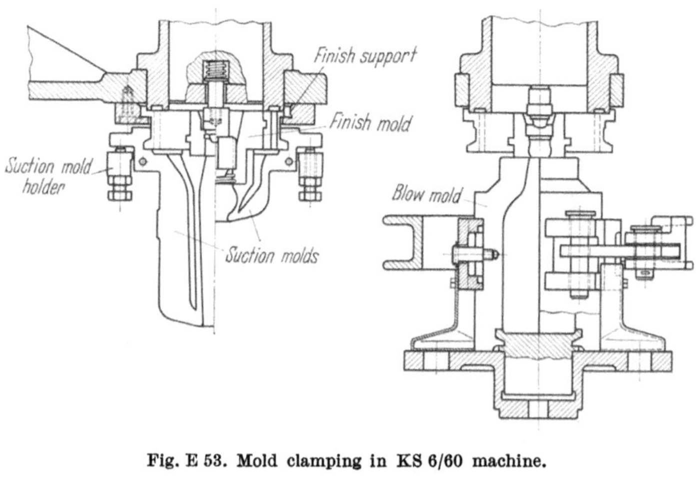

Fig. E 53. Mold clamping in KS 6/60 machine.

during operation. In this connection it should be noted that the proper design of the parison mold, and "after blowing" — an additional phase available in this machine — will largely eliminate shear marks.

In Fig. E 53 the conventional molds of the KS 6/60 automat are shown. The mold cooling air is brought from below. It is brought directly to the suction molds centrally through various machine parts provided with air channels and can be regulated for each station according to operational demands. The machine can work with two-part sleeve plungers. It is known that simple plungers, particularly in the fabrication of small bottles, may cause finish checks because of too rapid cooling. Sleeve plungers permit control of the dwell time of the sleeve on the finish so as to adapt cooling to prevailing conditions.

The manufacturer of the machine supplies the following production figures:

Single molds	Wt. (g)	Production (bottles/24 h)
	50	65,000
	100	50,000
	200	35,000
	400	25,000
	400	25,000
Double molds	20	114,000
	50	48,000

To exploit the productivity of the machine the tank must produce 5—10 t of glass/24 h.

The most important data for the machine and its auxiliary equipment can be found in Table E 20.

The entire power requirement (inclusive of auxiliary machines-vacuum pump, compressor, cooling blower) is about 30 kW.

Table E 20. *Data for automatic bottle unit Type KS 6/60*

Type	Number of stations	Energy requirement	Personnel	Net wt.	rpm of machine
KS 6/60	6	2 PS	1 machine operator	about 4500 kg	2—8 rpm stepless adjustment

Output of auxiliary units[1] (normally for air wt. of 1.2 kg/m³)

Vacuum pump	Compressor	Cooling air blower
178 m³/h at 90/97% Vacuum	about 100 m³/h 2,5 atü final pressure	6000—7000 m³/h about 800 mm WC

[1] Much lower output for bottles below 100 g.

References

[1] WENDLER, A.: Maschinelle Glasverarbeitung (Das Glas in Einzeldarstellungen Vol. 9). Ed. by G. GEHLHOFF and K. QUASEBART, Leipzig: Akad. Verlagsges. 1929, pp. 149. [Glastechn. Ber. 7 (1929/30) 550.]
[2] CREASER, J., L. G. CREASER, and F. W. HODKIN: Suction and feeder-fed bottle-making machines. J. Soc. Glass Technol. 37 (1953) 48—56.
[3] WYSS, A.: Neue automatische Einform-Saugmaschine. Glastechn. Ber. 27 (1954) 356.
[4] Anon.: The "AW 150" single mould suction machine. Glass 29 (1952) 451, 453.
[5] MOODY, D. M.: O'Neill machines. J. Soc. Glass Technol. 37 (1953) 44—47.
[6] BOCK, H.: Der Kleinglasautomat KS 6. Eine neue Saug-Blasmaschine für die vollautomatische Hohlglashütte. Sprechsaal 93 (1960) 305/06.

3.2.4 Suction-Blow Machines for the Fabrication of Thin, Seamless Glasses (Paste Mold Ware)

Bottle making processes are unsuitable for the manufacture of thin walled, seamless objects, since they cool the glass so much in the molds that it can no longer be blown out to thin enough shapes; and because seams cannot be avoided in the glass that touches the mold wall. The forming processes developed for this specific requirement closely follow the old mouth blow processes in which the blow pipe rotates and the mold is wetted. Only the take-out of the glass by suction (OWENS) or feeding corresponds with the bottle making methods.

The suction-blow machines originated under the impact of the Owens suction-blow machines after proper consideration of existing particularities. The Westlake machine came first and from it derived the Ivanhoe and later the Kutzscher rotation-suction-blow machine.

3.2.4.1 Westlake-Machine (By W. TRIER, Frankfurt). After the great success of the Owens machine attempts were made to apply its basic principles to the fabrication of thin walled, seamless objects such as lamp bulbs.

After initial difficulties A. KADOW succeeded in building a machine suited for the fabrication of bulbs, the so-called Westlake machine. This machine is a continuous round table machine with 24 blow pipes arranged in sets of two. Its working principles closely imitate the action of a glass blower (Fig. E 54). The gathering by suction was adopted from the Owens machine. The Westlake machine does not suction a true parison, rather a relatively compact gob. The glass is received by a pneumatically activated ram above the round table. The ram does not move with the machine, but only carries out a small swinging motion. At the front of the ram are two suction heads which descend obliquely into the furnace opening. To keep hot furnace gases away from machine parts, particularly suction heads, the suction area is covered by a hood. During suction, the suction heads with the ram move in the opposite direction to that of the machine so as to find fresh glass at the place of suction. After suction the ram moves back and drops the glass into the pipes underneath. The suction heads move a short while with the round table to facilitate transfer.

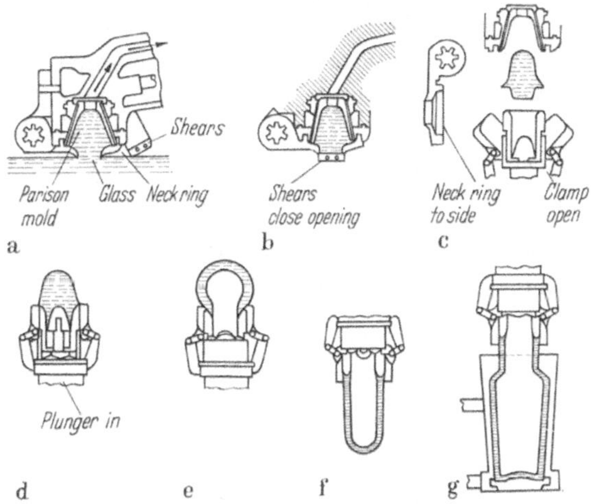

Fig. E 54a–g. Working process of Westlake machine.
a) Suction; b) Suction head retracts, shears cut; c) Transfer to pipe; d) Clamps close, plunger in; e) Parison blow; f) Swing and stretch; g) Blow pipe rotates.

The pipes of the Westlake machine are relatively long and, while turning first swing into a slightly inclined, then into a horizontal, and finally into a vertical downward position. KADOW believed that he could avoid kinking of the gob in the swing by using long pipes because this makes the swinging speed at head and foot nearly equal. But as development progressed it was recognized that much shorter pipes could be used (Sec. 3.2.4.2). The oblique position of the pipe during blowing together with its rotation favors equal glass distribution. The gob stretches under its own weight, particularly in the horizontal and the following vertical position of the pipe.

Blowing occurs in several, carefully regulated puffs. Only after the gob has attained its final length is the fixed mold closed around the gob that hangs on the rotating pipes. The molds are covered with paste on the inside and briskly sprayed with water before contact with the hot glass. The thin film of steam formed on blowing facilitates the sliding of the glass in the mold and prevents it from cooling too fast or unequally.

The finished bulb is thrown on a moving belt after the mold is opened by the holder claws. The glass part with the holder bulge is then fused off. The pipes swing back up and are ready for the next gob.

With the Westlake machine not only lamp bulbs, but also thin walled tumblers, goblet tops, etc. can be made. Today it has been mostly replaced by the Ivanhoe machine, which has been developed on much the same basis, but has a simpler and more sturdy construction. The productivity of the Westlake machine is, depending upon size, 40—70 bulbs/min.

3.2.4.2 Ivanhoe Machine by J. Goedicke, Berlin. The Ivanhoe machine — in the following called the IV machine for short — was developed from the Westlake machine and is a continuously rotating suction-blow machine for the production of thin-walled seamless hollow objects. It is used almost exclusively for the fabrication of lamp bulbs which have to meet the following specifications:

1. Uniform glass distribution over the entire bulb,
2. Wall thickness about 0.4 to 1.2 mm,
3. Wall thickness variation max. 0.3 mm.

The IV machine was introduced into Germany from America in 1928 and is now used in many countries.

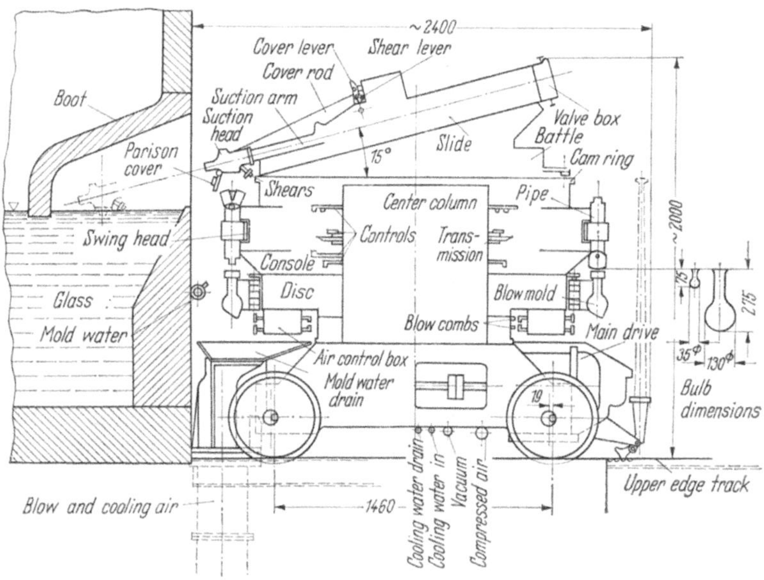

Fig. E 55. Arrangement and construction of a IV machine.

3.2.4.2.1 Operation. The dimensions of the machine are represented in Fig. E 55. To the right of the machine are shown in scale the smallest and the largest bulb that can now be made on it. The mode of operations is shown in Fig. E 56.

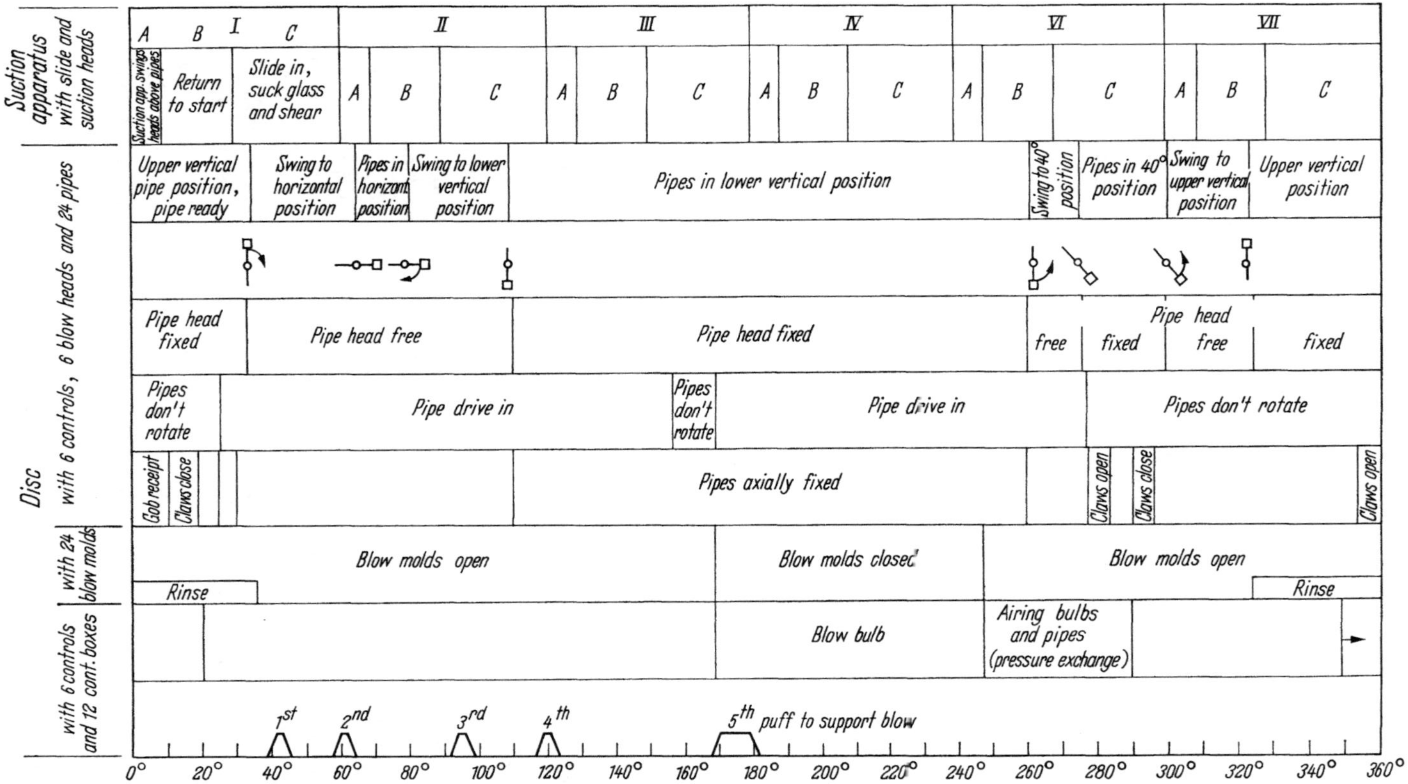

Fig. E 56. Working scheme of IV machine.

The driving motors and the drive of the machine are housed in the mobile base which permits height adjustment and movement by hand. On the cart or carriage is set a fixed center column of about 780 mm diameter which carries the suction apparatus and on whose circumference ere fixed the cams controlling the individual work phases. Around the column rotates continuously a hexagonal disc on each of whose six sides is attached one group of radially swingable blowpipes with their blow molds. The suction apparatus is attached to the column and consists of a group of four suction heads. To take out glass it is pushed towards

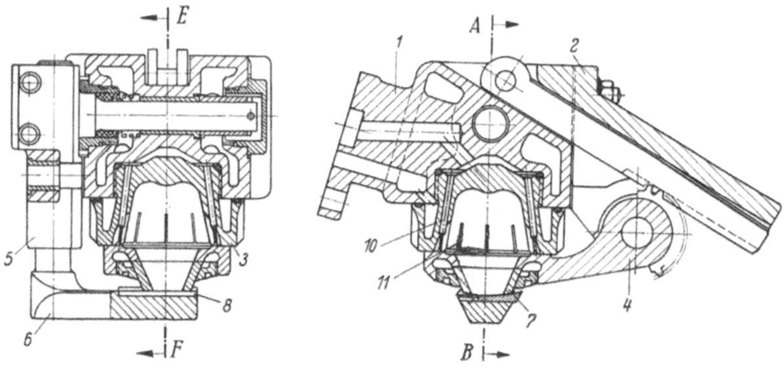

Fig. E 57. Details of vacuum neck ring.

the glass level at an angle of 15° by compressed air and tracted by compressed air after the suction is completed. In this starting position the glass is transferred to the pipes. This cycle is repeated six times during one revolution of the hexagonal disc.

The details of the suction process and the equipment used can be seen in Fig. E 57 and in E 65. The process corresponds with that of the Kutzscher-Sloan machine (Sec. 3.2.4.3). In the suction position the suction mold is sealed vacuum tight by a cover that is narrowed downwards. From the cover bore a sleeve protrudes which dips into the glass about 2 mm deep when glass is taken out. Through this bore suction air pulls liquid glass into the cavity formed by parison mold and cover. As soon as the suction equipment leaves the furnace the revolvable knives fastened at the suction heads swing past the sleeves, cut the tail formed during mold removal from the glass, and return to their original position. During the further lift of the suction apparatus to its upper position the parison mold covers open. The sucked gobs (Fig. E 58), held by vacuum, now keep hanging in the parison molds until the next group of 4 blow pipes has arrived under the suction heads. Now the suction apparatus swings about 8° with the revolution of the machine. During this time the switch from vaccum to compressed air releases the gobs from the parison molds and they fall into the mouth-pieces of the pipes in which they are held mechanically by claws (Fig. E 65b). The suction apparatus then swings back to its original position initiating a new working cycle.

The pipes have tips or needles which push a small cavity into the gobs and then retreat. The pipes, which now rotate around their own axles, also swing radially outward 90° to a horizontal position. After remaining in this position for a short time they swing another 90° to their lower vertical position in which the parisons are developed by several air puffs. The parisons then hang in the open blow molds until they are blown to bulbs in the closed molds. After the blow molds open, the pipes with the completely blown bulbs retract by 40° outward and are no longer rotated. The pipe claws are opened and the bulbs fall on a moving belt which conveys them to a cut-off machine in which the excess glass is fused off. The empty pipes swing another 140° to their upper vertical position and are ready for their next load. In Fig. E 58 the suction gob, the bulb leaving the IV machine, and the finished bulb are shown for a 60 mm diameter bulb.

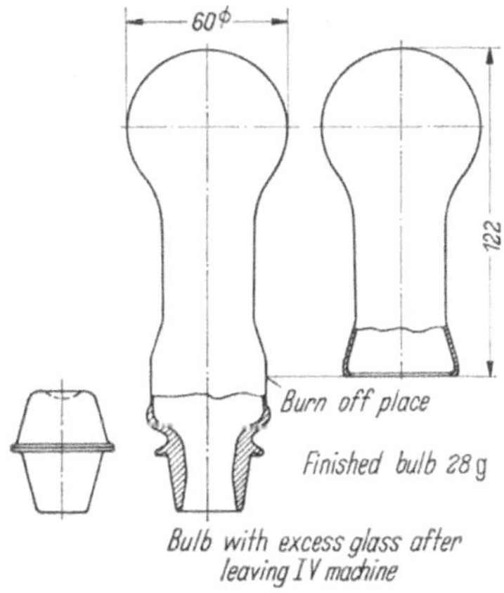

Fig. E 58. Manufacturing steps for an electric lamp bulb.

3.2.4.2.2 Machine Construction. The IV machine is driven by a main motor and a pipe motor which sit on the mobile base of the machine and are connected to the drive mounted on the cart by a friction clutch. Fig. E 59 is a schematic of the drive. The revolutions of the main motor can be controlled from a switch-board so that they can be adapted to glass quantity and glass temperature. The usual times for one revolution are 18 to 70 sec depending on the type of bulb.

The eccentric wheel axles of the carriage are connected by worm gear and permit lifting and lowering of the machine by 19 mm each by hand manipulation, to adapt the suction heads to the glass level. The removal of the machine from the furnace is also effected manually by means of a ratchet which engages a rack fixed in the machine base. The connections for blowing and cooling air are on the furnace side of the cart. On the center column are fixed, as mentioned before, all cams and the bearings for the gear wheel driving the pipes. The hexagonal disc rotates around this column on a bearing consisting of three ring discs. It is driven continuously by the main motor.

The disc carries underneath it 6 stations which control, by short cams, two air throttles each, with baffles and valves. They are called control boxes. The six sides of the disc are equipped with 4 blow molds each. At the six corners brackets carry the control stations. In each control station are found 3 baffles and one shaft for the control of the work phases which will be described. Baffle and shaft are activated by the cams attached to the center column, or, respectively, by the gear wheel positioned there. Their motions are transmitted by bevel gear

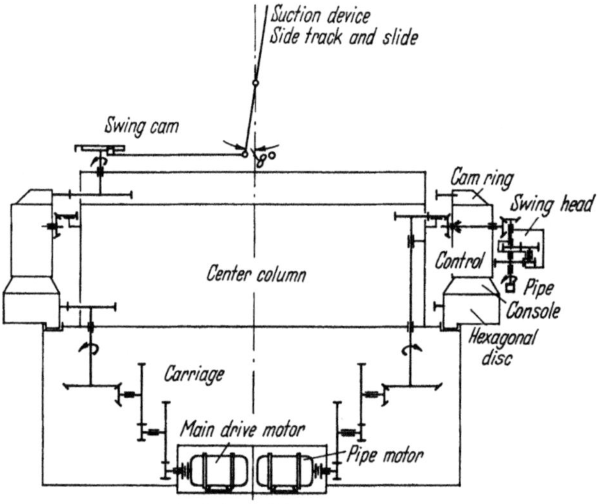

Fig. E 59. Drive of IV machine (schematic).

wheel to the swinging heads and their pipes. These heads are positioned at two adjoining control stations. These stations control the following work phases:

1. The opening and closing of the pipe claws at gob receipt, and at bulb ejection; and the pushing of the gobs over the ends of the pipes,
2. The swinging of the pipes from their upper vertical to their lower vertical position and vice versa;
3. The arresting of the swinging heads in the two vertical positions and in the bulb ejecting positions;
4. The pipe rotation.

The control stations also carry an adjustable cam ring which controls a shell slide and activates the advance cylinder of the suction apparatus (Fig. E 55). Two brake cylinders serve to brake the slide in its end positions. The entire suction apparatus, consisting of shell slide, advance and brake cylinder, and slide and slide bearing, rests on the center column and swings easily.

The swing, which has to take place six times during one rotation of the machine, and which covers 8°, is synchronized with the rotational speed of the disc, and is controlled by the swing cam. It sits on the same shaft as a gearwheel which in turn engages a gearwheel attached to the cam ring (Fig. E 59) with a transmission ratio of 6 : 1. The swing of the suction apparatus with the disc is required, as has been mentioned, for the proper transfer of the gob from the suction head to the pipe.

The swing cam holds further cams which control the valves in a box for the alternating supply of the suction heads with vacuum and compressed air. Valve box and slide are connected by a mobile conduit consisting of cylinder, piston and piston rod. At the slide the conduit forks into four branches which lead to the suction heads through the suction arms of the slide. The suction arms are in water tight connection with the slide and consist each of three tubes stacked into each other. The innermost tube alternately carries vacuum and compressed air. The exterior tubes serve to bring and take away cooling water for the suction heads. All 4 suction arms are interconnected at their ends by a manifold to which the suction heads are connected. Cooling water for suction heads, slide, slide bearing, as well as compressed air and vacuum, are brought from below through the hollow center column.

Blow and cooling air are brought to the disc through a duct that is cast into the standing cart and is continued to the height of the disc in the center column. Through a ring-shaped slot the air arrives at the disc which is gasketed by felt strips at the center column. The air for puff and final blow of the bulbs is led through the air control boxes, which are placed under the disc, and then goes through hoses to the pipes. To cool the parison the air streams through adjustable nozzles which are fastened to the upper side of the disc. Through other non-adjustable cooling nozzles the cooling air is blown against the pipes.

The lubrication of the machine is central, through two Bosch oilers, one of which supplies the suction apparatus, the other supplying all the other lubrication spots on cart and column. The machine has 24 blow pipes with a length of 200 mm from turning point to end of claw, which are carried at the swing head in groups of 4. In the same manner 4 of the 24 blow molds are always activated together. Pipes as well as blow molds can be exchanged in operation with just a few manipulations; but the suction heads only when the machine is pulled away from the furnace. The exchange of the suction heads requires 15 to 20 min.

A change from one type of bulb to another usually takes 2—4 hours, but takes 8 hours if one changes from the smallest to the largest bulb. By exchange of the accessible control cams, the preblow cams, and the suction time regulators, the IV machine can be adapted to operational conditions in a manner to attain optimum performance. Next to proper adjustment, the perfect condition of suction heads, pipes and blow molds is decisive for good production.

Fig. E 57 shows the cross section of a suction head. *1* represents the main body, *2* the cover bearing, *3* the parison mold cover, and *7* the sleeve. In the flange of body *1* are the entrance and exit slots for the cooling water, which flows through all the parts of the suction head. There is also a bore for the vacuum. The latter reaches to the parison mold. In the parison mold are cavities *10* which by narrow slots *11* are

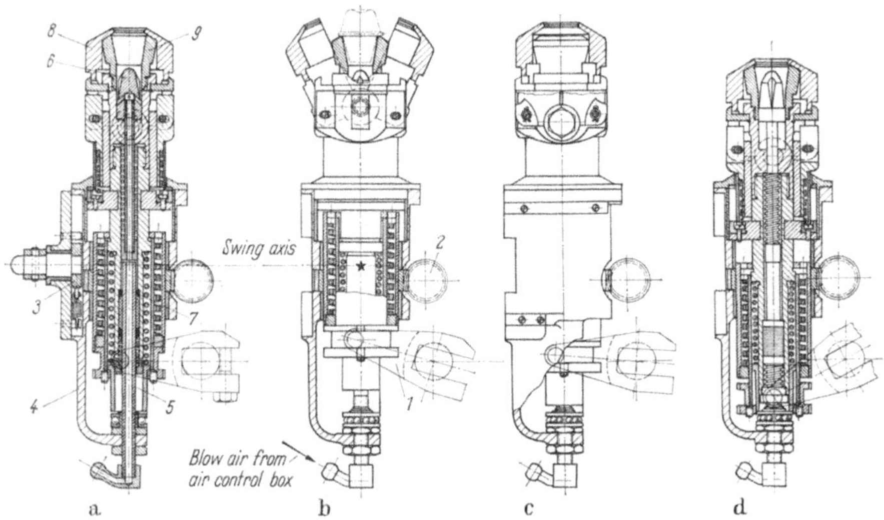

Fig. E 60a—d. Design and working principle of a pipe.

connected with the cavity into which the glass is suctioned. The geometrical shape and the size of this cavity must be determined with great accuracy for every type of product to obtain optimal results. The time of operation for properly mounted suction heads is between 80 and 140 hours.

Figs. 60a to d show the blow pipe in its different work positions. The forks *1* (indicated by thin lines) and the screw wheels *2* are parts of the swing head in which the housing *4* is fastened by the bolt *3*. The inner part of the pipe turns in the housing *4* around the blow tube *5* through which blowing air gets to plunger *6* and finally to the parison. The inner tube and the blow tube *5* are gasketed by sleeve *7*.

Fig. 60a shows a pipe in normal position with closed bolted claws *8*. The upward motion of fork *1* opens claws *8*.

Fig. 60b shows the pipe after receiving the gob, with claws still opened. If the fork *1* is brought back to its normal position (Fig. 60a) both claw halves close and are arrested. In this position they press the rim of the premolded gob airtight to the mouthpiece *9*. For the formation of an initial cavity plunger *6* is pressed into the gob. Next fork *1* draws down the entire interior of the pipe with the exception of plunger *6* (Fig. 60d). The plunger is supported by the fixed blow tube *5* so that the gob which is sustained between mouthpiece and claws is drawn over the plunger. At the same time the outer surface of that part of the gob which is in the mouthpiece is pressed firmly against its wall. Immediately afterwards the plunger is returned to its normal position (Fig. E 60a) by activation of fork *1*, and the blowing of the gob begins. The lifetime of pipes averages about 100 hours and is limited mostly by the leaking of the gasket.

The plunger wears most and must be replaced when it has become too pointed, because it would no longer be able to press the part of the gob which is in the mouthpiece firmly against the inner wall. If this happens the heat of the glass is no longer removed sufficiently through the mouthpiece and the point of the retracted plunger is heated so much by radiation that it reaches the sticking temperature so that at the next cycle it draws the glass back inside instead of forming a cavity.

The blow molds are opened and closed by baffles which are attached to the upper side of the disk. The forming surfaces of the grey cast iron blow molds, which have a thickness of about 5—12 mm, are painted with a paste consisting of a fast drying resin lacquer and charcoal dust, and are dusted with finest wood sawdust. The first gob blown into the treated mold burns the sawdust to charcoal. After every blow the mold is rinsed with water. The steam formed on blowing is carried away by small bores and slots in the mold halves. During operation the molds become only hand warm When the paste is worn they may be changed with a few manipulations while the machine runs. On the average about 1000 gobs are blown into a mold before it has to be cleaned and repasted.

An efficiency diagram of the IV machine is represented in Fig. E 61. Glass weight as well as production rate are plotted against exterior bulb diameter. The diagram shows clearly that for optimum production, i.e. between 50 and 90 mm gob diameter, the suctioned glass weight conditioned by the process (Fig. E 56) is about twice that of the finished bulb. The lower limit for the production of bulbs is given by the fact that too small amounts of glass in the parison mold cool too much and can no longer be formed. The upper limit is fixed by the dimensions of the machine.

Limits reached so far are:

	Minimum	*Maximum*
Bulb diameter	25 mm	130 mm
Neck diameter	18 mm	52 mm
Bulb length	75 mm	275 mm
Bulb weight	10 g	158 g
Suctioned amount of glass	30 g	270 g

Energy requirement:

a) IV machine

Compressed air 3.0 atm., 5 m^3/h eff.	7.2 kW
Vacuum air 85—90%, 75 m^3/h eff.	3.1 kW
Blow and cooling air 350 mm water column, 6000 Nm^3/h	10.2 kW
Cooling air for pipes 65 mm water column, 3400 Nm^3/h	1.1 kW
Blowing off of molds 100 mm water column, 4500 Nm^3/h	2.2 kW
Circulating cooling water	0.6 kW
Machine drive	0.8 k
Pipe drive	0.8 kW
	26.0

Machine cooling water 3 m^3/h

b) Cut-off machine

Burner air 1000 mm water column 60 Nm^3/h	0.3 kW
Machine drive	0.8 kW
Total	1.1 kW

City gas 4200 kcal 12—15 Nm^3/h

Oxygen 500—700 l/h

Servicing the machine requires one person continuously. In addition a mold lubricator and a smith are needed for three machines.

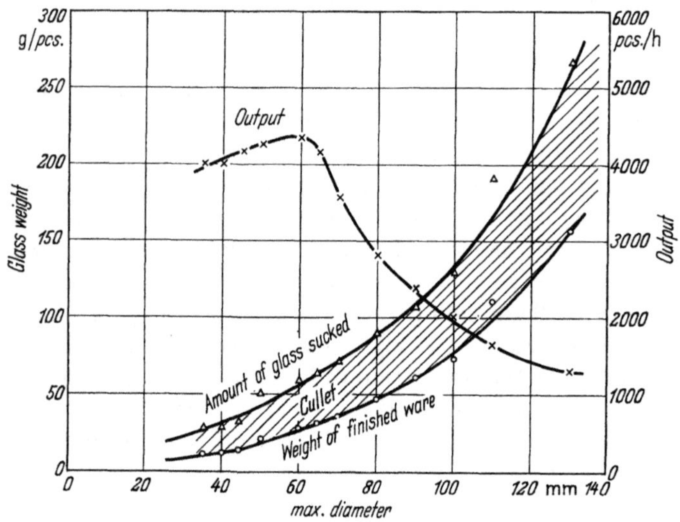

Fig. E 61. Production diagram of a IV machine.

3.2.4.3 Kutzscher-Sloan Machine (By W. GRUNDMANN, Mainz). This machine is intended for the production of thin-walled, seamless, hollow-glass ware of medium size with exacting tolerances for uniform wall thickness and surface smoothness, (e.g. tumblers, dewar flasks, special bulbs, glass inserts, etc.). It is constructed similarly to the Ivanhoe machine but has only 6 molds. The machine is particularly suited for low volume production.

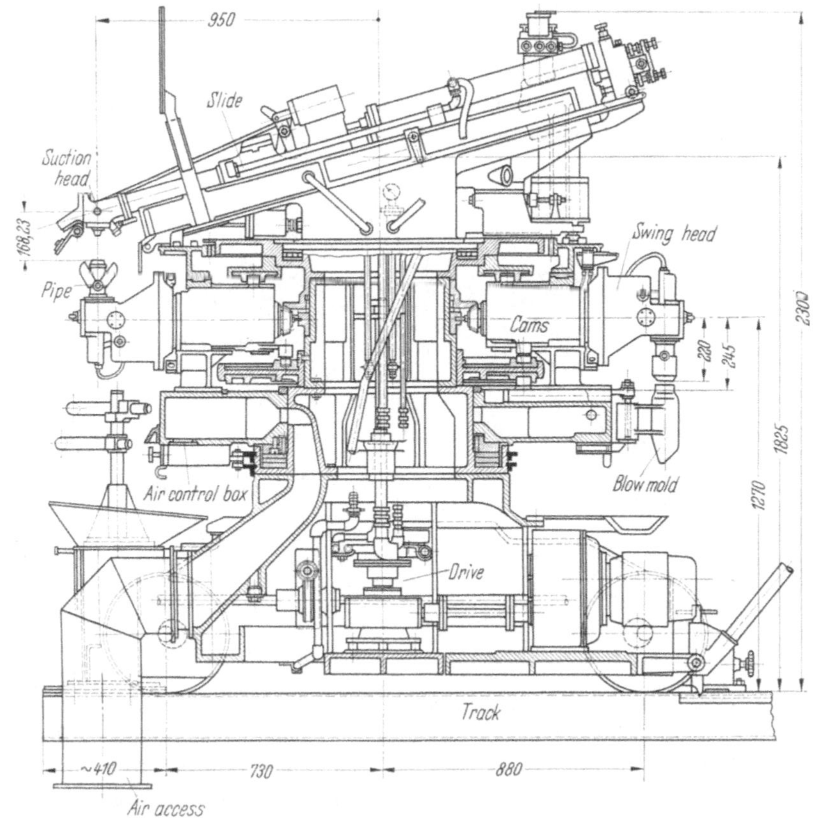

Fig. E 62. Construction of the Kutzscher rotating vacuum blow machine (Courtesy Kutzscher).

The construction corresponds largely to that of the IV machine (Fig. E 62). Its development, based on the IV machine, can be credited to the Osram company. Since the beginning of the fifties it has been built in Finland and Sweden and has become known under the name of the Sloan machine, model 184. Since 1955 the machine has also been built in Germany by the Kutzscher company in Kiel, under the designation Rotation-Suction-Blow Machine Type UM 6 S. In the course of further development Kutzscher also equipped it with 8 stations: type UM 8 S.

3.2.4.3.1 Operation. A pneumatically activated slide with suction head and parison mold enters the opening of the working end of the tank from above, in the usual manner, at an angle of about 15°. The working end is shielded by a hood or boot. By insertion of a swing cam "swinging back" can be effected also with the suction head, i.e. during suction. The working scheme is seen in Fig. E 63.

3.2.4.3.2 Construction. Drive and rotary table parts are similar to those of the IV machine. On the base plate of the carriage stands the main drive motor for the round table, and the pipe motor for the rotation of the blow pipes. Transmission is by friction clutch and a stepless drive

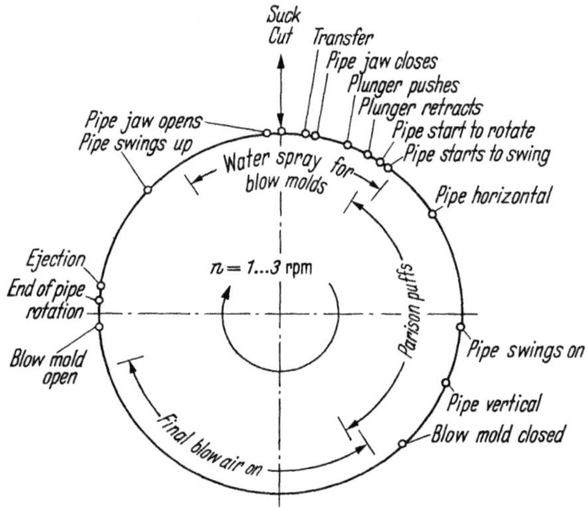

Fig. E 63. Operational diagram of the Kutzscher machine.

to the vertical drive shafts located in the cavity of the center column, and by pinion to the two gear wheels for the rotary table and pipe rotation (Fig. E 62). (1—3 rpm i.e. 6—18 cuts per minute.) The slide and the mobile parts of the rotary table are lubricated automatically from a central location.

Lifting and lowering (±20 mm) is by hand in the Sloan, and by motor in the Kutzscher machine using excentric cart axles. The machine is moved by a hand ratchet which engages a cog rod on the track. The connections for vacuum, compressed air, and cooling water which are brought through the free center column have to be kept flexible by short hose joints. At the tank side is the large air tank for blowing and cooling air. The flange of the lower tube is pressed against the machine flange by springs.

The slide with the suction head sits on the center column. Below the upper, fixed cover plate is the rotating turn table with 6 (or 8) swing heads, pipes and blow molds and the air control boxes. The cams for the functions shown in the process scheme are fastened to the center column. To change timing single cam segments or lugs are set with the machine stopped. The carriage housing and the lower part of the center column form a tube at the tank side, which conducts blow and cooling air by a slot in the center column to the table or, respectively, to the air control boxes. Above and below the air control boxes the slot is sealed against

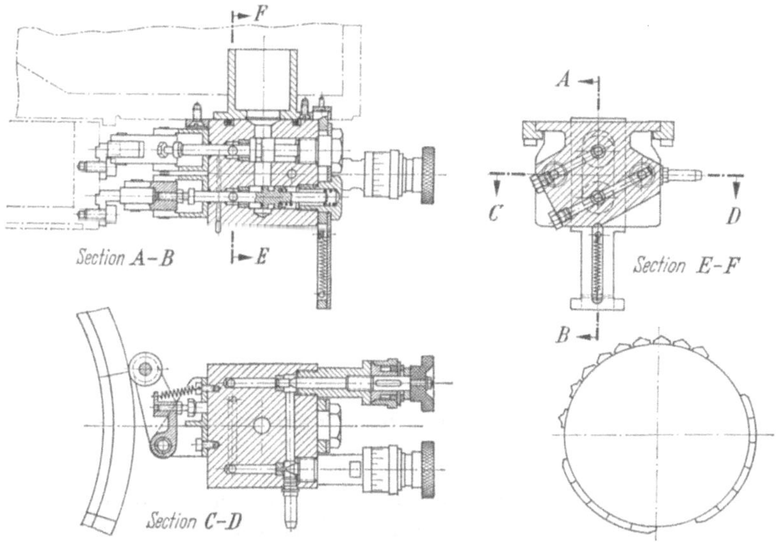

Fig. E 64. Construction of air control box.

the center column by a felt gasket. Parison and final blow air is led through a hose blow pipe via the regulating valves below the air control boxes. Adjustable lugs or cams fastened to the center column serve as controls. The exact dosing of premold air is of great importance for the formation of the parison so that much attention should be given the adjustment and maintenance of these air valves. Round valves with ring gaskets are used (Fig. E 64). Pressure adjustment during parison formation is by sight control. Parison and finish blow pressure is between 10 to 100, and 300– 600 mm water column respectively.

By adjustable nozzles the cooling air is blown to the lower edge of the pipes or the upper part of the parison (to influence glass distribution and lengthening). The suction head, the function of which can be seen in Fig. E 65, is constructed in correspondence with the required amount of glass or, respectively, the size of the parison mold in 4 different sizes. Suction head, slide and tracks are water cooled. It is recommended that a special circulation system with softened water be used. The parison mold is connected with vacuum and compressed air leads by fine bores and slots. Size and profile must be tuned exactly to glass requirements and the shape of the article. Two typical parison mold profiles are shown in Fig. E 66. Particularly for surface smoothness, high alloyed steel has been proven to be the best material (e.g. 24% Cr, 19% Ni). The suction head is always changed completely, at least at each job change. Lifetime

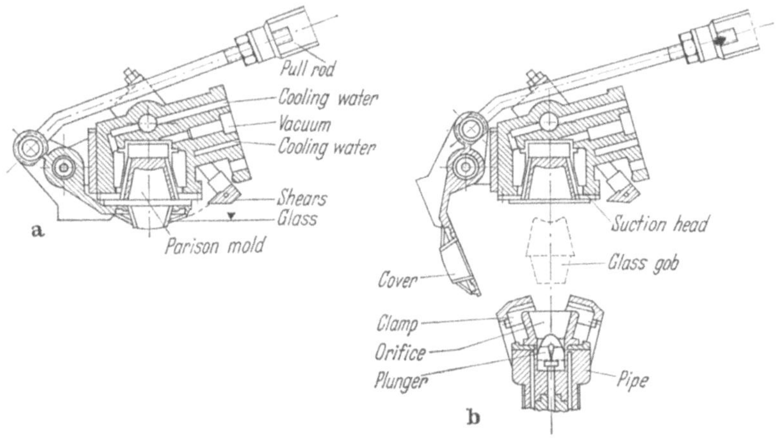

Fig. E 65a and b. Operation of vacuum head.

varies considerably, depending greatly upon operational conditions (hood temperature, water cooling) and maintenance (sealing, knife adjustment). On the average it is 100 h.

The pipes must be considered an ingenious but complicated construction since they have to permit not only rotation but also opening and closing of claws and plunger motion (Fig. E 60). Also, blow air must be brought through the rotating pipe without excessive loss of pressure. A change of the entire pipe can be made while the machine rotates. The adjustment of the rotation becomes necessary particularly if the articles differ much in wall thickness and profile. The blow molds are fabricated from regular cast iron and are generally hinged in mold holders. Needle bearings are preferred (Fig. E 67). For large diameter ware, molds with cast-on hinges are still in use. Opening and closing is effected by a cam through a shaft and a slide with a spring. To attain smooth glass surfaces the mold is painted with a special paste (e.g. charcoal and lacquer or linseed oil), which then is dried and burnt in. Sufficient ventilation during blowing, particularly of profile edges and on the bottom, must be provided. The lifetime of the blow mold depends much upon the shape

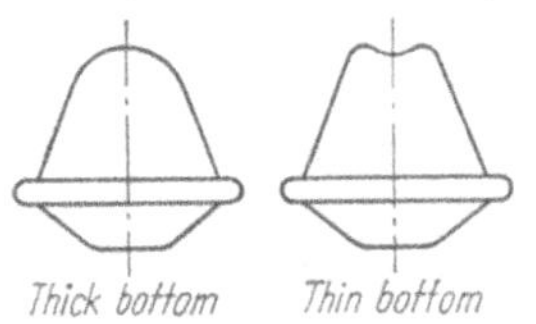

Fig. E 66. Parison shapes.

of the article, the required blow pressure, and the specified quality, and is about 500 to 1000 articles per pasting.

The machine is stoutly constructed and well adapted to the conditions of glass house operation. Interruptions can be minimized by good maintenance of suction heads and blow pipes. Pipes, valve box and blow molds are changed during operation. A suction head change requires 10—15 min. Sufficient spare parts (particularly for suction heads and blow pipes) must be available.

Time requirements for article changes are $^1/_2$ to 2 h, depending upon available skilled personnel. Personnel requirements are: 1 machine fore-

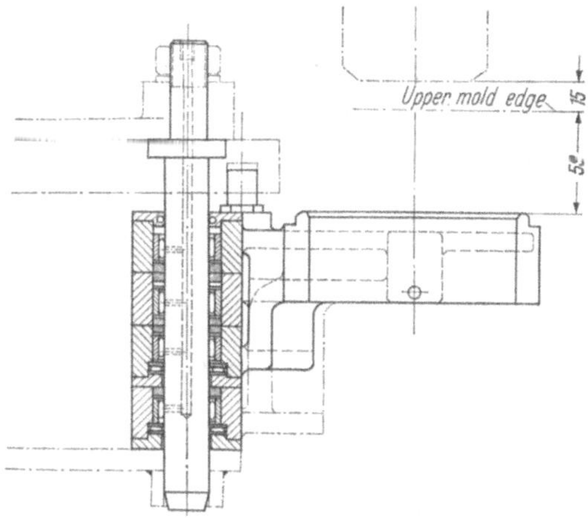

Fig. E 67. Mold carrier bearing.

man per shift; 1 machinist who, for one machine, can maintain suction heads, pipes and molds; and 1 auxiliary operator at the cut-off machine, if transportation to the machine and the lehr is automatic.

3.2.4.3.3 Technical Data. Limits of production:

Diameter	40—175 mm
Length	100—375 mm
Suctioned Wt.	60—350 g

About 50% of the suctioned glass must be allowed for the waste glass at the top.

Productivity:

Suctioned Wt. in g.	Indicated pieces/min.	Operational Data
to 100	12—18	
100—150	10—14	10—12 for 150 g
150—200	8—12	10—10.5 for 170 g (Tumblers)
200 and above	6—10	7—8 for 250—300 g (Dewar flasks)

These are target data since the influences of glass composition and of the type of article are particularly large for free formed parisons. The UM 8 S has a productivity about 33% larger corresponding to the use of 8 molds, since the time for feeding by one suction head is still sufficient. The energy requirements are given in Table E 21.

Table E 21. *Energy requirement*

	Sloan		*Kutzscher machines*			
			UM 6 S		*UM 8 S*	
	m^3/h eff. or Nm^3/h	kW	m^3/h eff. or Nm^3/h	kW	m^3/h eff. or Nm^3/h	kW
Compressed air 3.0 atü (m^3/h eff.)	33.0	13.0	20.0	8.9	30.0	12.5
Vacuum 85–90% (m^3/h eff.)	100.0	5.0	120.5	5.5	120.0	5.5
Blow and cooling air (600 mm WC m^3/h)	1500.0	4.6	800.0	2.5	1200.0	3.7
Cooling water:						
Suction head and slide (3 atü)	2.0	0.8	2.0	0.8	2.0	0.8
Mold cooling (1 atü)	6.0	0.8	8.0	0.8	8.0	0.8
Drive (kW):						
Main drive	—	2.2	—	1.5	—	1.5
Pipe drive	—	1.1	—	1.1	—	1.1
Height adjustment	—	—	—	0.5	—	0.5
Total power req.		27.5		21.6		26.4

The preceding data are given by the supplier of the machine. The larger reserve of compressed and cooling air in the Sloane machine provides some security.

3.3 Feeder Machines

During recent years the use of the feeder process has greatly increased. The exact control of glass temperature and of glass level permitted by the feeder has improved productivity and quality decidedly. In addition, the parison fabrication by feeder permits stronger preblow, resulting in better glass distribution. Feeder machine produced bottles are generally lighter for equal volume. Another advantage of feeder machines is that more machines can be placed at the tank than in the suction process. The revolving pot is not used and the feeders when arranged radially at the working end allow a larger distance from tank to machine. For tanks with high melting efficiences this fact is of great importance. In the following the most important feeder machines and their individual features will be discussed.

3.3.1 Lynch Rotary Table Machines

By E. Schulze, Essen, and H. Ganns, Essen
(with the assistance of H. Frieg, Essen)

In 1910 the Lynch corporation of Anderson, Indiana, USA, brought out the "No Boy Machine", so called because it saved manpower. Further developments of this machine are the Lynch machine types 6, 8, 10, 16 B,

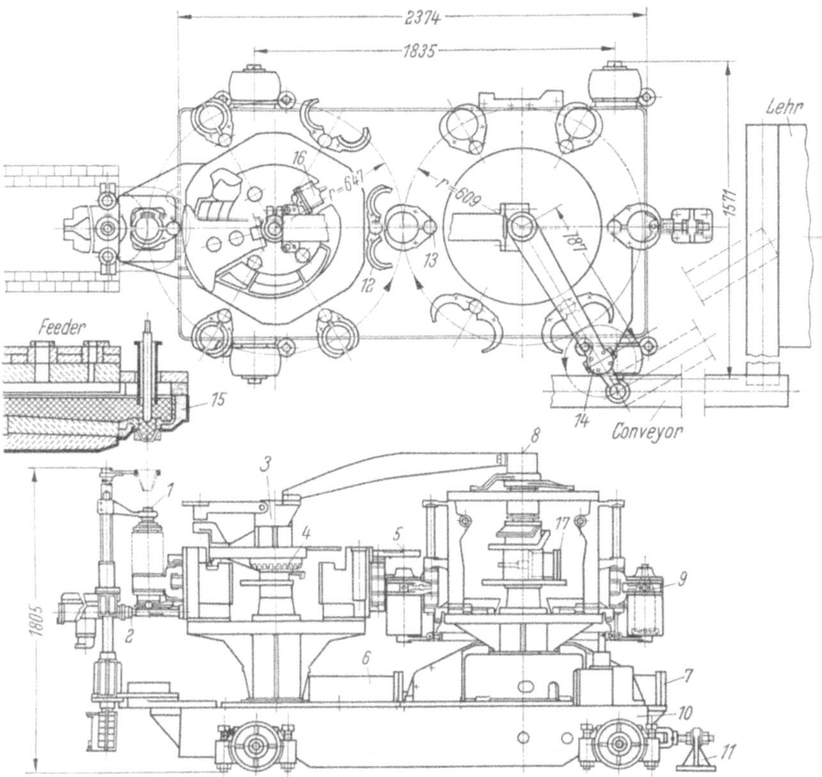

Fig. E 68. Schematic of a Lynch 10 machine.
1 Gob station; *2* Vacuum baffle; *3* Parison mold-table center; *4* Gear segment for turn over mechanism; *5* Transfer; *6* Drive cylinder; *7* Drive cylinder; *8* Blow mold center; *9* Blow mold station; *10* Machine base; *11* Device for fixing machine; *12* Parison mold holder; *13* Blow mold holder; *14* Jaws; *15* Feeder port; *16* Opening cylinder; *17* Closing cylinder.

and, recently, 44. The most widely used current Lynch machine is the model 10.

3.3.1.1 Lynch 10 Machine. This type is a compressed air driven, intermittently working, two table machine with feeder operation (Fig. E 68). The machine can be run with single or double gobs and 6 to 12 molds (Fig. E 69).

3.3.1.1.1 Process. The functioning of this machine, which is typical for all feeder process machines, is seen from Fig. E 70.

The gob, cut by the feeder shears, falls through a short tube and a funnel into closed two-part parison mold which hangs mouth down. The funnel is placed on the parison mold shortly before the gob falls.

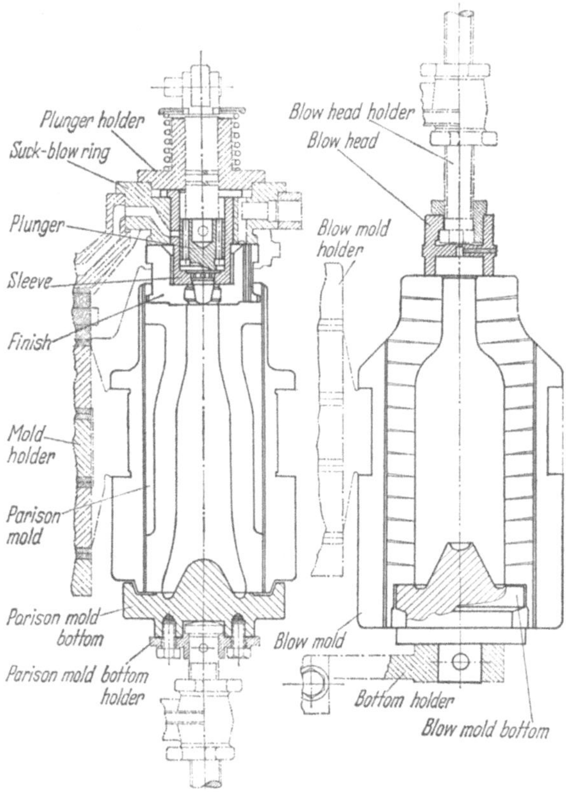

Fig. E 69. Diagram of blank and blow mold of a Lynch 10 machine.

After gob delivery the settle-blow head swings over the funnel. By compressed air impact the glass in the parison mold is pressed against the interior mold wall and through the neck, narrowed by the plunger, against the neck or finish ring. The air between mold and glass must be carried away between plunger and plunger sleeve. For high speed operation the glass is sucked into the mold by vacuum. In this mode of operation the finish is formed in the neck ring at the first station. After settle-blow head, funnel, gob ring and plunger have been withdrawn and the baffle arm with parison baffle has swung in, the parison mold gets to the second station, the counter blow station. Here the gob is expanded from the finish end through the plunger opening. On the way

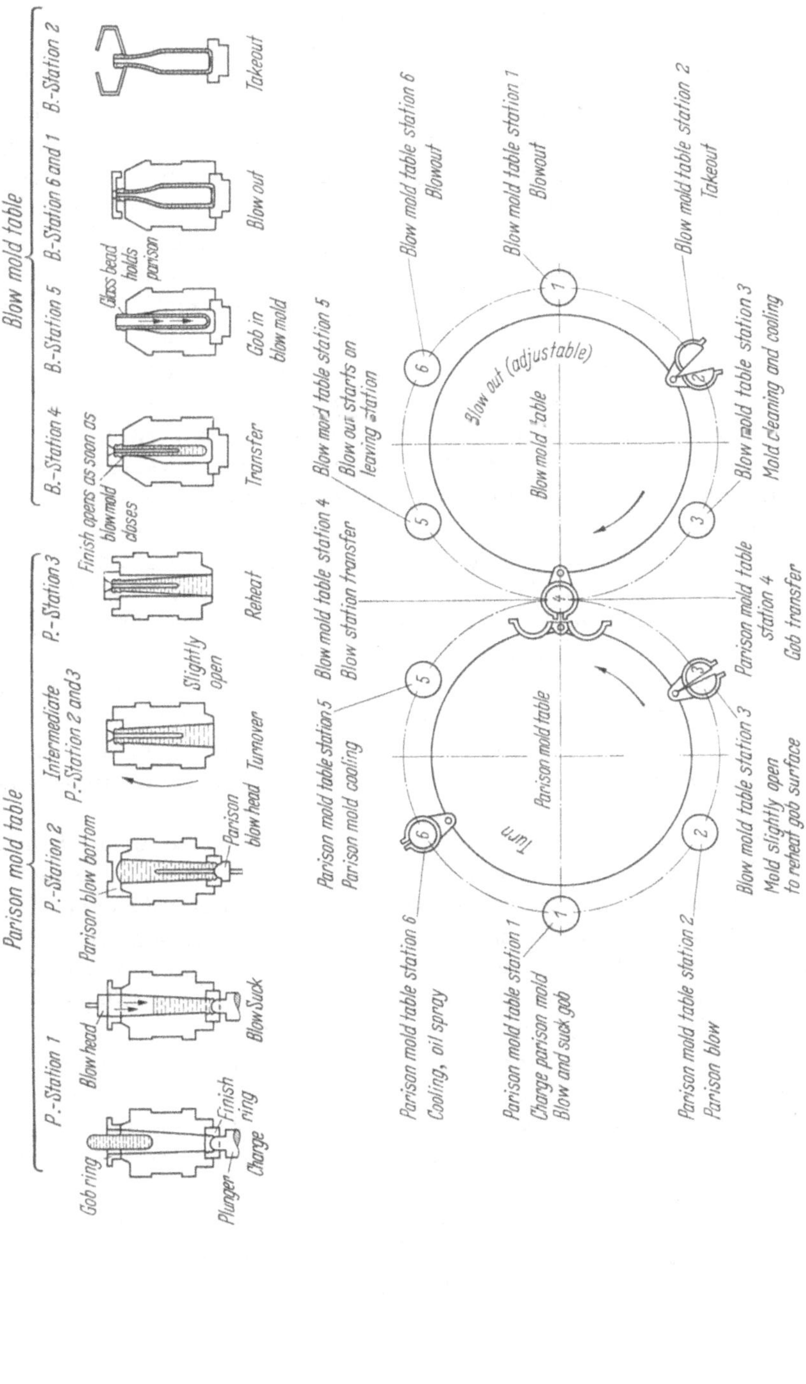

Fig. E 70. Working principle of a Lynch 10 machine.

from the second to the third station, holder and bottom swing back and the parison mold is turned by 180° so that in the third station, the turnover station, the parison mouth comes up. The neck ring still encloses the finish while the two parison mold halves already open a little. This permits reheating the parison surface by heat flow from the glass interior to the exterior. To prevent excessive stretching of the parison through the parison bottom opening it is pushed back by the parison cylinder by means of a bottom plate. Station *4* is the transfer station (from parison to blow mold). While the parison still hangs in the neck ring the parison mold halves open wide enough to allow the open blow mold halves to take it. Simultaneously with the closing of the blow mold the neck ring opens and liberates the parison. At this instant the parison falls about 1 to 2 mm, corresponding to the distance between the upper edge of the blow mold and the lower edge of the parison mold neck ring.

At the next station the parison mold is closed again and turned over to become ready for the next gob. On the way from the first to the second station, the blow head swings in above the finish and, upon passing the second station, sets on the mold and, at the third and fourth stations, blows the parison into a bottle through the neck opening. At the fifth, or withdrawal station, the blow head has already swung back. The blow mold opens while the bottle stands on the bottom plate. The take-out tong seizes the bottle at the finish and swings it by 90° above a dead plate where it is air cooled to prevent sagging. Special burners for fire finishing may be built into the take-out tong. After one or two more 90° swings the bottle is deposited on a conveyor belt which brings it to the lehr through a heating or cooling zone. In this way the Lynch 10 machine is characterized by the following:

(a) A mechanism permitting the parison mold to open slightly, shortly before it is inverted, so that the parison mold wall is detached from the parison and the skin may reheat from the interior before transfer to the blow mold. In Germany the English term „cracking" is used for this procedure. It improves the appearance of the glass article since higher glass temperature in the surface zones during blowing permits more uniform distribution in the finished bottle.

(b) Internal cooling after final blowing causing better stability of the ware when it is withdrawn. This cooling is achieved by allowing excess blowing air to escape through openings in the blow head without interfering with the blow pressure. The circulation of blow air cools the glass on the inside.

(c) The possibility of forming finishes by compressed air as well as by vacuum. For difficult finishes both procedures are applied simultaneously to make the process as brief as possible without the danger of incomplete finish formation.

(d) Equipment for the fabrication of push-up bottoms for brandy bottles. This equipment provides a push-up by a cone shaped baffle at the second station. It is blown into final shape by means of the properly shaped bottom plate in the final blow (Fig. E 69). In this process it becomes necessary to water spray the baffle in the parison mold after every cycle to cool the push-up to the extent that the bottle can be transferred without sagging.

3.3.1.1.2 Construction. The synchronization of the operation of the machine is provided by a valve timer connected with the feeder mechanism. It directs the control cylinder of the machine, the swing arm of

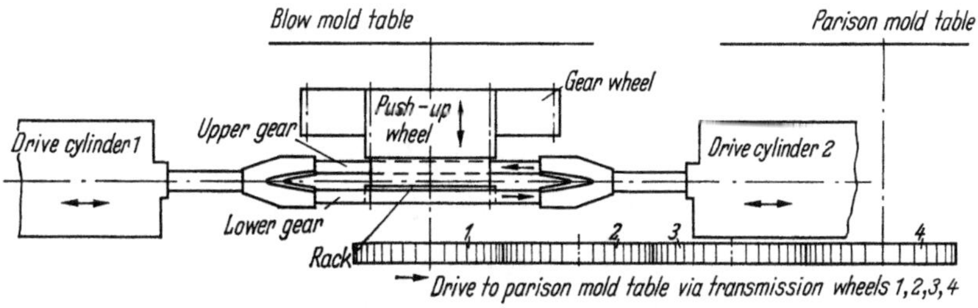

Fig. E 71. Side view of table drive (see Sec. B 1.2.3).

the parison mold blow head, and the conveyor in front of the lehr. An additional timer is the so-called 10 valve timer which controls the following work phases: plunger cylinder up, plunger cylinder down, vacuum for finish on, vacuum off, swing cylinder down, swing cylinder up, and let-in control valve. The 10 valve timer has a shaft with adjustable cams. The shaft is driven by the transmission wheels of the feeder-timer. The compressed air for the control impulses released by the cams is carried by hoses and tubes.

The Lynch 10 machine has two driving cylinders (Fig. E 71) one of which alternately drives the machine while the other brakes it by air cushion. The machine possesses two mold tables, one carrying six parison molds, the other six blow molds. The tables progress intermittently by 60°. The work cylinders drive two gear racks lying on both sides of the switch wheel which turns in the same sense. The gear wheels are arranged at different heights so that only one gear rack engages the switch wheel at any given moment. Cylinder *1* (Fig. E 72) shoves the lower gear rack to the right, cylinder *2* brakes; the switch wheel with the flanged-on push-up wheel is pressed to the height of the upper gear rack by the push-up cylinder, cylinder *2* shoves the rod rack to the left, and cylinder *1* brakes. By adjustment screws the air may be throttled and the working rate of the drive cylinder changed.

All motions of the machine are controlled by cams or air cylinders. For motions that need no adjustment fixed cams are used. The desired machine part motions are effected by rolls running over these cams. For motions requiring adjustments valve or baffle controlled air cylinders are used. The parison mold table is braked by a jaw brake arranged at the table shaft.

The loading station is mounted on the machine foundation. Its center column serves as a piston shaft for the swing arm of the settle blow head.

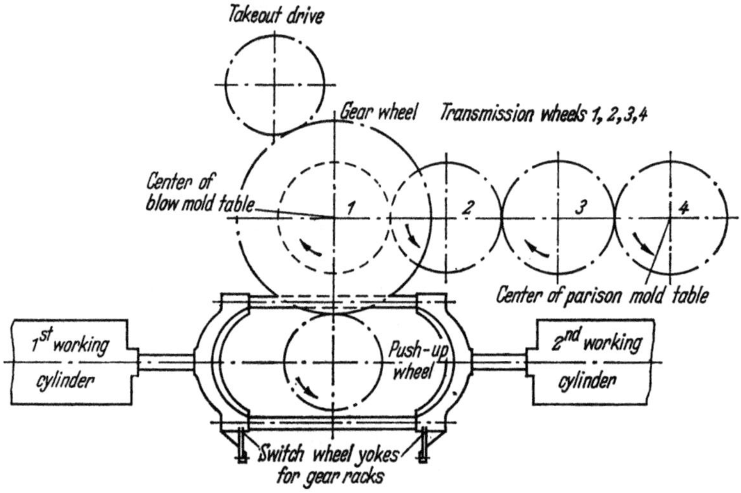

Fig. E 72. Top view of table drive (see Sec. B 1.2.3).

The funnel holder (E 68) sits at another column. The plunger cylinder at the machine foundation presses the plunger holder up and the plunger into the parison mold as the latter arrives below the loading station. The parison mold stations consist chiefly of the turnover mechanism which turns the parison mold by 180°. The parison mold holder receiver is supported in the turnover bearing housing and is rotatable. A gear wheel at the parison mold holder receiver runs through a segment gear at the table. Parison mold holder, neck ring holder, blow ring holder, and plunger holder turn with the turnover mechanism. The parison mold baffle arm at the housing of the turnover bearing does not rotate by 180°. The neck ring opening cylinder is located above the parison mold table at the position of transfer of parison to blow mold. After the closure of the blow mold the piston rod presses against the neck ring opening valve and opens the neck ring holder. The parison mold open-and-close mechanism is controlled by cams. The parison cylinder is mounted on the machine foundation. The blow mold stations consist of blow mold holder, blow mold bottom, plate mechanism, and blow head. The blow mold holder runs in a slot and is connected with the shaft of the blow head mechanism. The motions of the blow head and bottom plate are controlled by cams. After withdrawal of the finished bottle the bottom plate swings down to drop possible glass fragments. Blow mold open and close cylinders are controlled by valves.

The machine take-out is driven from the main gear of the machine (Fig. E 72). It has four arms each with a grip cylinder which is controlled by the control cylinder on the take-out shaft. By proper adjustment the bottle can be placed on the conveyor belt after a turn of 180° or 270°, depending upon the relative position of the machine to the lehr. 270° is more advantageous since it permits longer holding of the finish in the grip heads. If one has to deposit at 180°, in the double gob process, the grip head is turned another 90° by an additional air cylinder at the grip arm, to align the bottles on the conveyor.

3.3.1.1.3 Control Plan (Fig. E 73). The machine drive is controlled by compressed air of 2.8 atm. It arrives at the piston by gear rack valve *4* and shoves the switch wheel to the height of the upper gear rack. The safety valve *3* is pressed down by the switch rods. An impulse from the timer shoves the piston of the chief control valve *1* to the left. Compressed air goes through *1* via *3* to the main drive control cylinders *7* and shoves the control piston to the left. Compressed air flows through *7* to the main drive cylinders moving the pistons to the left; the gear rack turns the machine via the switch wheel. The reversing switch *6* is changed to position *6a*. Now the air presses the control piston *1* from *4* through *6a* to the right. By the reversing of *1* the air shoves the piston of the brake cylinder outwards and stops the parison mold table. A check valve *10* permits the air cushion of the brake cylinder to escape through the main valve. By a stop the piston of the gear rack valve *4* is pressed to the left. The incoming air pushes the piston of the switch wheel cylinder down and the safety valve *3* is pushed up by the clutch rods. A new impulse from the timer moves the piston of the main control valve *1* to the left. Now control air for the working cylinder gets to the left side of the main drive control cylinder *7* by *3* and presses the piston to the right. The piston of the brake cylinder is controlled downwards by *1*. The brake releases and the table starts turning. The reversing valve *6* reassumes the position drawn and the course of the control phases starts over again.

The control of the cycles occurs at the funnel cylinder and is synchronized with the feeder. Both are supplied with air by the control cylinder *2*. Reversing of the control cylinder *2* is by control valve *1* and cam valve *5*. The plunger cylinder is controlled from the 10 valve timer by control valve *2b*. Control of the swing cylinder is optional; either from the feeder timer through valves *1* and *1a* (cocks *a*, *b*, closed; *c*, *d* open) or from the 10 valve timer by valve *1a* (cocks *a*, *b* open, *c*, *d* closed).

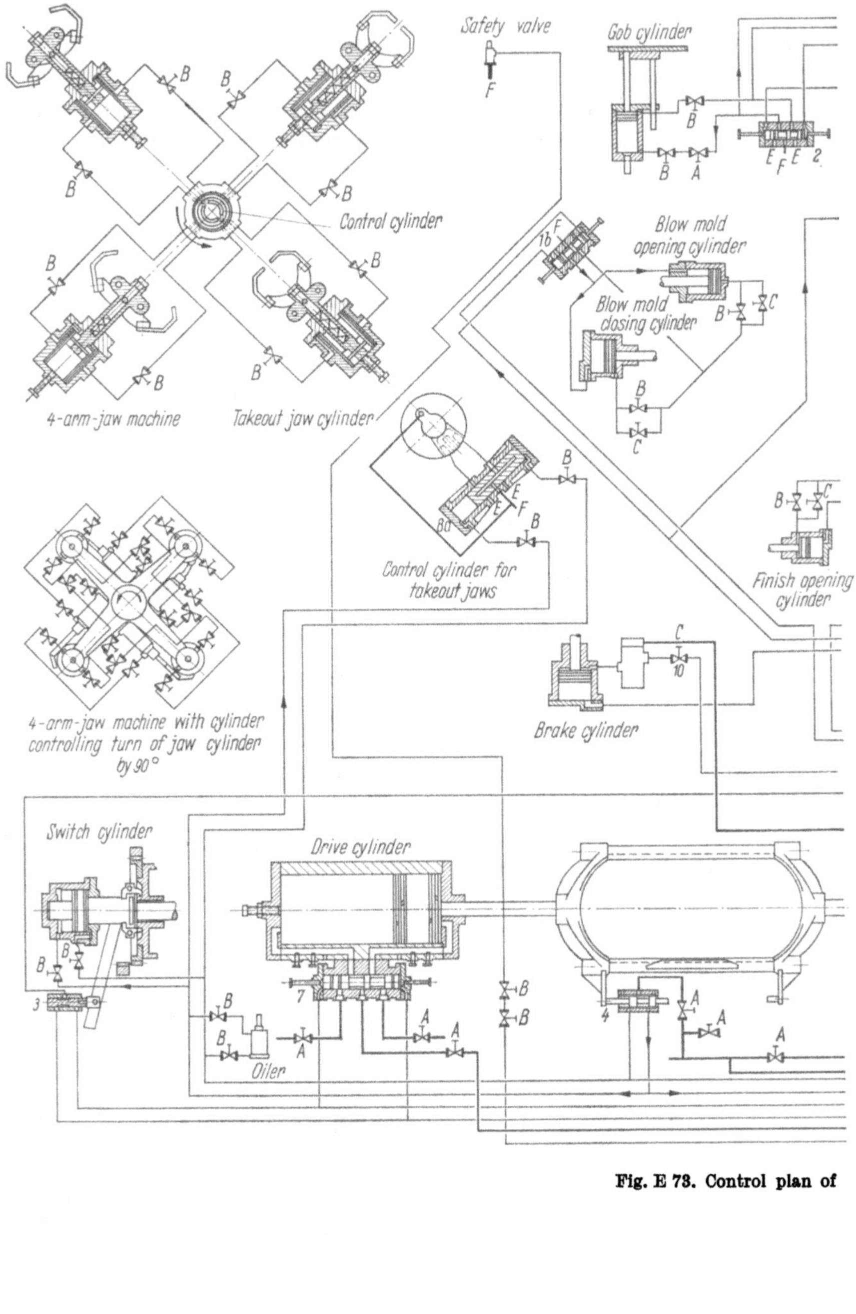

Fig. E 73. Control plan of

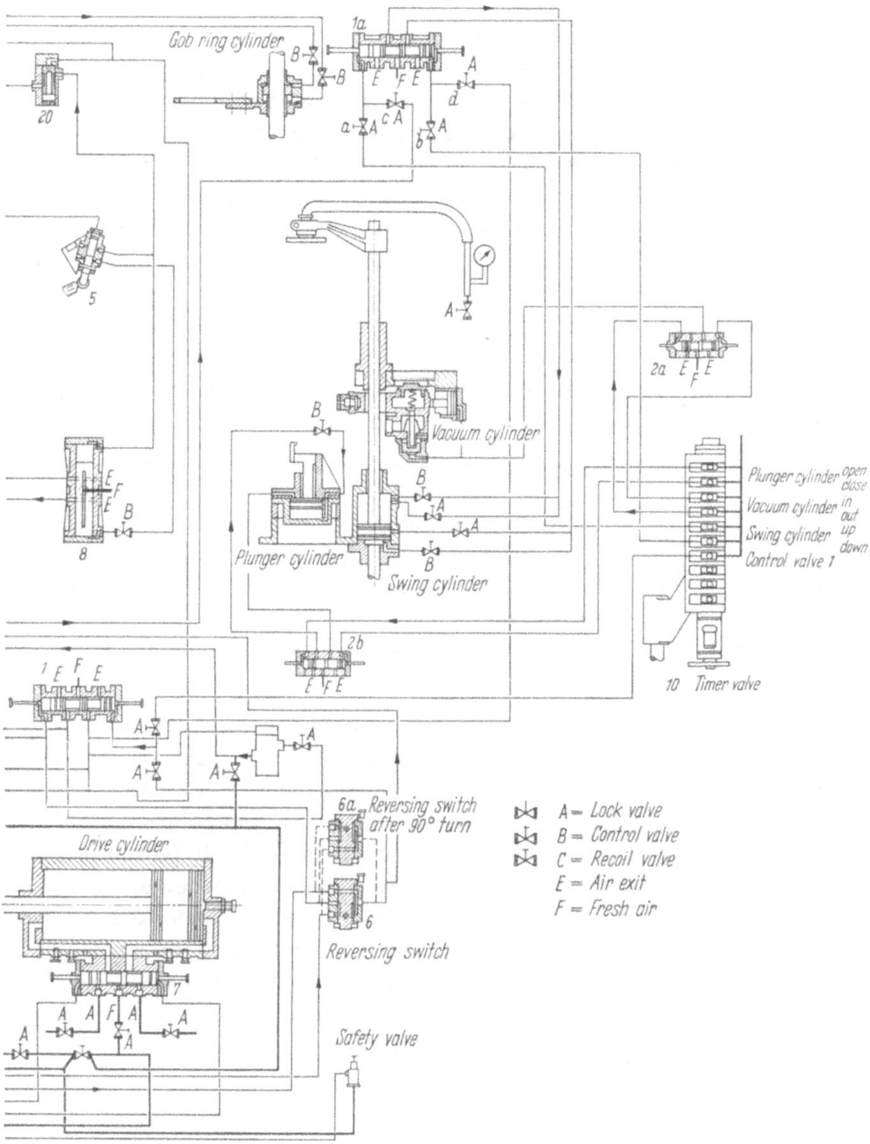

the Lynch 10 tandem-machine.

The vacuum cylinder is controlled from the 10 valve timer by valve *2a* and the vacuum is released; reversing is by spring pressure. The rack ring opening cylinder is controlled by cam valve *5* and vacuum valve *8*. Cam valve *5* below the parison mold table is connected to reversing valve *6*. The blow mold open and close cylinders are served from the reversing valve *6* through valve *1b*. Adjustment of cylinders is by needle and one way valves. The gear rack valve *4* sends an air impulse to the vacuum valve *8a* which releases the compressed air to the control cylinder. Depending upon the selection of the control sleeve in the control cylinder, the bottle is deposited upon the conveyor belt 180° or 270° after withdrawal. The impulse for turning the take-out by 90° is supplied to the additional air cylinders from the control sleeve of the control cylinder. The central oiler is connected with the impulse line of the gear rack valve.

3.3.1.1.4 General Machine Data. Personnel requirement: One machine foreman is required. The changeover times in minutes for a group of five mechanisms are: (ET = single gob, DT = double gob process)

Process	Changeover of		Startup	Production loss of changeover of	
	job	mold	time	job	mold
(ET) (articles same height)	20	20	15	35	35
(ET) ("different")	25	20	15	40	35
(DT)	30	25	20	50	45
from ET to DT	80	25	20	100	45
from DT to ET	80	20	15	95	35

Operational experiences: The cut numbers given by the supplier (50/min for ET, 86 min for DT) are achievable for certain articles. The operation of the machine requires great care to assure uniform acceleration and deceleration. Records about gob shape, temperature, rate of production, adjustment, cooling air, parison and finish blow pressure and control, are valuable tools for the machine foreman when he produces the same or similar articles at some future time. Dimensions of the machine, energy requirements, mold requirements, and dimensions of articles can be seen from the following summary. Dimensions, weight, energy requirement according to supplier [2]:

Base area	2.80 × 1.70 m
Height	2.00 m
Wt.	6.8 tons
Glass level	2.50—4.00 m (ideal: 3.20)

Target numbers for

Drive and control air	2.8 kp/cm²	8.5 m³/min
Cooling air	300 mm water col.	400 m³/min
Vacuum	660 mm water col.	3.5 m³/min

Mold requirements:

1. Single gob process:
6 each parison molds, parison mold baffles (bottoms), blow molds, blow mold bottoms (bottom plates), neck rings, plungers; 4 pairs of grip claws, 1 funnel

2. Double gob process:
12 each as above, 4 pairs of take-out tongs, 2 funnels.

It is recommended that twice the required number of molds, and sufficient auxiliary parts be stored to assure the smooth exchange of dirty parts.

Limiting dimensions of hollow glass ware producible by this machine in the blow process are given in Table E 22. Table E 23 summarizes

Table E 22. *Maximum dimensions of glassware* (mm)

	Single gob	Double gob
Diameter	114	72
Height under finish	368	254
Finish diameter	46	41

production figures obtained in a German glass house. With one mold the single gob process yielded an average of 450,000, the double gob process 300,000 bottles.

3.3.1.2 Lynch 16 B Machine. An 8-station machine with parison and blow mold tables is the outcome of further development of the Lynch 10 machine for the fabrication of bottles with narrow to wide mouth, long to short neck, for single and double gob processes. Its productivity in the double gob process is equal to that of two Lynch 10 machines in the single gob process. Important characteristics are simple adjustment, quick mold changeover, improved lubrication, and better safety devices. The machine is driven by electric motor. Each table has a Geneva drive. The machine has pressurized oil lubrication, and the oil is filtered in circulation. A 24 valve timer controls cycle and speed in synchronization with the feeder. A particularly long forehearth (6 m or more) with a deep spout is needed. Gob delivery can be direct or off-set. The yield in single as well as double gob processes is 92% or better according to the supplier.

A mold set for single gob process comprises 8 each parison molds, parison mold bottoms (baffles), blow molds, blow mold bottoms (bottom plate), 8 pairs of neck rings, 8 each blow heads and plungers, 4 pairs of take-out tongs and 1 funnel.

Table E 23. *Effective total output of a German glass house*
Color: b = brown, g = green, c = colorless

Article	Capacity (cm³)	Wt. (g)	Finish diam. (mm)	Article diam. (mm) ca.	Ht. below finish (mm)	Color	Pcs./min	
							Single	Double
Cordial sample bottle	25	38	11 mm screw	27	87.5	b	—	85
Container bottle	100	95	22 mm screw	45	102	c	42	74
Cleaner bottle	125	125	26 mm screw	49	98	b	38	70
Pocket bottle	105	150	18+22 mm comb. screw + cork	61×29	105	c	36	65
Container bottle	205	210	22 mm screw	63	143	c	—	62
Screw neck bottle	265	225	25 mm screw	57	197	c	32	59
Half pint bottle	297	295	26.7 mm crown cork	61.4	184	g	34	60
Water bottle	350	340	26.4 mm crown cork	61.5	218	c	32	58
Brandy bottle	365	315	26 mm band ring cork	61.5	218	g, c	34	58
Lemonade bottle	265	400	26.4 mm crown cork	60.5	191.5	g	28	48
Water bottle	520	460	26.4 mm crown cork	69.5	238	c	30	46
Milk bottle	535	470	26.4 mm crown cork	70	233	c	32	48
Beer bottle	520	520	punct.	72	233.5	b	27	42
Brandy bottle	720	640	band ring cork	78/74	292	g, c	28	—
Wine bottle	720	620	E 29 band ring cork	76	285.5	g	29	—
Med. wt. wine bottle	1020	650	E 29 band ring cork	89.5	302.5	b, g	24	—
Wine bottle normal	1020	780	E 29 band ring cork	89	320	g	23	—

Table E 24. *Feasible dimensions of glass articles of different manufacture*

	With Normal Equipment	With Special Equipment
Finish max. diameter	140.0	—
Article diameter	190.5	216.0
Art. height under the finish, min.	152.5	—
Art. height under the finish, max.	317.5	368.5

Table E 25. *Productivity of American glass works according to supplier of machine* [2]

Description	Capacity (cm^3)	Wt. (g)	Finish form	Pcs./min Single	Pcs./min Double
Ketchup bottle	355	270	Vacuum		64—68
Utility vinegar bottle	296	277	CT		68—70
Chili sauce bottle	355	290	Vacuum		64—68
Export beer bottle	355	348	Crown cork		66—70
Seven-up bottle	207	355	Crown cork		66—70
Pint bottle	473	358	CT		64—68
Coca-Cola bottle	177	397	Crown cork		66—70
Water bottle	296	425	Crown cork		64—66
Drinking bottle	710	440	Crown cork		60—64
Utility vinegar bottle	591	453	CT	38	60—64
Purex bleach bottle	946	466	CT	33	52—56
Pepsi-Cola bottle	296	466	Crown cork		60—62
Whisky fifth bottle	769	474	CT	33	54—60
Beer bottle	946	482	Crown cork	34	54—60
Wine bottle	769	496	CT	34	54—56
Milk bottle	946	496	Milk	33	
$^1/_2$ Gallon bottle	1892	830	CT	28	
Quart soda bottle	946	850	Crown cork	26	
$^1/_2$ Gallon milk bottle	1892	908	Milk	28	
70 mm wide neck bottle	3785	1115	CT	20—22	
120 mm wide neck bottle	3785	1120	CT	17	
89 mm wide neck bottle	3785	1143	CT	19—20	
Coca-Cola gallon bottle	3785	1220	CT	24	
Quart soda bottle	946	1303	Crown cork		40—44
132 mm Wide neck gallon bottle	3785	1360	CT	16	

Specifications:

Motor	5.5 kW	1800 rpm
Compressed air	2.8 kp/cm^2	8.5 m^3/min
Cooling air	360 mm WC	400 m^3/min
Vacuum	660 mm Hg	3.5 m^3/min

The mold holders are constructed in a way to permit the use of Lynch 10 molds on the Lynch 16 B machine. Servicing the machine requires one machine foreman. Possible dimensions of produced items are shown in Table E 24. Productivities in American glass houses are summarized in Table E 25.

Dimensions, weight, energy requirements of the 16 B machine according to supplier:

Base area	2.8 × 2.45 m
Height	2.15 m
Wt.	14.5 t.

References

[1] Anon.: The Lynch-Overmyer Glass clinic. Glass Ind. 34 (1953) 542/43.
[2] Prospectuses and data of Lynch Corporation, Anderson, Indiana.
[3] Anon.: New bottle machine by Lynch Corporation. Glass Ind. 12 (1931) 113, 119.
[4] Hunter, F. W.: O'Neill and Lynch machines. J. Soc. Glass Technol. 37 (1953) 76—79.
[5] Wardley, T. A.: Hartford feeder, Lynch machines and the Hartford I.S.-machine. J. Soc. Glass Technol. 37 (1953) 32—40.
[6] Anon.: New Model Lynch 16. Glass Ind. 21 (1940) 491.

3.3.2 Roirant R 7 Machine

By W. Giegerich, Heilbronn

The precursor of the now widespread R 7 machine was the 8 arm Roirant M 8 for feeder operation (1928). It was one of the first feeder machines with continuously rotating table in which the parison molds were arranged above the finish molds. Since at that time the feeder operation was not exact enough to achieve a precise fall of the gob, and since the synchronization of machine and feeder drive created difficulties at high speeds, it was preferred to feed the parison mold with the table at rest. E. Roirant at this time had the idea of achieving relative rest of the parison mold on a continuously rotating machine by certain construction measures. W. Miller had made a similar proposition earlier (1920). For this purpose the entire structure of the machine was placed on an oscillating turntable. At the moment at which the parison mold arrives below the gob ring of the feeder the turntable turns with the same speed in the opposite direction. After filling the turntable oscillates back. In this manner a short standstill of the parison mold during filling is achieved in a continuously rotating machine. The second innovation was the use of vacuum for the blowing of the gob in the finish mold. Ashley had proposed this procedure in 1889 (DRP 52,208, March 2, 1889), but only the Roirant M 8 solved the problem. The Roirant M 8 machine has prevailed for many years, particularly in France and Belgium. SEVERIN [1] gives a productivity of 22.8 one liter bottles/min.

From this machine the R 7 machine originated between 1949 and 1954. It is a continuously rotating one-table machine with 7 arms driven and controlled mechanically. In the topmost row sit the parison

Fig. E 74. View of a Roirant R 7 machine (Courtesy Hanrez).

molds with funnels, under them swing the neck ring molds and in the bottom row are the blow molds (Fig. E 74). The operation of the machine proper is seen in Fig. E 75, its construction in Fig. E 76. The total construction is seen in Fig. E 77.

Before filling, the parison mold is lubricated by two high pressure spray nozzles. While the gob falls into the funnel and parison mold suction begins and continues until the plunger rises. The funnel in the meantime has been set upon the parison mold. After the retraction of the plunger, counter blowing starts, next the parison mold is opened and the gob is swung. This procedure had already been used by GRAHAM, SEVERIN, AFTERGUTH and ROIRANT, in the M 8 machine. Roirant was one of the first to apply vacuum air exclusively in the parison and blow molds of his machine. The stretching and reheating of the gob thus occurs in part while it swings in free air and in part in the closing and closed blow mold before the final blow begins. In the blow mold blow-out occurs by vacuum after opening the neck rings; it can be adjusted

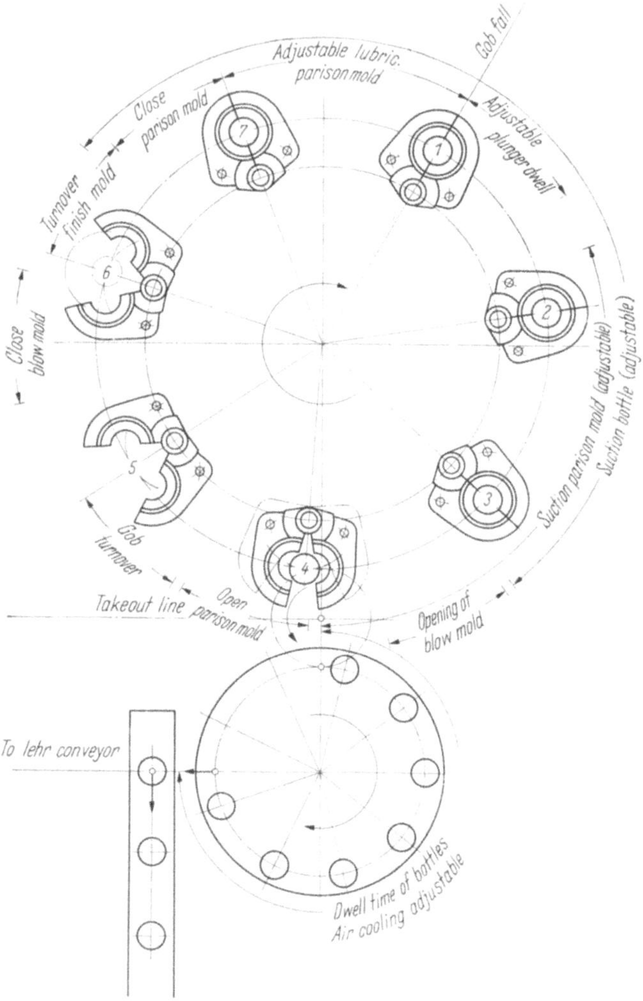

Fig. E 75. Function of a Roirant R 7 machine.

by individual stations and can be carried out with gradually increasing intensity. Because of the use of vacuum for blowing the otherwise required blow head and its moving mechanism can be omitted. Therefore parison and blow molds can be arranged in close vertical array and in consequence the swinging of the gob is simplified. Before the blow molds are opened the vacuum of the blow mold is turned off and the transfer tongs reach into the opened blow mold to set the bottles on the cooling table. In contrast to the machine the latter works intermittently. The bottles cool on the grills, which supply cooling wind from below, until the bottles are shoved to the conveyor belt. On the cooling table an internal bottle

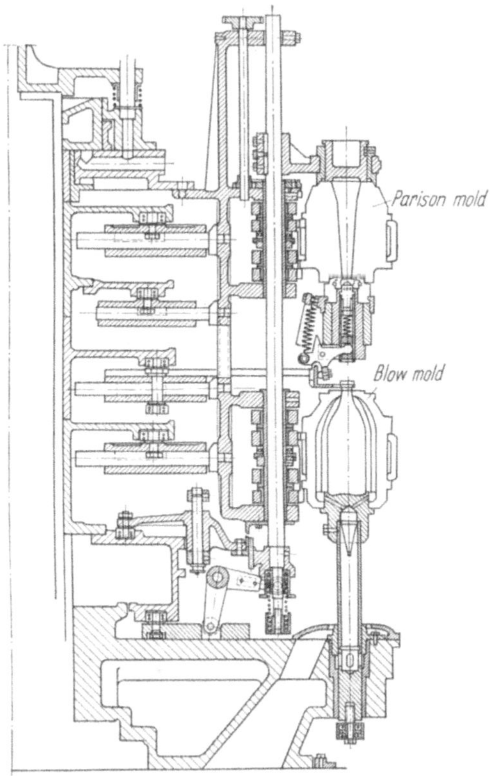

Fig. E 76. Cross section of a station of Roirant R 7 machine.

cooling mechanism or a glazing operation may be attached. Thus the R 7 machine performs loading by vacuum suction, parison blowing by compressed air and finish blowing by vacuum. Basically the R 7 machine uses no novel processes in any fabrication step; it is simply a very skillful arrangement and combination of older ideas using present fabrication and process techniques. This has resulted in its present importance and widespread use; next to the IS machine it is one of the most highly developed machines.

The cooling of molds is accomplished as usual by a blower placed in the basement, the pressure outlet of which is connected with the hollow

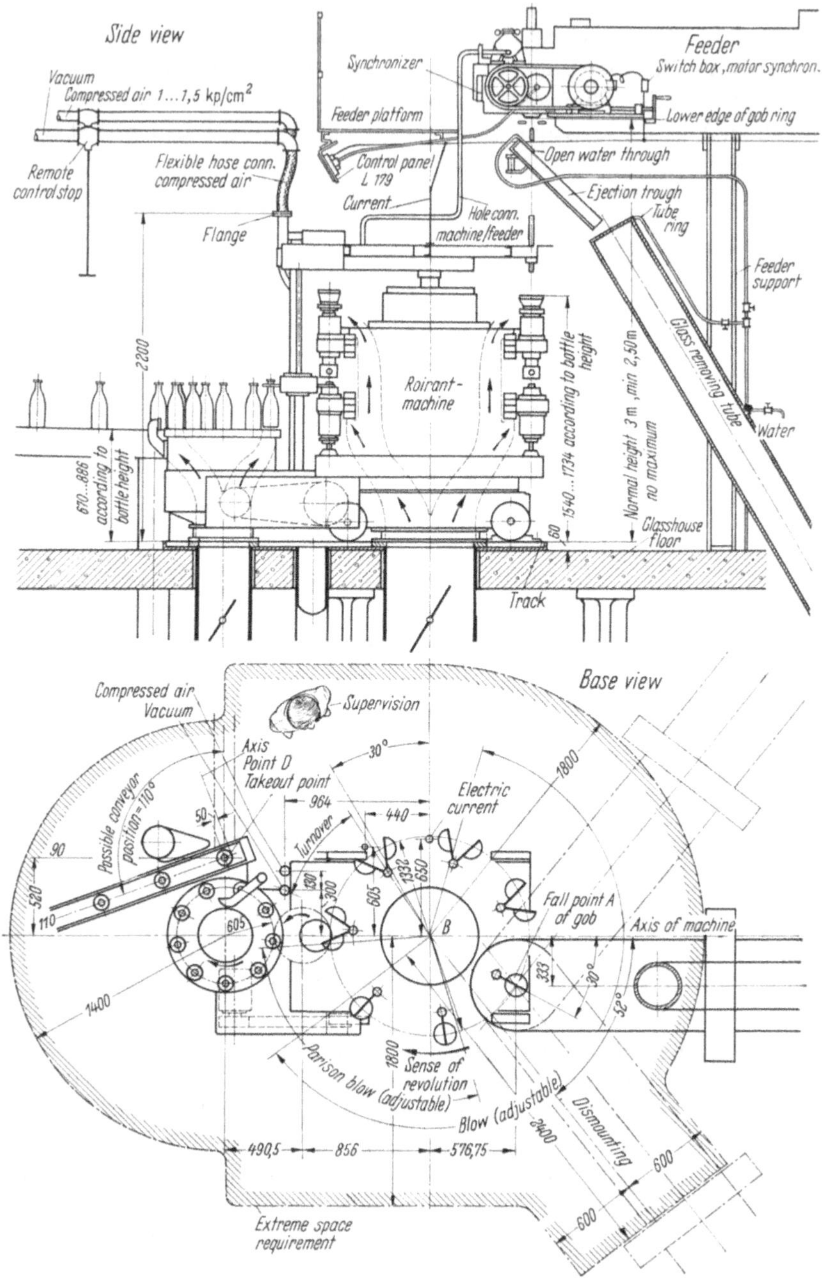

Fig. E 77. Plan for arrangement of the machine.

center axle of the machine. From there the air is delivered to the parison and blow molds through adjustable nozzles.

The R 7 machine can be operated under any type gob feeder. But feeders with mechanical control are preferred to those with pneumatic control because synchronization with the machine is easier. However, pneumatically controlled feeders have been used. The pneumatically driven feeder in connection with a mechanically driven machine can easily deviate from synchronized operation by minute fluctuations in the compressed air supply. The feeding of a running machine demands by far more exact synchronization (between feeder and machine) than the older feeding of the machine at rest. At a cutting rate of 35 bottles/min corresponding to $35 : 7 = 5$ rpm the peripheral speed for a mold circle diameter of 1.33 m is:

$$1.33 \times \pi \times 5/60 = 0.35 \text{ m/sec.}$$

The falling time of the gob from ring to upper parison mold edge (1.4 m) is:

$$\sqrt{2 \times 1.4/9.81} = 0.53 \text{ sec};$$

the falling time to lower mold edge (1.7 m) is:

$$\sqrt{2 \times 1.7/9.81} = 0.59 \text{ sec.}$$

While the gob falls to the upper mold edge the machine proceeds $0.53 \times 0.35 = 0.185$ m = 185 mm. The gob must be cut 185 mm before the parison mold arrives. For the fall of the drop through the parison mold (300 mm length) $0.59 - 0.53 = 0.06$ sec is needed. During this time the parison mold moves $0.06 \times 0.35 = 0.021$ m = 21 mm. To avoid the contact of glass and mold wall, the inner diameter of the parison mold must be 20 mm larger than the outer diameter of the gob, and this can be done easily. But the calculation shows also that the accuracy of the fall of the gob must be high in relation to the speed of the machine to avoid its deformation. The distance from lower edge of feeder ring to upper edge of machine track should be about 3 m.

The correct operation of the mechanism that synchronizes feeder and machine is necessary for proper functioning of the machine. For this reason special equipment was developed for the R 7 machine: type L-1 for the Hartford feeder 144, and type 12 for the Hartford feeder 81. There is no fundamental difference between the two types: they differ only in their application to the feeder structure. Space requirements for this synchronizing equipment are: Ll: $1.3 \times 0.85 \times 0.6$; 12: $1.05 \times 0.67 \times 0.57$ m. The weight of either type is also about 260 kg, the output of its drive motor 1.6 kW (current about $^1/_2$) and the requirement in compressed air 250 liter/min with an overpressure of 1 to 1.5 kp/cm^2.

The time tolerance for the fall of the gob is 0.01 sec according to the supplier. For a machine speed of 50 bottles per min. this corresponds with a motion of the parison mold by about 5 min arc on the circle of mold centers. The synchronizing equipment developed by the Roirant company is connected with the R 7 machine pneumatically, by a rubber hose. An air impulse from the machine activates the friction clutch to the feeder; after one revolution of the feeder cam the clutch is released; next comes a brief stop of the feeder with the plunger in the upper position and the shears open; then follow: reclutching, plunger downstroke, shear cut, plunger upstroke etc. Thus the continuous rotation of the machine proper provides intermittent rotation of the feeder cam. In this construction feeder and machine have separate drives, and disturbances in one or the other have no mutual effect.

The fabrication of different bottles at the same time is possible only if they have equal weight. The R 7 machine allows the fabrication of bottles differing in height by about 30 mm and in bottom diameter by about 10 mm. For changeover to another type one requires about 1—2 h depending upon the kind of bottom (flat, hollow), kind of finish and the difference in weight (ring, plunger), provided a special changeover group of 4—6 mechanics and skilled operators is available.

Lifetime is hard to estimate: in general, 800,000 to 1,000,000 bottles might be made in one blow mold. Parison molds should be cleaned and polished every 6—10 days, neck rings at shorter intervals.

Table E 26. *Energy requirement, dimensions of bottles and output of Roirant 7 machine*

a) *Energy requirement*

Compressed air	2.2 atü	100—144 m³/h	7.0 kW
	2.1 atü feeder	30— 84 m³/h	2.5 kW
Vacuum	82% on machine	960 m³/h	18.5 kW
Cooling air	270 mm WC	21600 Nm³/h	32.0 kW
	(in reality 265 mm WC, 25370 Nm³/h)		
Drive			2.0 kW
Drive for connecting device			2.0 kW
Feeder drive			0.5 kW
	Total energy requirements		64.0 kW

Heat consumption feeder 43500—126000 kcal/h (city gas 4000 kcal/m³)
Lubricating oil requirements ca. 15 l/day

b) *Limiting dimensions of bottles*

Height including finish	115—375 mm flat bottom
	115—341 mm hollow bottom
Outside diameter max.	125 mm
Finish diameter max.	58 mm
Puncture max.	40 mm
Plunger diameter max.	60 mm
Weight	150—1200 g

The space requirements for the machine are about 2.8 × 1.9 × 2.4 m and its weight without molds about 6000 kg. Energy requirements, bottle dimensions and production data for the Roirant 7 machine are given in Table E 26. For some time attempts have been underway to

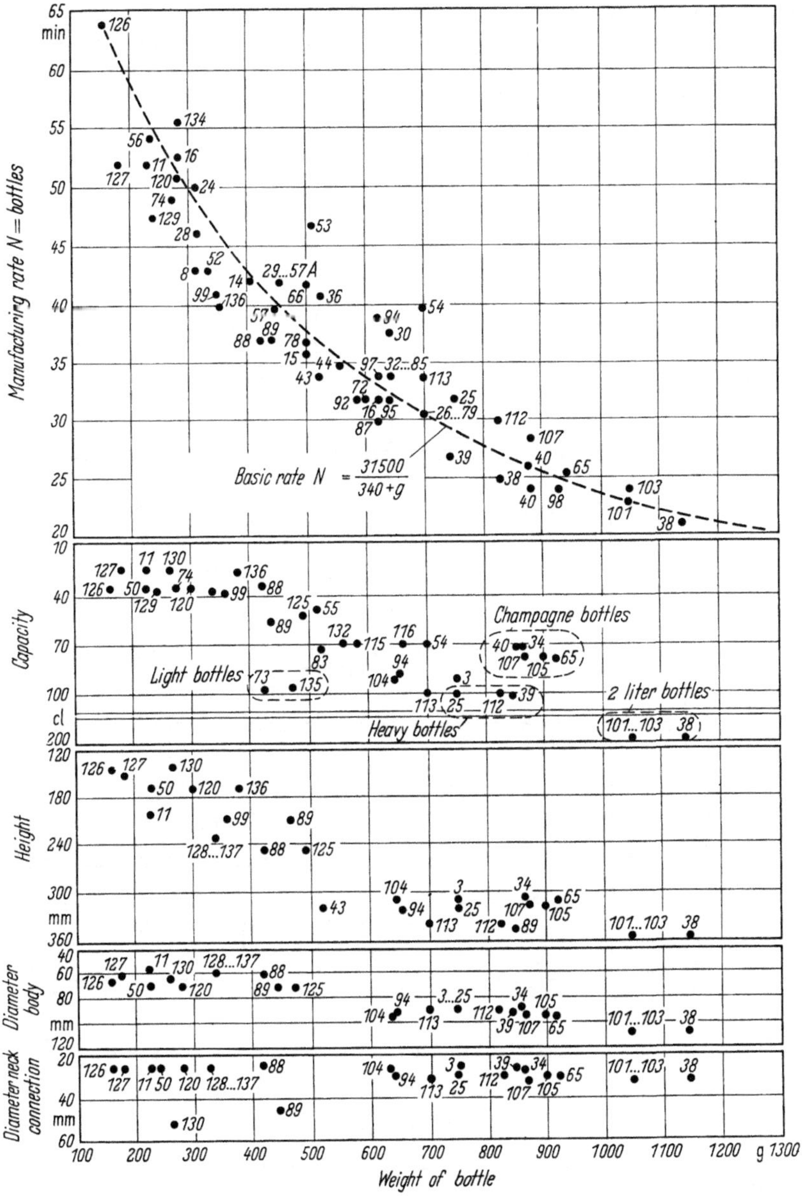

Fig. E 78a. Productivity of a Roirant 7 machine (single gob process).

work the machine in the double gob process; production data for single and double gob process can be seen in Figs. E 78a and E 78b.

Rates for single pieces in degrees as well as seconds for various machine speeds are tabulated in Table E 27. A comparison with other machines was made in Section E 3.1, p. 204 to which reference is made for detail. Table E 28 contains mean data for various necks, as given by the supplier.

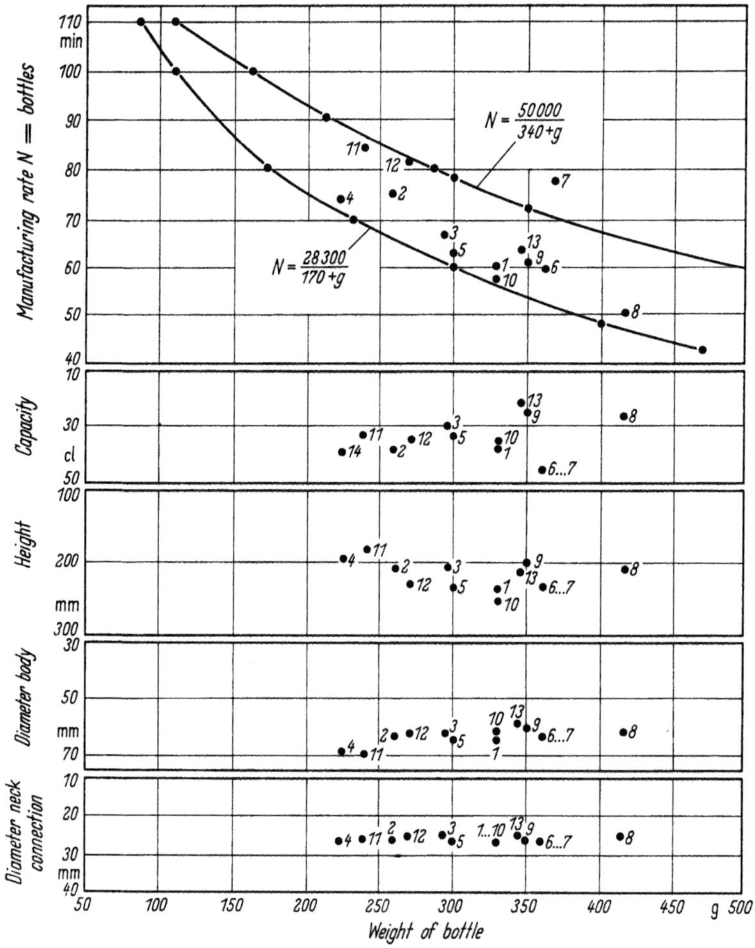

Fig. E 78b. Productivity of a Roirant 7 machine (double gob process).

References

[1] Severin, H.: Die Entwicklung der Roirant-Maschine A 6. Glastechn. Ber. 20 (1942) 65.

[2] Information of manufacturing firm (Ateliers Hanrez, Monceau, Belgium).

Table E 27. *Fabrication times for Roirant 7 machine*

	Angle (°)	Time in seconds for speed (bot./min) (e)			
		50	45	40	35
Gob in: start (a)	0	0	0	0	0
Gob received: end (a)	12	0.28	0.31	0.35	0.40
Suction starts (parison mold) (b)	55—65	1.51	1.55	1.60	1.84
Suction ends (parison mold)	145	3.38	3.75	4.23	4.83
Opening starts (parison mold)	153	3.57	3.92	4.46	5.10
Parison starts rotation	170	3.97	4.40	4.96	5.68
End closing (blow mold)	245	5.72	6.34	7.15	8.20
Suction starts (c)	305	7.05	7.88	8.90	10.15
Next gob starts (d)	360	8.40	9.30	10.50	12.00
End suction (blow mold)	473	11.05	12.20	13.75	15.75
Start open blow mold	483	11.25	12.47	14.05	16.10
Start takeout	510	11.90	13.20	14.80	17.00
Start dwell time on table	562	13.10	14.50	16.40	18.80
End ejection, conveyor	920	21.40	23.80	26.80	30.70

(a) Zero pt = start of entry of gob in funnel. At 12° the funnel starts to be moved away.
(b) Start of suction adjustable (e.g. 55° for large, 65° for small bottles).
(c) Adjustable, depends on glass type and machine type. The given value represents an average.
(d) 360° corresponds with one revolution of the machine.
(e) These times are cumulative and correspond with the angles in col. 1.

Table E 28. *Mean adjustments for various finishes* (in °)

	Mach. speed (Pcs./min)	Vacuum before gob falls	Parison mold end. Gob falls	Dwelling time plunger	Beginning of blow
Basic adjustment		−7	+29	31	46
Limits		0—7	23—41	25—43	44—95
Smallest		1	3	3	3
Finishes					
Crown cork	34	−7	+29	31	56
other thin walls	46	−7	+35	37	62
Champagne	24	−7	+29	31	53
Sprinkler finish	34	−7	+35	37	59
other heavy walls	46	−7	+41	43	62

3.3.3 Hartford IS Machine

By H. Riede, Düsseldorf

3.3.3.1 Description of the Machine. The Hartford IS machine is an in-line machine which is built with 2, 4, 5 or 6 sections. The name IS derives from the initials of the words "individual section", indicating that each section is an individually functioning hollow glass machine.

The machine is a fully automated machine for the fabrication of narrow neck as well as wide mouth containers which allows a wide range of operation and the corresponding cycles. A plan of the location of the machine with respect to the feeder, showing the main dimensions, is given in Fig. E 79 which represents a 5-section IS machine with a left-

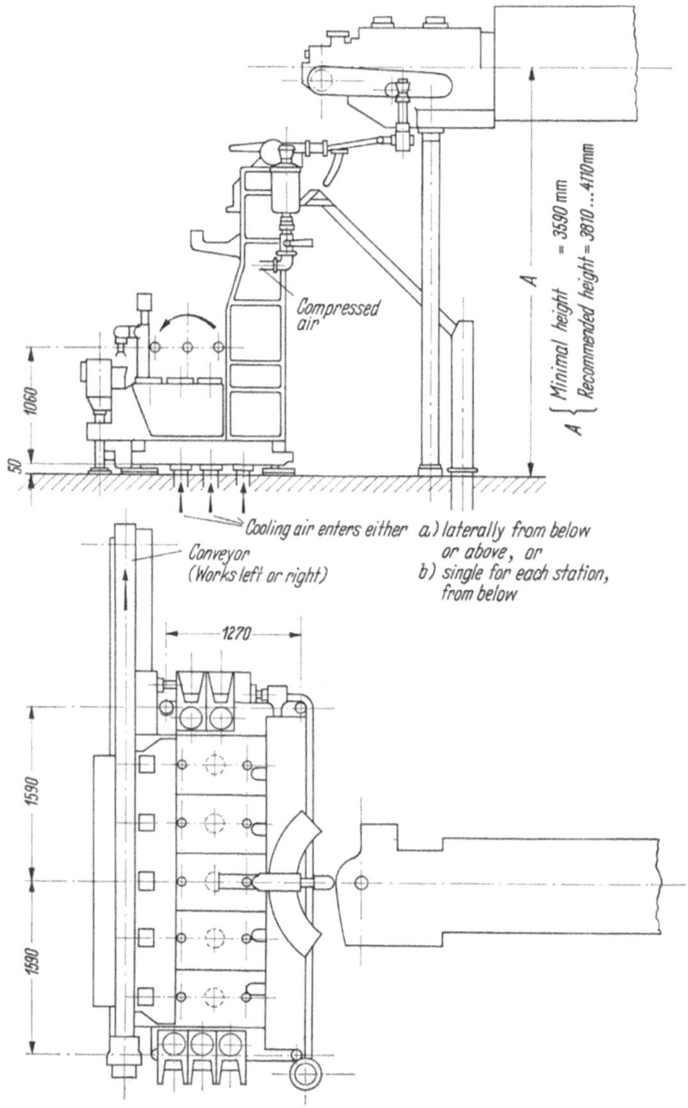

Fig. E 79. Hartford IS machine, position of machine in relation to feeder, critical dimensions etc.

hand conveyor. The control of the machine and feeder requires one operator. The construction and the control of the machine permit the back to back placement of two machines under one feeder. If one works with double gob molds the machine must be at a right angle to the feeder unless special equipment allowing to angle the feeder shears is available. The motions of the machine are pneumatically activated. The chief phases are:

1. Activation of scoop,
2. Activation of plunger,
3. Opening and closing of molds,
4. Swinging of funnel, parison mold bottom (baffle) and blow head,
5. Inverting the parison by 180°,
6. Opening and closing of neck ring,
7. Takeout of bottle from blow-mold.

The control of these phases, including the necessary blowing operations, is also pneumatic by means of an ingenious control mechanism. Each station has a control drum with 19 valves. Adjustable buttons on this drum release impulses and work phases in a pneumatic valve block above the drum. Fig. E 80 shows schematically the construction of an

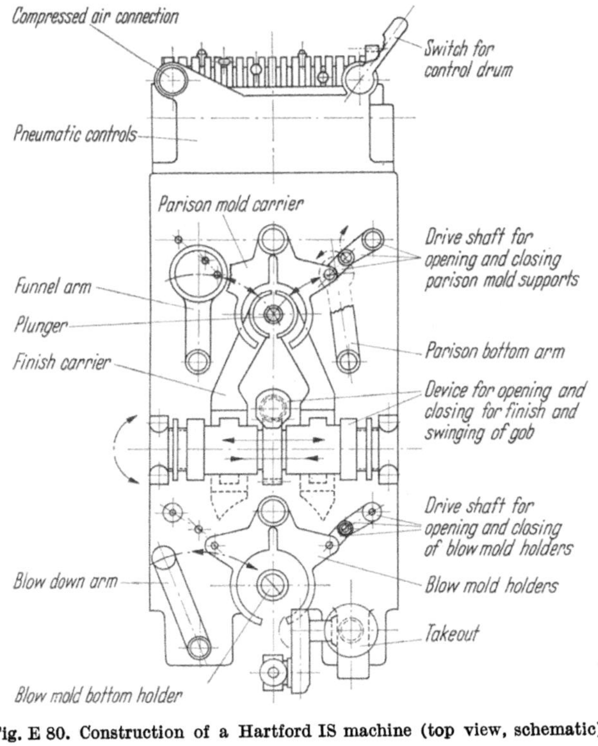

Fig. E 80. Construction of a Hartford IS machine (top view, schematic).

IS station with the parts and motions required for production. It is a characteristic of the IS machine that the various production processes are carried out by separate independent assemblies. These assemblies which are self-contained, are screwed and dowelled to the section frame or the base plate, and are connected to the pneumatic valve block by corresponding tubes and special hoses. One may speak of the building block principle. On the basis of this building method a fast change of any assembly is possible in the case of a breakdown while the rest of the sections continue to operate. The control drums are driven mechanically by gears from a common shaft. In order to be able to engage and disengage the control mechanism independently at all stations a pneumatic clutch (see Fig. B 21 in Sec. B 2) is inserted between the gear drive and the timing drum. The drive scheme of the Hartford IS machine in connection with the feeder is shown in Fig. E 81.

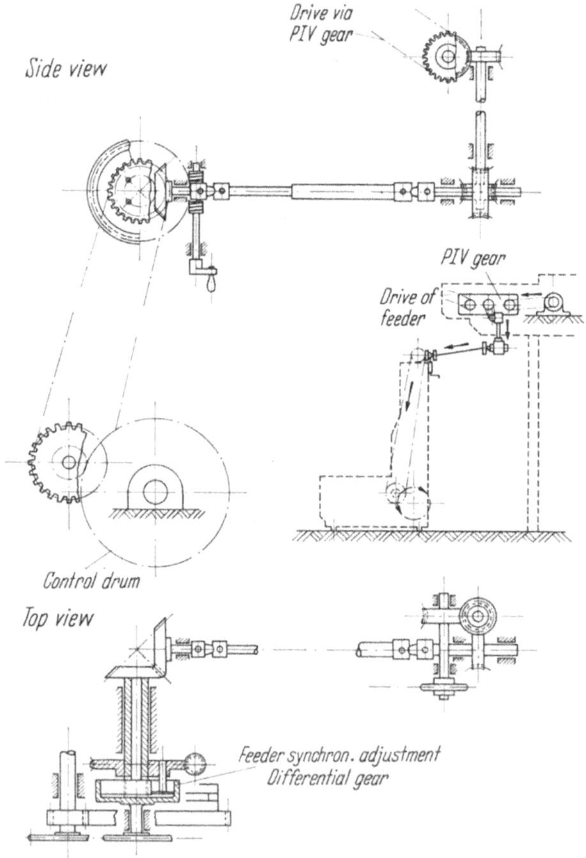

Fig. E 81. Machine drive of the Hartford IS (schematic).

The gobs from the feeder are distributed to a system of troughs. The gob is intercepted by a curved receiving trough after a brief vertical fall. It is directed into a straight intermediate trough which slants downward. From this point the glass is directed downward by a third trough or deflector and finally falls vertically into the parison mold. Each station is equipped with the delivery system just described. The scoop is fastened to the piston rod of a pneumatically operated cylinder. The forward and backward motion of the scoops is effected by this well known control mechanism. The scoop of the just fed section always retracts from below the feeder opening. In multiple section machines various gob sequences are possible so that bottle distances can be changed for left and right handed conveyors in connection with belt speed. Every multi-section machine may be operated with less sections, even just one, if it has to be, e.g. for samples or small orders. For equal weight and rate the IS machine may fabricate items of various shapes and sizes. The first Hartford IS machine was built as a 4 section blow machine according to H. INGLE's U.S. Patents in 1925. Five years later the development of the additional press-blow equipment started: the "62" process.

3.3.3.2 Production Phases

3.3.3.2.1 Blow-Blow Process. The production of a narrow neck container is represented schematically in Fig. E 82. By the delivery system just described the feeder-formed gob is brought to the parison mold. To make the gob fall if possible concentrically, and without touching the upper mold area, it is led through a funnel which sits on the parison mold for a short period (Fig. E 82). In a further phase the gob is forced into the lower part of the parison mold by a brief compressed air impulse. This phase is called settle blow. The settle blow should not be longer than absolutely necessary, since the formation of the well known "settle wave" depends, in part, upon settle blow time (b). After the settle-blow the plunger is retracted. Before air is blown into the gob through the opening formed by the plunger, the finish may be given the opportunity to heat internally for a short period (reheating). The duration of parison blow time (counter-blow) and the amount of parison blow pressure depend largely upon size, shape and weight of the bottle (c). The parison blow air simultaneously cools the plunger. After the gob is blown to a parison, the parison mold opens and the parison is inverted to the blow mold side while being held at the finish (d). During this time and until final blow the surface glass of the parison heats up. This diminishes temperature gradients within the glass (reheating). At the same time the parison stretches in the blow mold (e). By the final blow in the blow the container obtains its final shape. Here too, final blow pressure and mold time can be varied, again depending upon shape and weight of the bottle. The blow head of the Hartford IS machine possesses an additional part which permits internal cooling during final blow. This internal cooling gives increased productivity (f). The final blow time must be long enough for the glass container, immediately after withdrawal, to be stable enough to suffer no deformation when set upon the dead plate. The blow mold is opened, take-out tongs of the take-out equipment seize the finish and set it on a dead plate (g). From there the glass container is shoved onto the conveyor belt and transported to the lehr.

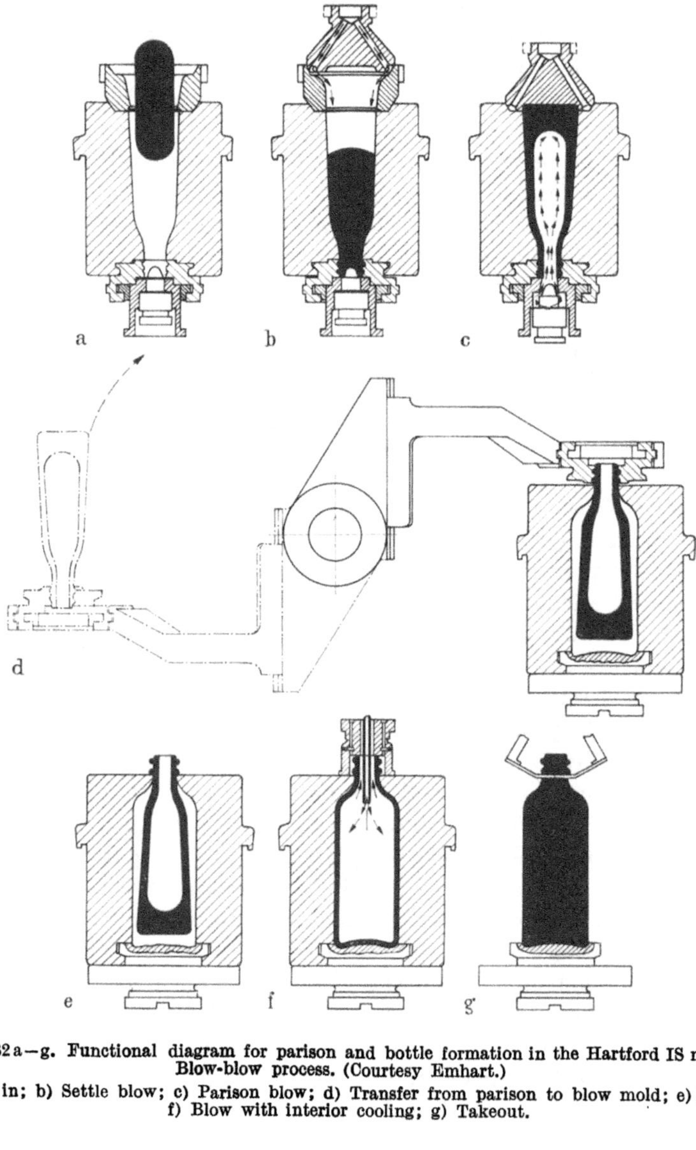

Fig. E 82 a–g. Functional diagram for parison and bottle formation in the Hartford IS machine. Blow-blow process. (Courtesy Emhart.)

a) Gob in; b) Settle blow; c) Parison blow; d) Transfer from parison to blow mold; e) Reheat; f) Blow with interior cooling; g) Takeout.

3.3.3.2.2 Press-Blow Process. The press-blow process is used for the fabrication of wide mouth containers. To convert from blow-blow to press-blow process the plunger cylinder is exchanged for a press cylinder operated by compressed air in both directions. Hartford designates this press cylinder plus accessories, as well as the method, as the "62 process" (Fig. E 83). The gob falls through the parison mold (open on top, with or without funnel) on the plunger top which is in loading position (b). Now the baffle closes the parison mold opening and the glass is pressed to a parison by the rising plunger (c and d). Pressing time can be varied through the timing drum. After the plunger has been withdrawn (e) the parison is blown from the inside to obtain solidification of the glass near the rim. Next the parison and the neck ring are removed to the blow mold side after the parison mold is opened (f). From here on the cycle proceeds as in the blow-blow process. While the parison swings to the blow mold side and the empty neck ring reverts the plunger has reached its lowest position (a). Shortly before the gob falls into the parison mold the plunger rises by an exactly measurable amount to the so-called loading position. This is to avoid onesided squeezing of the gob between plunger and finish which would cause bad production conditions. Since the plunger is exposed to severe heat it is cooled by compressed air from within (d).

3.3.3.2.3 Production Limits and Rates. Apart from the general capability of producing narrow and wide mouth containers the IS machine has remarkable versatility. Narrow neck containers can be produced from a few cm^3 to 4 liter contents. Experience teaches, however, not to schedule the extremes as a rule. When the smallest containers are made, even in double gob molds, machine and mold wear becomes very large due to high speeds. If orders cover one week's production or more double gob molds are nevertheless recommended. The increase in productivity through double gobbing is about 60—70%. The changeover of a 5 station IS machine from single to double gob or vice versa takes about 5—6 h. The exchange of equipment for wide mouth "62" production from single to double gobbing or vice versa needs more time because of the change of press cylinders. The necessary adjustments in the tubing and air piping systems can be carried out while the molds are being changed. The production limits as to dimensions and outputs of Hartford IS machines, depending upon glass weight, are shown in Table E 29. These are average total values referred to speeds of the machine.

General Production Experience

3.3.3.2.4 Control Times. Production Phases. When the machine was described (3.3.3.1) it was mentioned that simple adjustment of control buttons in the timing drum permits changing production phases and corresponding motions. Important as this feature is for production, too

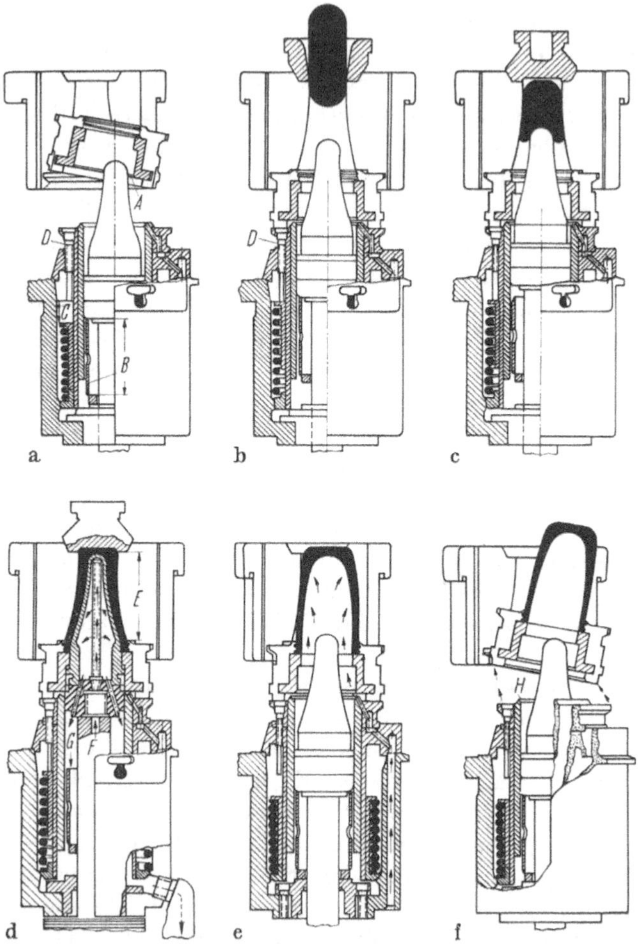

Fig. E 83a–f. Function of gob forming in the Hartford 62 press-blow process.

a) Return of finish mold to the parison side to accept gob. Lower plunger position fixed by part *B* which must have the proper length to assure the required reversal shown in *A*. The large spring is compressed in the lower plunger position. In most cases *B* can be found by subtracting the length of the plunger (*E* on the figure) from 6.5″. b) Closure of parison mold Spring activated upward motion of plunger to charge position. This motion is effected by turning off the air supply that pushes the plunger down. The upward motion *C* of the plunger is controlled by the screw length *D* which is obtained experimentally to assure charge conditions illustrated in Fig. c). The plunger should be high enough to keep the glass from the finish mold, yet admit the upper part of the gob below the parison mold bottom connection. An initial screw length of $2^1/_2$″ is recommended. This brings the plunger tip about $^3/_4$″ above the finish mold line if *B* is correctly adjusted as described in a). Subsequently the screw should be gradually lengthened or shortened to assure proper charging conditions. c) At any rate the parison bottom must be adjusted for rapid connection after charging. The pressing action of the plunger must start immediately afterwards. The illustration shows the plunger at the start of the pressing stroke. d) The plunger is in the pressing position. The cooling system is indicated by the air entrance at *F* and the exit holes at *G*, and in the lower right. Air cooling is continuous. The plunger has a stroke of $^1/_8$", more than sufficient to compensate for weight variations. f) The parison mold is wide open. The gob swings to the blow mold side. The plunger head as shown serves finishes to 70 mm For finishes from 70 to 90 mm the insert *H* is removed so that a larger plunger head can be mounted.

Table E 29. *Production range and output of Hartford IS machine* (Limited only in maximum dimensions. Narrow neck containers of just a few cm^3 can be made)

Production range		Single mold	Double mold
Narrow-neck (blow-blow process)	Max. height without finish	324.6 mm	225.4 mm 273 mm (2)
	Max. body diameter	165 mm (178 mm) (1)	76.2 mm 90.4 mm (2)
	Max. finish diameter	89 mm	53 mm
Wide mouth (press-blow "62" process)	Max. height without finish	241.3 mm	114.3 mm 220 mm (2)
	Max. body diameter	165 mm (178 mm) (1)	76.2 mm 92 mm (2)
	Max. finish diameter	120 mm	68 mm to 83 mm (2)

(1) By means of special devices.
(2) Extension tool, $4^1/_2''$ center distance.

The output for a 5-section IS machine is given in the following table. (Rounded averages according to new Emhart information.)

Wt. (g)	Pcs./min (av.)			
	Narrow neck		Wide neck ("62")	
	Single molds	Double molds	Single molds	Double molds
to 30	73	132	—	—
50	61	108	66	125
100	53	96	63	110
150	47	85	56	104
200	45	83	53	88
300	42	70	46	—
400	36	56	41	—
500	35	—	35	—
750	26	—	28	—
1000	24	—	24	—
1500	15	—	15	—

many adjustments may cause trouble. Often inexpert adjustment of control buttons during sudden upsets or temporary defective production causes mechanical collisions that are hardly visible but which lead to excessive mold wear in a short time. Two general things should be kept in mind as to control time: matters concerning fabrication, and matters concerning motion security i.e. the avoiding of mechanical collisions of cooperating mechanisms. As to this latter condition there are available exact tables for button settings for the entire production, worked out by the supplier of the machine. The remaining possibilites for improvement or increase of production are based on experience data depending on weight and shape of the glass articles. To improve efficiency it is advisable to plot corresponding control and function times in suitable diagrams for proper evaluation.

3.3.3.2.5 Changeover and Lifetime of Molds. The single line production and the independence of its sections in IS machines favor rapid job changes. Job change times depend largely on the deviation in dimensions and weight of the article to which one wishes to change. If they are great, e.g. in changing from a $^1/_4$ to a $^7/_{10}$ liter, bottle mold changes usually must be accompanied by changes in the machine. In that case delivery equipment, mold holders, funnel holders, etc. must be replaced and button settings altered completely. If several IS machines are available, intelligent production scheduling can avoid extensive and wasteful mold and machine changes. Short changeover times demand skilled personnel, systematic operation, and exact readying of molds and machine parts close to the machine. A changeover group consists of 4 men including the operator in charge. The changes to the feeder also are the responsibility of the changeover group. A simple mold changeover on a 5-station IS without change in delivery equipment, holders, etc. can be executed in about 30 min. For a bigger changeover involving rings, delivery equipment, feeder, etc., two or three times this period has to be allowed. Lifetime data on molds for the IS machine are not significant, since under certain conditions lifetime can be very long, while it can be considerably reduced by negligence. Conditions for an optimum include:

(a) Thorough training of operating and supervisory personnel in the construction and operation of the IS machine,

(b) Conscientious testing and adjustment of control times, correct height adjustments on adjustable mechanisms and on the delivery system,

(c) Control of motions in their rate of operation through corresponding valves,

(d) Dimensions of molds,

(e) Continuous maintenance and care of the machine and execution of repairs.

During changeover of molds, changing molds in operation, lubricating of molds, and maintenance of the machine, certain safety measures are required to prevent accidents.

3.3.3.2.6 Energy Consumption and Productivity. Production data given in Table E 29 are average total values related to 5-section machine data. In Table E 30 energy requirements of a 5-section IS machine are

Table E 30. *Energy requirement of a Hartford 5-station IS machine (compressed air, cooling air, vacuum drive)*

		Blow-blow process		Press-blow process ("62")		Remarks
		Single molds	Double molds	Single molds	Double molds	
Compressed air 2.1—2.3 kp/cm²	Working air	590 m³/h	590 m³/h	730 m³/h	680 m³/h	Average total requirement: Blow-blow: 900—950 m³/h Press-blow ("62") · 1250—1300 m³/h
	Inner cooling	170 m³/h	—	170 m³/h	—	
	Blow down air	86 m³/h	86 m³/h	—	—	
	Plunger cooling	—	—	300 m³/h	430 m³/h	
Cooling air (pressure 380 mm WC)	Mold cooling and conveyor cooling* (average 18800 m³/h)					* Conveyor with double plates: +20%
Vacuum**	170 m³/h (at 650—700 mm Hg)					**Only for special device on parison side
Drive requirement	Machine, feeder, sometimes carry-in max. 3.5 kW					

For machines with different number of stations, energies are proportional to these numbers.

summarized. The data coming from the supplier are average values. Compressed air consumption in a machine shortly before overhauling is higher than in new or overhauled machines. There also are differences in compressed and cooling air consumption for the production of large and small articles.

A description of the IS machine would be incomplete if, besides the possibilities of double mold production and of "62" process wide mouth fabrication, other special equipment was not mentioned. If certain bottle finishes require fire polish this can be achieved easily by the take out heads. In the parison mold vacuum operation is also possible. Furthermore there are accessories permitting the fabrication of thin walled seamless containers. These accessories do not convert the IS machine to a perfect tumbler machine but they do contribute to its versatility.

References

[1] Prospectuses and information of Emhart Mfg. Co.
[2] Ingle, H. W.: U.S.A.-Patent No. 1843100.
[3] Holscher, H. H.: Feeding and forming. Handbook of Glass Manufacture, Ed. by V. F. Tooley, Vol. I, New York: Ogden Publishing Comp. 1953; particularly Section G: The individual section IS-machine, pp. 326–332.
[4] Anon.: The Hartford-Empire I.S. machine. Glass 24 (1947) 143.
[5] Riede, H.: Sonderheiten über die IS-Maschine (Abstract). Glastechn. Ber. 24 (1951) 142.
[6] Wardley, T. A.: Hartford feeder, Lnych machines and the Hartford I.S. machine. J. Soc. Glass Technol. 37 (1953) 32–40.

3.3.4 Lynch 44 Machine

By E. Schulze, Essen, and H. Ganss, Essen[1]

The Lynch 44 produces hollow glass ware by the blow-blow as well as by the press-blow process. The basic idea in its construction was the creation of a production unit permitting the rational fabrication of hollow-glass ware of high quality in short runs.

The Lynch 44 is the first machine not requiring time consuming changeover work for the transfer from blow-blow (bottle) to press-blow (mason jar) process, or from single to double gob operation.

The mechanisms that have to be exchanged in conventional machines are all directly built into this machine so that a fast change from press-blow to blow-blow process is possible.

In comparison to other machines the following cost-saving features are stressed: reduced down-time, high mold efficiency, and reduced operating cost.

[1] The authors where supported by Mr. P. Borgböhmer, Essen.

3.3.4.1 Construction. The Lynch 44 is a multipurpose hollow glass blowing machine. Like the IS machine it receives the gob from synchronized troughs following the well proven Hartford feeder principle. It consists of single block tables which can be assembled in groups of 2 or 4 units at will and which work independently. In larger, e.g. 6 unit machines difficulties in gob delivery are avoided by arranging rows of 3 block tables each back to back (Fig. E 84a). The total arrangement as seen

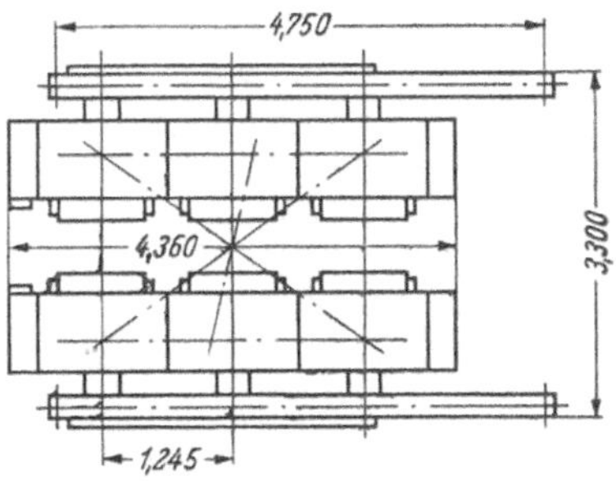

Fig. E 84a. Mounting sketch of a 6 station Lynch 44 machine.

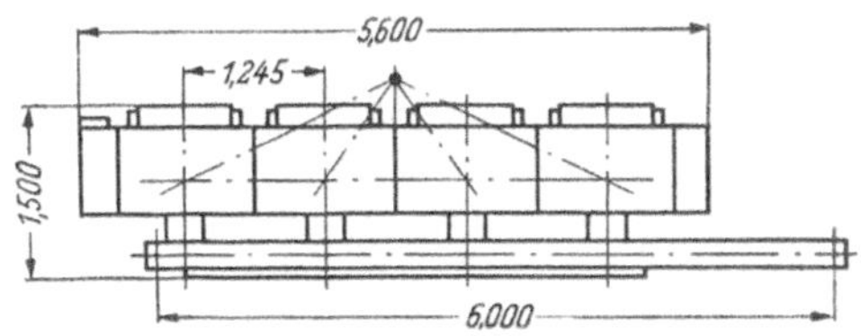

Fig. E 84b. Mounting sketch of a 4 station Lynch 44 machine.

Fig. E 85. Total view of a 3 station Lynch 44 machine (Courtesy Lynch Corp.).

from the control side is shown in Fig. E 85. Each unit has a vertical rotating column with three finish or neck ring arms separated by 120° which serve the parison mold, blow mold and takeout stations (Fig. E 86). The activation of all work phases is hydraulic using high and low pressure

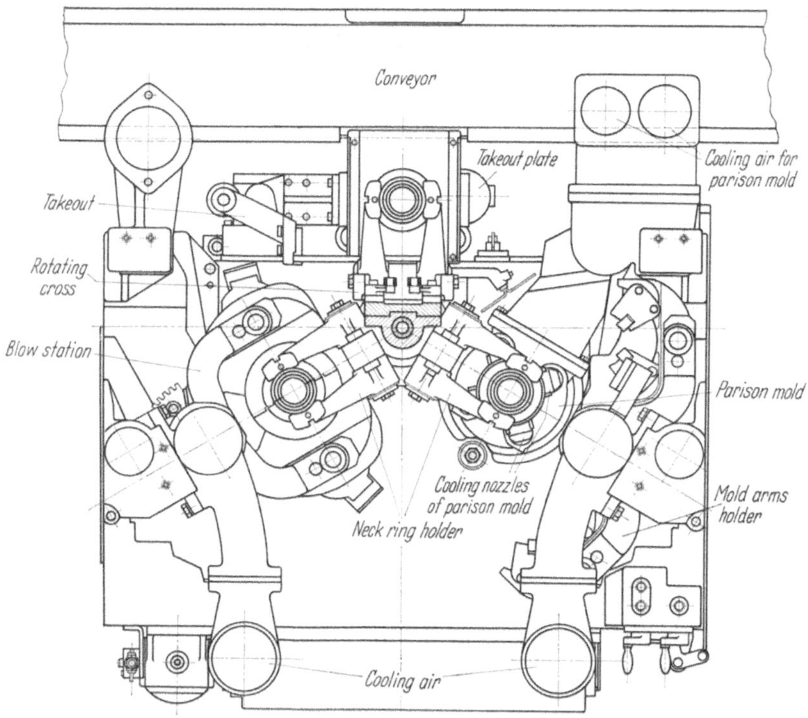

Fig. E 86. Arrangement of revolving column with parison mold, blow mold, and takeout station.

oil circulation. The neck ring arms are synchronized with the feeder. The high pressure oil activates the working pistons of the machine, the low pressure oil serves to return the piston to its starting position when the operation has been carried out. For high and low pressure oil circulation one tank for each (with oil reservoir) provides constant oil pressure. High and low pressure pumps circulate the oil. Each block table possesses its own control shaft with adjustable cams to regulate all work phases. A lift table with cooling plate receives the finished article, and a stripper shoves it on the conveyor.

3.3.4.2 Main Dimensions. The space requirement of a station of the Lynch 44 is relatively large. A 4-station machine has a length of about 5.6 m. The depth of the table is about 1.5 m (Fig. E 84b).

3.3.4.3 Blow-Blow Process. The machine receives the gob from the delivery system which feeds it through a funnel and into the parison mold. The neck ring arm is below the parison mold and is in the loading position (Fig. E 88a). The settle blow occurs through the swung-in blow head and, if necessary, is assisted by vacuum through the plunger (Fig. E 88b). The parison is then blown against the bottom plate with the plunger withdrawn (Fig. E 88c).

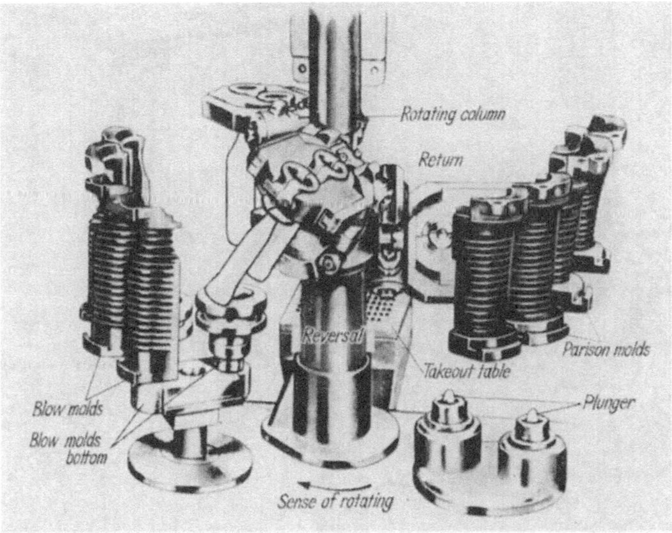

Fig. E 87. Working principle of a Lynch 44 machine using the blow process.

After opening the parison mold the neck ring arms are turned into the blow mold station (120° on) and the finish holder with the parison is swung by 180° so that the finish is on top (E 88d).

This is followed by the closure of the blow mold and the swinging in of the blowhead with subsequent final blow (Fig. E 88e). The swinging of the parison is shown clearly in Fig. E 87.

When the blow mold opens, the neck ring arms proceed by another 120°. Next the neck ring arms open and release the article which is received simultaneously by the lifted cooling plate. The cooling plate drops to the level of the conveyor belt so that the bottle can be shoved on to it (Fig. E 88f).

After completion of this operation the mechanisms go on to the starting position and the neck ring is ready to receive the next gob. In the meantime the two other neck ring arms are activated in the same cycle but with a 120° phase shift so that with every 360° turn of the column there are three complete cycles.

Through this far reaching overlap of single work phases high productivities are achieved.

3.3.4.4 Press-Blow Process. The changeover of the machine from blow-blow to press-blow operation requires only the exchange of directly article-oriented parts. The supplier emphasizes that changeover times

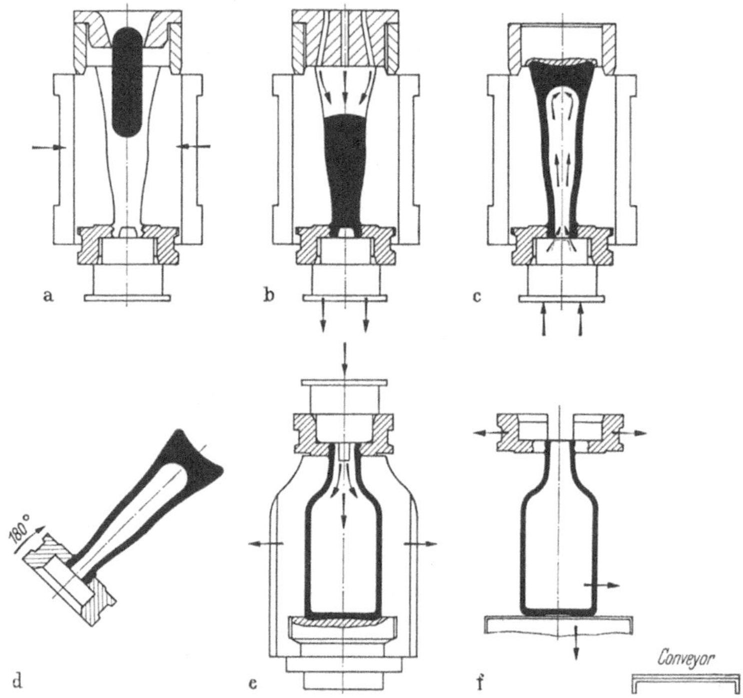

Fig. E 88a–f. Blow-blow procenss o the Lynch 44 machine.
a) Gob falls; b) Settle blow; c) Parison blow; d) Turn parison; e) Blow; f) Takeout.

are comparable to those from single to double gob operation. The gob is guided into the solid parison molds. The neck ring now is placed above the mold (Fig. E 89a), in contrast to the blow-blow operation. Then the plunger penetrates the mold through the neck ring and presses the glass up along the plunger into the neck ring. Before the plunger retracts the finish has been completed (Fig. E 89b, c).

Next the plunger moves upwards out of the parison mold which drops, the neck ring turns 120° with the prepressed parison, but without the swing characteristic of the blow-blow process, and enters the open blow mold (Fig. E 89d).

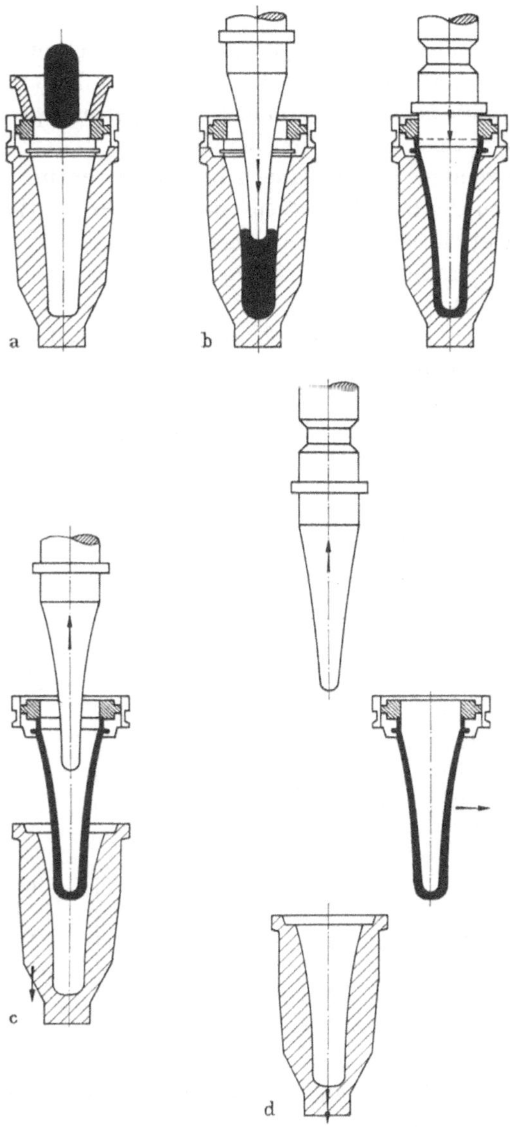

Fig. E 89a–d. Press-blow process of the Lynch 44 machine.
a) Gob falls; b) Plunger down; c) Plunger up; parison mold down; d) Takeout.

All further phases — such as closure of blow mold, final blow, cool table transfer, stripping to conveyor — correspond with the phases of the blow-blow process.

3.3.4.5 Special Characteristics of the Lynch 44. Through the construction of the turning arm for the neck ring it is possible to provide a fixed mounting for parison and final molds. This permits perfect control of all operations by the control mechanism.

The finish remains in the same ring for the entire cycle. The ring opens only when the finished articles are deposited on the cooling plate. Thus mechanical checking of the finish in operation is prevented.

The control cylinders and valves work similarly to those in pneumatic control systems. But since instead of compressed air a non-inflammbale liquid is used, more precise controls and higher speeds become possible. The incompressible liquid results in quiet operation.

Table E 31. *Output of a station of the Lynch 44 machine*

Dimensions		Bottle capacity	Bottle wt.	Pcs./min	
Ht. (mm)	Bottom diameter (mm)	(cm³)	(g)	Single gob	Double gob
		a) *Press-blow process*			
181	92		311		22
177.8	95.25	907	311	12	
128.6	77.8	453	198.5	14	
114.3	76.2	482	170		26
177.8	95.25	907	311		21
92	82.5	340	226.8		26
131.8	79.4	482	269		25
117.5	76.2	397	155		30
		b) *Blow-blow process*			
233.4	88.9	937	410	9	
268.3	80.9	722	425		16
298.5	161.9	3785	1190	5	
144.5	68.3	340	142	15	
209.5	63.5	340	255		20
284.2	74.6	722	368		15

Each station can be adjusted, served and maintained independently. A station needing repair can be replaced within one hour according to the supplier. By far-reaching overlaps, due to the total construction of the machine, the maximum production per station can be as high as 25/min. Production data according to the supplier are given in Table E 31.

Maximum dimensions of articles producible on the Lynch 44 can be seen in Table E 32.

Table E 32. *Maximum dimension of articles for the Lynch 44 machine* (mm)

	Ht. under finish	Article diam.	Finish diam.
	Press-blow process		
Single gob	248	165	120
Double gob	216	102	89
	Blow-blow process		
Single gob	368	165	120
Double gob	305	102	89

3.3.5 Other Feeder Blow Machines for Bottles

By L. Muschalek, Bad Godesberg

3.3.5.1 Pöting Feeder Blow Machines Type F 2 and F 14. The bottle blow machine Type F 2 which works according to the feeder process resembles the Hartford IS machine and has 2 sections (Fig. E 90). It

Fig. E 90. Pöting bottle blow machine F 2/55 (Courtesy Pöting).

is similar to the Hartford machine and has the same work phases using fixed parison and blow molds as well as inverting neck ring arms. The machine owes its origin to the demand for fully automated rational fabrication of relatively small quantities of bottles with a minimum of molds and of changeover time.

The F 2 machine differs from the IS machine by avoiding the delivery of gobs through curved channels and instead bringing the two parison molds below the feeder outlet by horizontal shifts. The treatment of the gob is much simplified by its direct fall into the parison mold; the gob does not suffer deformation and is not cooled on one side by the delivery channel. Besides, the distance machine floor-metal line can be reduced to 3.2 m from the 4 m in the IS machine. These ideas led to the construction of a mobile parison mold side and a fixed blow mold side which is restricted to two juxtaposed stations.

The production range of this machine is smaller than that of the IS machine. The motions are pneumatically controlled by two control shafts, separately for each station. A third shaft, at the side, regulates accessory motions for both stations, such as those of slide and dish table, or of the take out tongs. The control shafts are driven by an electric motor through a variable speed drive which is coupled to the feeder timer. A pneumatic clutch allows stopping the machine without interrupting the operation of the feeder. The adjustable cams on the control shafts act in the known manner upon compressed air valves which actuate the individual motions. The "slide table", which is driven by a compressed air cylinder with 300 mm stroke, is braked in its end movement on each side by a brake cylinder. It runs in slide tracks and is held in the middle by a supporting shaft which diminishes the sliding pressure. The sidewise motion of the parison mold does not create an additional time requirement since the entire equipment for work phases on the parison mold side moves along and does not interrupt any function during the slide. The bottles are deposited from the blow molds on the cooling table and are exposed to cooling air to avoid deformation. They are then shoved-together with the motion of the parison mold side from the left or right by means of push outs to the belt which runs at a right angle to the machine.

The most important data for the machine are: maximum length 3000 mm, total height 2200 mm, maximum width 2000 mm, wt. 4 t, distance glass level-machine floor 3200 mm.

Power requirement: Machine and feeder: 2.2 kW; vacuum about 30 kW (for difficult mouths, additional); compressed air about 7.4 kW; cooling air about 8.1 kW.

Air requirement: vacuum 80 m^3/h; compressed air 330 m^3/h for 2.5 atm; cooling air 3600 m^3/h for 300 mm water column.

Production limits: content 150—2000 cm^3, minimum height without mouth 150 mm, maximum height 350 mm, maximum diameter 170 mm, maximum mouth diameter 60 mm.

Production data:

Contents (cm^3)	Pieces/min
150	17
250	15
500	13
700	11
1000	9.5
2000	8

The machine works with any feeder.

A further development of the F 2 is the F 14 machine. It works with 2 parison and 4 blow molds. Each station thus has one parison and two blow molds. The dwell time of the bottle in the blow mold thus can be made relatively long. The working times of the parisons are made as brief as possible.

As in the F 2 the parison molds sit on a horizontally movable carrier. A special delivery system for the gob thus can be omitted. The blow molds are also set on a horizontally movable carrier in a further development of the system.

The production range of the machine comprises bottles of the following dimensions:

	Double Molds	Single Molds
Contents	100—250 cm^3	250—2000 cm^2
Mouth diameter max.	28×20 mm high	60×35 mm high
Body diameter max.	60 mm	100 mm
Body height without mouth max.	200 mm	350 mm

The productivity is given as 12—18 pieces/min depending upon bottle height. The glass requirement of the machine is 5.5 to 14 t/24 h.

3.3.5.2 Putsch SBK 22 Feeder Blow Machine. A similar feeder fed blowing machine comes from Lindental, Voitsberg, Austria. It was developed during the years after 1950. Lindental solved the supply of the 2-station machine, and also of a 4-station machine, in a simple manner by letting the gob falling from the feeder drop into the center of a segment which he then swung or turned (Fig. E 91 a). The axis around which the segment is swung and turned is vertical. Particularly simple is the mechanism supplying the 4-arm machine. Here the above mentioned segment is turned by an intermittently rotating shaft arranged vertically in the center of the machine. A certain disadvantage in the 4-arm machine might be the arrangement of the four blow molds in a square. This causes the 4-arm machine to have a short conveyor belt at each side and to be somewhat inaccessible. Much better is the arrangement of all machine parts in the 2-arm machine (Fig. F 91 b). The construction of this machine was taken over a few years ago by H. Putsch, and Co., Hagen, which

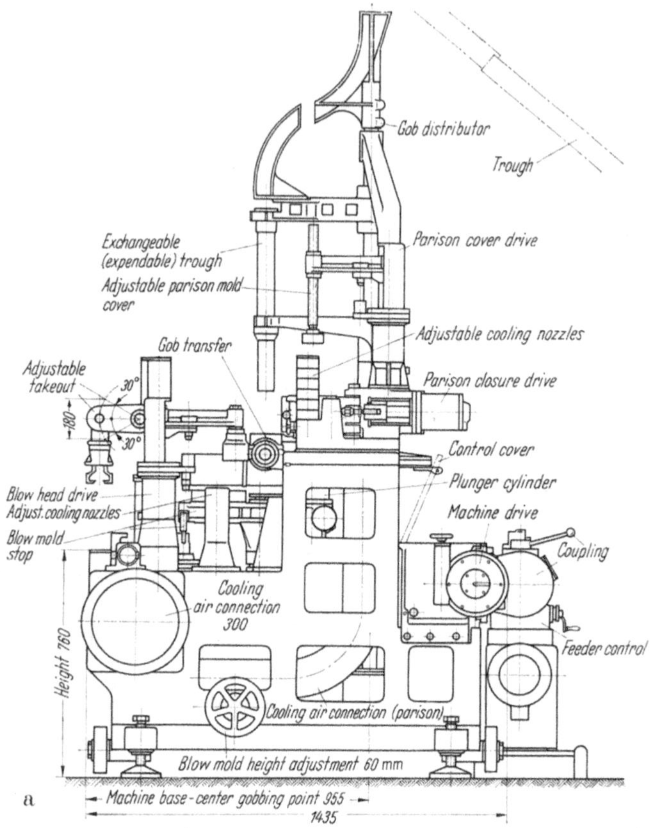

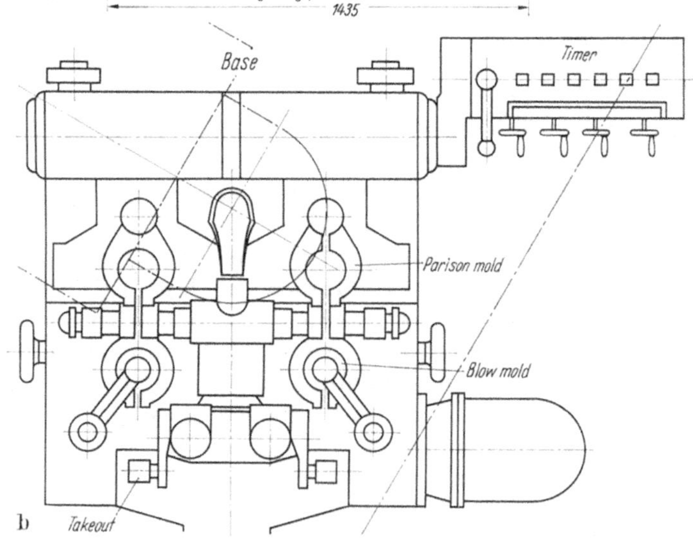

Fig. E 91a and b. Putsch SBK 22 machine (Courtesy Putsch).
a) Elevation; b) Base.

developed from it the two arm SBK 22 and the four arm SBK 14. The machines are driven pneumatically and use a control shaft with adjustable cams to activate the individual valves. In contrast to many feeder machines it does not blow down the gob falling in the parison mold but sucks it into the mouth parts by vacuum (Fig. E 92a). It turned out that this mode of operation yields very good parisons. The final blow in the blow mold is accomplished by the utilization of atmospheric air pressure. The suction channels worked into the blow mold halves are subjected to vacuum so that the air streaming through the mouth blows the parison to a final shape. The procedure is not new; it was first used by Roirant in his type F machine and it is still used successfully in machines A 6, B, and R 7. Since the conventional blow head here becomes unnecessary it is used for internal cooling which contributes to a significant

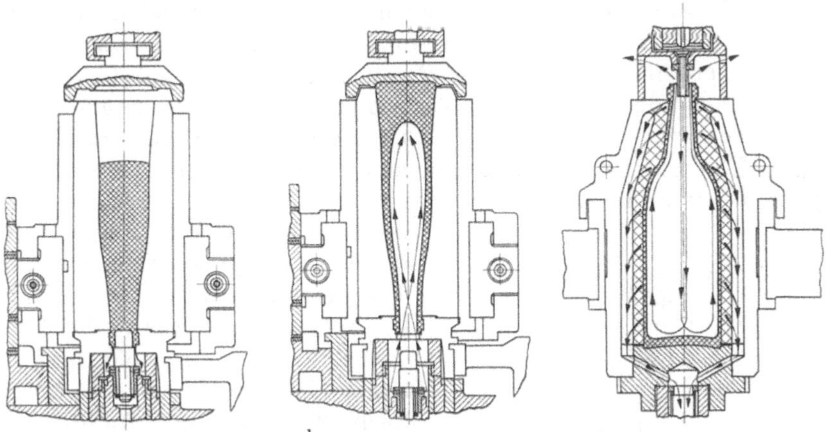

Fig. E 92a–c. Working scheme of the Putsch SBK 22 machine (Courtesy Putsch).
a) Suck; b) Parison; c) Blow out, suction, internal cooling with compressed air.

shortening of blow mold times (Fig. E 92c). The author of this chapter verified production numbers for 0.5 liter bottles weighing 505 ± 10 g at 13.2 pieces/min for the 2-arm machine, i.e. at 19,000 pieces/24 h. The supplier indicates the following numbers:

Wt. (g)	Pieces/min/station
150	8–9
300	6–7
600	5–6
900	4–5

3.3.5.3 Mitchell Feeder Blow Machine [1, 2]. This machine was developed by the Mitchell Glass Co. of Bradford, England, and has been marketed since about 1945. Like the Mitchell Press-Blow Machine it possesses 6 or 8 arms and represents a continuous one-table machine.

The closed parison mold is rotated 180° with compressed air. Vacuum air, and final blow air can be hand controlled by valves from a distributor head. The bottles are blown through the neck rings. After a turn by 450° withdrawal is mechanical. The production range is: height to neck piece 25—192 mm; mouth diameter max. 50.8 mm; maximum external bottle diameter; 102 mm. The total production reportedly is between 6 and 54 pieces/min.

Besides this round table machine Mitchell developed a machine designated B. V. which is constructed somewhat like the IS machine. Details about this machine are not known.

3.3.5.4 Knox WD Bottle Blow Machine [3, 4]. The WD machine is a one-table rotary machine with 10 sets of molds working according to the blow-blow process. As in all known fully automated machines the parison is transferred from a parison mold to another mold in which the bottle is given its final shape. The gob is delivered into the continuously rotating machine by the feeder. A guide tube for the gob follows the parison mold during the fall of the gob at a certain distance and in full synchronization. Then it snaps back to be ready for the next mold (Fig. E 93). The time for cutting the gob and the motion of the guide tube can be adjusted while the machine is running and delivery continues. The machine has 10 equal stations each forming a complete production unit. The parison molds are arranged on the rotary table below the blow molds. The parison mold circle is larger to permit the gobs to fall in freely. The machine is, for instance, capable of running with just one section producing one bottle per revolution while the other nine sections idle.

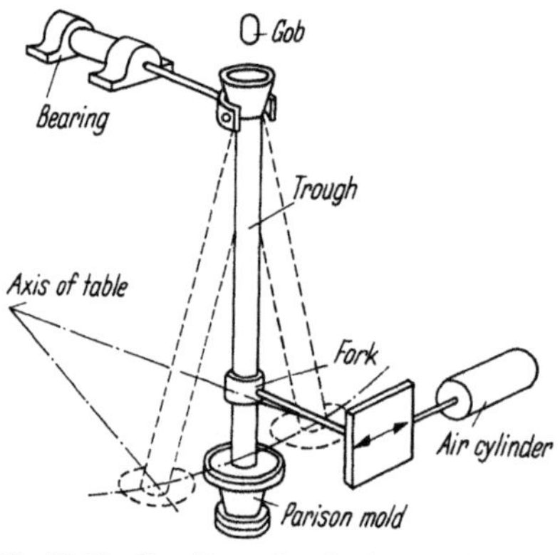

Fig. E 93. Function of gob trough in the WD machine.

It can also produce in one revolution ten bottles of different shapes but equal weight, provided the difference in shape is not extreme.

The machine is driven electrically. Activation cams, and rolling and guiding notches are attached to the center column and to the stationary framework. Cams and guides can be adjusted in operation.

The activation processes for the single stations are pneumatic for safety. Molds are opened, closed and locked pneumatically, the parison is transferred from parison to final mold pneumatically, and the parison mold baffle is also controlled pneumatically. Plunger and blowhead are dropped and lifted by compressed air.

The electrical drive of the machine makes possible a very exact sequence of operations; and the pneumatic processes simplify the construction of the machine while preventing interferences. The machine is represented in Fig. E 94.

3.3.5.4.1 Construction. Mold units are interchangeable. Blow mold carriers and parison mold carriers can be interchanged. Each mold holder has a complete constructional unit that contains the following parts:

Fig. E 94. Knox WD machine (Courtesy Knox).

carrier arms, closures, compressed air cylinders, and a joint shaft mounted on a stand. The whole thing is controlled by cams at the stationary column. The valve touches the cams only when the mold is closed. Control is stationary and accessible to the machine foreman. At present the machine is made and sold only for single gobbing.

3.3.5.4.2 Parison Transfer. The parison is not transferred by its finish as in all other hollow glass machines — but by the bottom plate. The latter is perforated and can be switched to vacuum. It is lowered onto the closed parison mold and contacts the parison bottom as early as possible. The parison is blown by vacuum and the parison mold is opened. The bottom plate with the parison swings up 180° and into the blow mold. After closure of the blow mold the vacuum is released. The parison is as long as the finished bottle, which favors the manufacture of lightweight bottles. The blowhead can be put on as soon as the parison is in the blow mold. Depending upon the reheating time desired it can be dropped earlier or later. A safety gadget prevents collision with the ejector. The execution and guidance of the neck ring is also novel and unique. It is shoved into the parison mold where it is protected. Yet it can not slip from its location while sitting in the parison mold. It can be changed in the machine. This construction safeguards the neck ring from damage during mold changes and faulty moves during fast machine runs. There are no neck ring holders, which could be bent or broken, — a unique advantage of the WD machine.

The plunger is activated pneumatically by a number of guide rails fastened to the base plate of the machine. One rail serves to lower and raise the plunger, one to open and close the cover. These guides can be adjusted in operation. Results can be seen immediately and corrected. The adjustments are self aligning.

Parison blowing starts as soon as the plunger is withdrawn, and it may last until the cover starts to lift. The cover is lifted as soon as the unloader is free. This long duration of parison blowing permits a high working speed of the machine and makes it possible to make very light bottles. This is the main advantage of the machine.

According to experts the take out is the best ever designed for a continuous machine. It has 5 units i.e. half as many as the machine has molds. It is driven by the same drive as the machine. When the bottle has been grasped the blow mold opens and the tongs, with the bottle, rise to take it off the plate.

The bottle may be taken out at any take out unit, from 90° after take-out to 45° above the machine axle, i.e. over a range of 135°. This will satisfy all requirements for transport from machine to lehr.

The unloader control valves are put on the upper portion of the rotary table and are activated by adjustable cams. A position once chosen is firmly fixed. Adjustment can be made in operation. Hand throttle valves are provided for the control of the speed of the single units and permit opening as well as closing of grips. The ejector does not use rubber hose but small copper tubes.

3.3.5.4.3 Safety Equipment. The main motor has an effective dynamic brake. It is push-button controlled. The machine can be stopped instantly by pushing one button which is generally placed at the feeder column. Another push button is found near the unloader to permit immediate switch-off if the operator desires.

To avoid accidents two protective fences are provided. One is in front of the gob guide channel column, so that nobody can be dragged between the column and the machine. The other is in front of the unloader to prevent dragging of people between unloader and machine. These fences are connected with the same circuit as that connecting the pushbuttons stopping the motor. A safety device controlled mechanically and electrically is attached behind the transmission. It has the effect that the funnel tube can be activated only if the bottom is in the blow mold.

Another device assures that the synchronizer can only be activated if the baffle is on the parison mold. An additional safety device is found behind the blowhead. If the blowhead should not rise the machine is switched off before it could touch an ejector unit.

A planet wheel drive permits synchronizing feeder and machine. The feder continues to be driven even if the machine is switched off. Devices

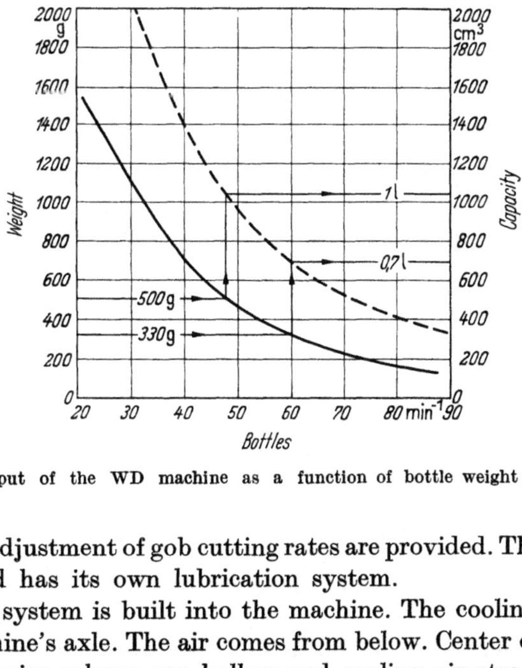

Fig. E 95. Output of the WD machine as a function of bottle weight and capacity.

for precision adjustment of gob cutting rates are provided. The mechanism is sturdy and has its own lubrication system.

A cooling system is built into the machine. The cooling air conduit is in the machine's axle. The air comes from below. Center column, main table, and carrier column are hollow and cooling air streams through. The parison molds possess adjustable nozzles. An accessory external cooling system can be supplied.

3.3.5.4.4 Production Data. The production range of the WD machine is between 59 cm^3 and 3.8 l bottles. Minimum cooling air quantity is 283.2 m^3/min at 354 mm water column. Production rates can be taken from Fig. E 95 and Table E 33. The relatively low bottle weights are noteworthy.

3.3.5.4.5 Installation and other Technical Data. Space requirements (m): length 3.43; width 3.56; height 3.05; wt. 15,000 kg (packaged for export: 16,780 kg); compressed air requirements: 14 m^3/min at

3.16 atm (25% blow, 74% control, 1% other (plunger cooling etc.) air.
Vacuum: 13 m^3/min at 645 mm Hg.
Drive: Main 5 PS; auxiliary for feeder 1.5 PS; lubrication 3/4 PS.

Table E 33. *Capacity, weight and production rate of some bottle types*

Capacity (cm^3)	Bottle type	Wt. (g)	Pcs./min
946	Knox mason	383	55
375	Red wine	255	65
473	Oval wine	298	66
473	Oval wine	269	65
762	Red bulb neck wine	412	61
946	Eastern wine	482	51
1893	$^1/_2$ gallon red jug	745	40
946	One way beer bottle	504	49
355	Tall tavern beer	241	72
355	One way beer bottle	255	72
355	Export beer	340	61
236	Jeris	227	65
946	Bleach	454	50
3785	1 gallon jug, round	1150	29
946	Oil bottle	426	52
355	Round packer	198	78
423	Round packer	234	78
946	Round packer	397	57
473	Pinch	305	$61^1/_2$
500	Standard round bottle	255	79
990	Standard round bottle	383	58
990	Standard round bottle	412	55
473	Forest	312	66
473	Amber peroxide	298	72
118	Evenflow nurser	125	80
236	Evenflow nurser	193	72
236	Standard round bottle	163	79

3.3.5.5 O'Neill Type 55 Bottle Blow Machine. The O'Neill Machine Co. Ltd., Burton Road, England during the past few years has developed a bottle blowing machine under the designation "Machine 55" which works with a feeder. The basic construction resembles the IS machine. It consists of several production units which are arranged in a series. The efficiency of these new O'Neill machines is not yet known. There are available only data from the suppliers' prospectus. The following report is based on such data.

The units work independently and possess separate control systems. As in the IS machines they can be combined in series up to 6 and supplied with glass by a feeder (Fig. E 96). The use of two parison molds and one finish mold per unit is noteworthy. The two parison molds sit on a turning head which rotates intermittently around a horizontal axis. The finish mold is fixed at the framework (Fig. E 97). The gob falls from above into the parison mold which is in filling position. After suction, blow-out and parison blow, the turning head is rotated by

Fig. E 96. Front view of O'Neill 55 (3 station machine) (Courtesy O'Neill).
1 Cylinder for blowhead; *2* For gob ring; *3* Parison bottom; *4* Gob ring; *5* Parison mold; *6* Plunger head with finish suction; *7* Drum; *8* Compressed air duct; *9* Cooling air duct; *10* Beam for trough; *11* Jaws and jaw cylinder; *12* Ejection; *13* Takeout cooling table; *14* Cullet chute; *15* Jaw turn cyl.

180° while the parison mold is closed. Next the parison is dropped vertically from the open parison mold and set into the open blow mold below (Fig. E 98).

This method of operation avoids swinging the free-hanging parison and all subsequent consequences. Each parison mold has its own neck ring. Both molds have a combination suction and blow head; the blow mold also has a blow head. The gobs are delivered to each parison mold from the feeder through troughs as in the IS machine system. At the back of each unit is a control drum which rotates around a vertical axle through worm gears. The built-in double valves control numerous compressed air cylinders which are needed for the work phases. So called "self-damping" work cylinders are used which are reported to effect a very constant rate. The work pistons are moved by compressed air from both sides. One must assume that a certain pressure difference between rear and front is established as in the Lynch 44. But the Lynch

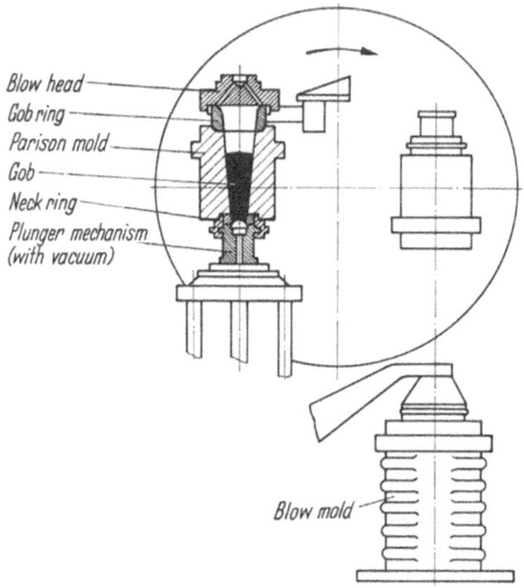

Fig. E 97. Function of turning drum.

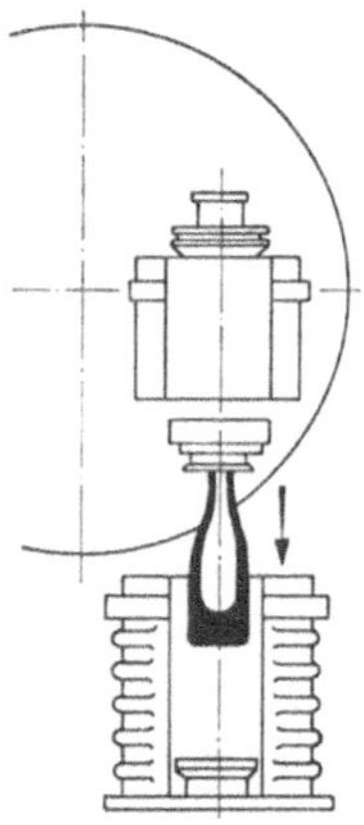

Fig. E 98. Lowering of parison into blow mold.

44 does not use compressed air, but rather hydraulics. A self lubrication system uses the waste air of the work cylinder for an automatic lubricant spray. A planet wheel drive is inserted in the drive shaft between machine and feeder and allows exact synchronization of feeder and machine.

It is characteristic of the working cycle of this machine that the blow mold has almost no idle time. As soon as the finished container has left the mold a new parison is set into the open mold from above. The blow mold in this process takes a heavy heat load and therefore is provided with strong ribs for better heat removal. The two parison molds are not heat loaded as badly.

Unfortunately the producer's data do not include production rates. The working principle used suggests that the productivity of one unit of the 55 machine corresponds to about that of an IS unit. It may even be assumed that the greater utilization of the blow mold permits slightly higher rates. The width of a station without auxiliaries is given as 69 cm, the depth without conveyor as 192 cm. The height varies from 220 to 250 cm depending upon the number of stations (channels). The cooling air requirement of a station is given as $2-3$ Nm³/min.

References

[1] MITCHELL, W. B.: Mitchell feeder-fed glass-forming machines. J. Soc. Glass Technol. 37 (1953) 41–43.
[2] PLATT, T.: Mitchell machines. J. Soc. Glass Technol. 37 (1953) 71–75.
[3] Anon.: The Knox WD bottle machine. Glass Ind. 27 (1946) 130.
[4] SCHUBERT, K.: Die WD-Flaschenmaschine. Silikattechn. 412/13.

3.3.6 Feeder Machines for the Production of Thin-Walled Seamless Hollowware

By E. SCHULZE and H. GANSS, Essen
(with the support of H. W. HORTIAN, Essen)

Thinwalled, seamless glasses such as tumblers, goblets and, in part, also bulbs, are now also made on feeder machines. The machines developed for this purpose show considerable deviations from the feeder machines described so far.

3.3.6.1 The Hartford Empire Machine Type 28 of the Emhart Mfg. Co. This machine provides fully automatic fabrication of drinking and packer's tumblers. It is a 12 station feeder machine and works with the press-blow single gob process. The product leaves the machine as an almost closed hollow object. The upper part ("finish") must be severed. The 28 machine therefore is synchronized with a burn-off machine, e. g. the 30 spindle burn-off machine of the Eldred Co., of Columbus, Ohio. Because of the high capitalization of the entire system (about $ 400,000) it is placed only in large glass houses (Fig. E 99).

3.3.6.1.1 Process (Figs. E 100 and 101). The short gob formed by the feeder falls into the swinging delivery of the loading station, glides through the intermediate trough, which is inclined downwards by about 120°, and down through the lower through into the parison mold. Parison mold and plunger move up and down, respectively. The plunger enters the loaded parison mold trough guide ring and neck rings. The glass is pressed up along the oil-sprayed wall of the parison mold and then into guide ring and neck ring. The parison is held in the neck ring during further working. Pressing time depends on the kind of tumblers made. With longer pressing times the bottom is pressed out better, and the parison is more rigid. With shorter pressing the parison stretches along its entire length before blowing out and the bottom becomes thinner. In the further course of fabrication, after about two stations, the two paste mold halves close, as does the bottom, while the free hanging parison rotates. At the same time the blowhead swings on top of the ring and the parison is preblown with a pressure of about 250 mm water column. Since the distance from neck ring to mold is about 13 mm, a

Fig. E 99. Hartford paste mold machine (Courtesy Emhart).

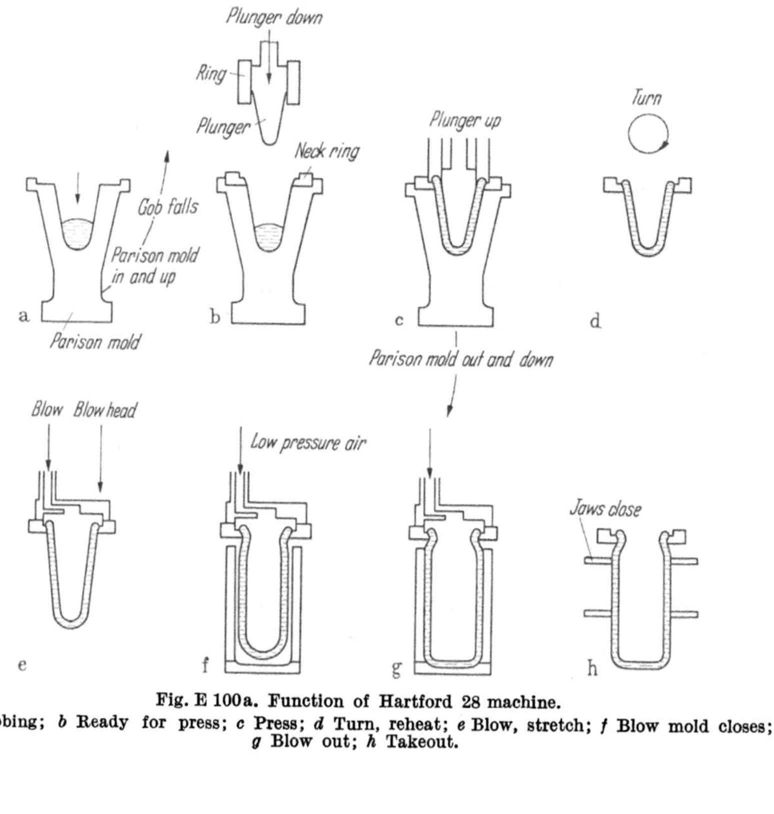

Fig. E 100a. Function of Hartford 28 machine.
a Gobbing; *b* Ready for press; *c* Press; *d* Turn, reheat; *e* Blow, stretch; *f* Blow mold closes; *g* Blow out; *h* Takeout.

bulge forms at the non-enclosed wall which permits observation of the blowing process.

A continual regulation of blow pressure is needed since it must be adapted to wear of the paste molds. Preblowing occurs while the machine

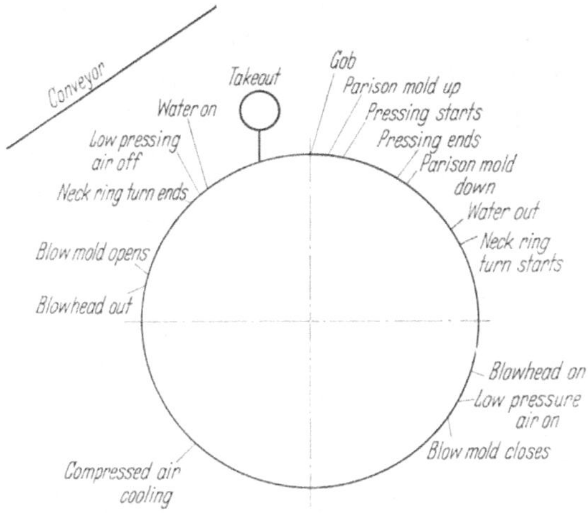

Fig. E 100b. Sequence of functions.

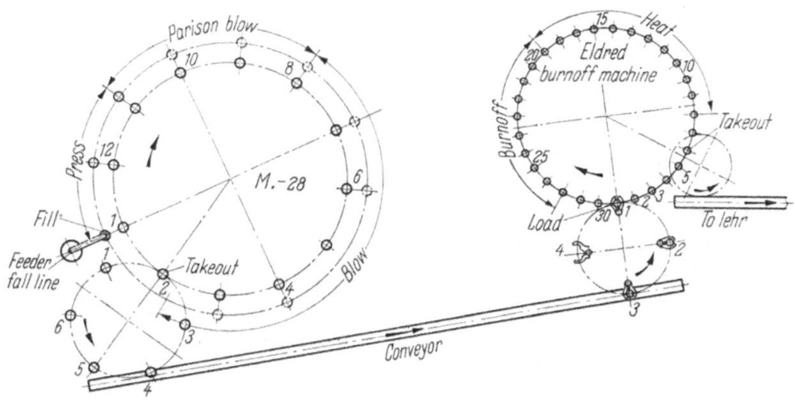

Fig. E 101. Arrangement and function of Hartford 28 machine.

turns 4 stations. Then air of about 1250 mm water column pressure is blown in. The glass solidifies enough to permit take-out without danger of deformation. After a total run of 11 stations the air is switched off. The blow mold opens, and the take-out tongs with open jaws come under the article. The four neck ring segments open, the article falls between the jaws of the take-out arm which transfers it to the conveyor leading to the burn-off machine. Cooling air is blown through the take out and cools the tumbler or other article from below through holes in the bottom plate. At the end of the conveyor the pair of jaws of the burn-off machine seize the tumbler below its mouth, turn it 180°, and transfer it bottom-up to the burn-off machine. A suction spindle with suction head and suction cup adapted to the object descends on the tumbler. The tumbler is held by vacuum, and the loading jaws open. The suction spindle lowers, and the ring burner encircles the tumbler at the set height. By adding oxygen to the burner gas flame the tumbler is burned-off while it passes along 10 stations. The oxygen is then turned off and the fire-polishing of the tumbler edge is completed through 6 stations. The suction spindle is elevated about 30 mm during the progression. Thus a bead forms at the edge which characterizes fully automated tumblers. The tumbler is moved over the transfer and the vacuum is cut off from the suction cup. From the transfer the tumbler is deposited on the asbestos conveyor. The conveyor brings the tumbler to the lehr loader which sets it into the lehr by double tong arms. The burned-off caps are caught by flat plates, the cap holders, and carried to cullet collectors.

3.3.6.1.2 Construction (Fig. E 102). The 28 and Eldred machines are driven by a 7.5 kW motor. The drive works through a PIV and a differential gear to the feeder mechanism, or by a universal drive and ring gear on the body of the 28 machine. The gears drive the delivery station, the take-out mechanism, and the conveyor belt. The burn-off machine is driven by a bevel gear and a universal shaft (Fig. E 103).

3.3.6.1.3 Details. The 28 machine consists of a fixed drum around which a ring of sections rotates continuously. On the ring 12 sections are placed. They work independently and may be switched off individually. The parison mold is supplied with glass by a 3 part delivery trough which is synchronized with the parison mold for a short time. It may be swung by about 30°. The upper trough part may be flipped back by an air cylinder which is operated by a valve through a cam at the parison cylinder. In this manner the gob can be directed to the cullet container. Each station consists of the following assemblies:

(*a*) *Parison Mold Mechanism*

A cylinder glides, by means of 3 atm compressed air, on a piston and a piston rod. At the cylinder housing sits the parison mold receiver. Controlled by a cam it swings inward below neck ring and plunger as it moves upwards. The working air is adjusted by a needle valve.

(*b*) *Press Mechanism*

It resembles the parison mold mechanism. The cylinder housing carries the receiver for the plunger. The plunger is inserted elastically by means of a spring which permits sensitive adjustment of pressure (from 2.2 to 3 atm depending upon the product). The plunger head is cooled by compressed air which is conducted by an inner tubing.

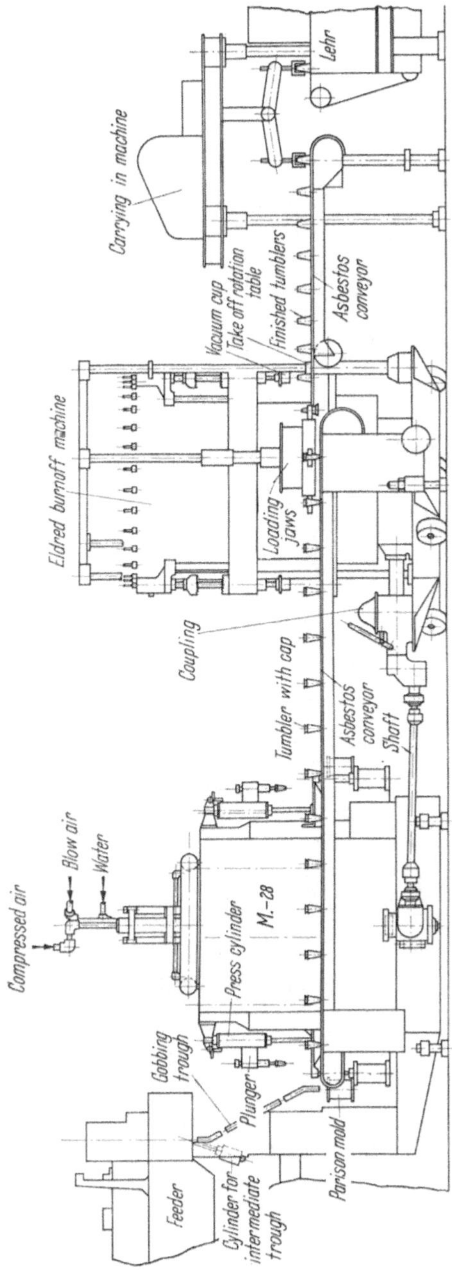

Fig. E 102. Side view of layout.

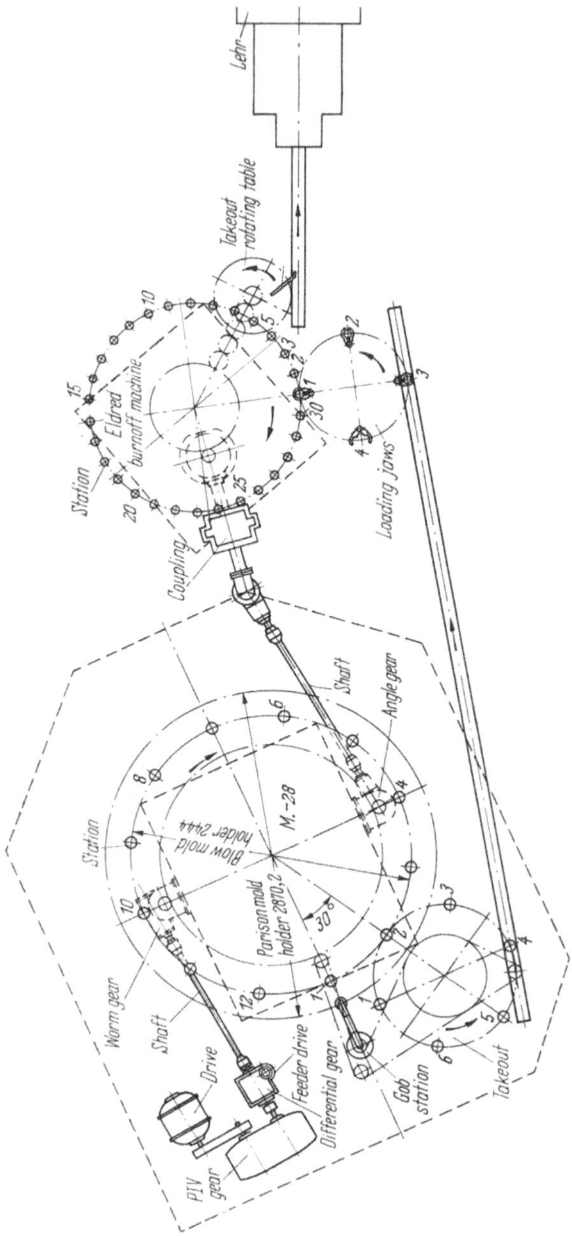

Fig. E 103. Top view of layout with drives.

(c) *Neck Ring Mechanism*

At an elevation half way between the plunger and the parison mold, the rotating neck ring holder with neck ring and guide ring is built into a table. The neck ring consists of 4 segments which open by lever and cam and which transfer the product to the take-out grip. The neck ring mechanism is driven by a gear placed on the fixed machine drum with bevel gear and chain. A clutch controlled by a cam makes sure that the neck ring holder is stationary during the press and release operations.

(d) *Blow Mold Mechanism*

This works below the neck ring mechanism. The closing of the blow mold halves is accomplished by a tension spring, the opening by cam, rods, and bevel gear. The opened blow mold is sprayed with water for cooling and for moistening of the paste. The water is switched on and off by a valve controlled by a cam.

(e) *Blow Head Mechanism*

This is located at the left side of the section and is swung over the ring and pressed on it by a tension spring. Through it the puffing and final blow air streams into the parison. Opening and closing of air ducts is by cam controlled valve. The blowhead is swung out by means of cam and rods. On the fixed machine drum are cam segments for the following control phases: parison mold cylinder up, press cylinder down, press cylinder up, parison mold cylinder down (cam segments are connected together on an adjustable bracket which can be moved during operation by rods, to control pressing time), parison blow air, parison mold open, blowhead down, final blow air, neck ring clutch, and water valve. The take out, a six arm rotating cross, is driven simultaneously with the delivery system. The conveyor belt consists of 8 mm asbestos weave. The belt from the "28" to the burn-off machine is chain driven by a sprocket from the take out. The Eldred burn-off machine has 30 stations since burning off and fire polishing of the tumblers take more time than their production in the "28" machine. A clutch is inserted between the two machines. The bevel gear wheel drive of the Eldred machine also drives the transfers (Fig. E 103). The 4 grip heads are opened by air cylinders and closed by tension springs. Strength and speed of grip jaws is regulated by air valves. The aluminum jaws have graphite inserts which are adapted to the profiles of the tumblers and hold them at the bulge below the finish. The grip head is swung 180° by cam and gear rack so that the tumbler arrives at the burn-off machine bottom-up. The stations of the Eldred machine consist of a suction and a cutoff device. At the lower end of the suction device the suction tube receiving the tumbler is screwed on; the suction spindle is turned by a chain. The suction tube is connected with the vacuum chamber by valve and hose. The valve consists of a cylinder and a piston with hollow axle, which is shoved up by a cam at the transfer point and frees the connection from suction tube to vacuum chamber. When the article is put in, with the suction tube closed, at the bottom, the vacuum enters through the piston bore and sucks the piston up against a compression spring. If no tumbler is placed into the suction cup the valve piston is pressed down again after having passed the cam. This station does not draw in air while running, so no hot air can get from the burner to the chambers and pumps. The vacuum chamber around the machine center axis consists of a fixed and a rotating half. The lower fixed part is connected with the vacuum pump and has a channel from the height of the loading station to just before the transfer table. The upper rotating part of the chamber has bores and connections for the 30 stations. The bores move over the channel of the lower part and are supplied with vacuum. At the transfer point there is a bore in the place of the channel. Air is blown in through this bore and the tumblers released from the suction head to the transfer. Below each suction station is a receiver for a ring burner with connections to gas and oxygen which come together in a mixing nozzle just before entering the burner. The burner is held by two lugs and can be easily changed in operation. The gas is piped in from the gas chamber and is regulated by valves. The oxygen comes from a distributor resembling the vacuum chamber. Below the burner tables receiver plates are running which hold the falling waste in horizontal position so that one sided burn-off is avoided.

At the end of the burn-off process the separated waste is directed to the cullet collector. The shaft of the aluminum transfer is driven by the gear of the Eldred machine through another gear.

The following accessories are required: blower for cooling, molds, neck rings, vacuum pump for suction in the Eldred Machine (20—70% vacuum), gas reducer for ring burner supply (0.2 atm), and oxygen plant for burn-off process (0.25—0.30 atm).

Personnel requirements are: 1 foreman, 1 burn-off operator, one helper for lubrication and mold change, and one helper for pasting of blow molds serving 3 machines.

Mold equipment comprises 12 each parison molds, blow molds with bottoms, neck rings (4 part), plungers with sleeves, cooling tubes and rings, 6 pairs of grip jaws, 4 pairs of graphite jaws, and 30 suction cups. For molds and neck rings cast iron is used; for the plunger, structural steel with wearproof coating; for the grip jaws, copper strip; and for the suction cups aluminum.

The blow molds are brushed with lineseed oil, varnish and then pressed into finely ground cork meal, so that a thin layer of cork meal adheres. This paste is fired for 30 minutes at 400 °C and assumes a firm velvet-

Tumbler					Pcs./min	
Capacity and job	Form	Diameter (mm)	Ht. (mm)	Wt. (g)	Theoretical after Emhart	Production attainment
0.12 l, tapered tumbler		52	82	68	63	71
0.17 l, tapered tumbler		61	80	98	66	66
0.25 l, wine tumbler		78	94	133	—	50
0.25 l, tapered tumbler		67	110	140	58	63
Barrel tumbler with foot		66	80	118	58	60
0.25 l, beer		65	143	193	—	49
0.25 l, beer tumbler with thick bottom		63	140	200	44	52
0.5 l, beer tumbler with thick bottom		81	175	331	40	42

Limiting values of tumbler dimensions and weight

Diameter	35—100 mm
Height	51—165 mm
Capacity	0.03—0.5 l
Weight	30—500 g

like surface which is smoothed with a gob of hot glass and rubbed with a 2:1 mix of graphite and talcum. This treatment makes the mold ready for use. The mold change in the "28" and Eldred machines takes about 70 min, the run-in about 35 min., assuming a change group of 5 mechanics. The change involves parison and blow molds, neck rings, guide rings, plungers, sleeves, cooling tubes, takeout grip jaws, transfer grip jaws, suction heads with suction plates and suction cups, and ring burners. For major changes gob weights and delivery equipment must be changed,

Table E 35. *Dimensions, weights and energy requirements (data by supplier)*

Hartford "28"-machine	
Base	4.60 × 4.00 m
Height	2.80 m
Weight	14.5 t
Diameter of mold centers	2.444 m
Glass line	2.44—2.70 m
Driving motor	7.5 kW
Compressed air	20 m^3/min 3.2 atü
Cooling air	85 m^3/min 300 mm WC
Cooling water	1.9 m^3/h 2.1 atü
Eldred burnoff machine	
Base	2.80 × 2.40 m
Height	3.10 m
Weight	12 t
Burner diameter	1.727 m
Vacuum	5.7 m^3/min 20—70%
Gas	Quantity according to heating value 0.2 atü
Oxygen	2.5 m^3/1000 tumbler 0.25—0.30 atü

too. The average life of the parison mold (until change because of soiling) is 6; that of the blow mold (until change because of paste burnup), 4 h. A complete set requires double to triple mold equipment. The feeder height should be about 300 mm from the upper edge of the trough to the orifice ring. The yield is about 90—95% of cut (Table E 34), depending upon the type of product. The waste is about 20% of the gob weight, varying with type of product. The output of a mold is from 200,000 to 250,000 pieces. Dimensions, weights and energy requirements of the machine are listed in Table E 35.

3.3.6.2 The Machine Type M 16, Produced by Jerome and Bonnefoy, France. (Sec. E 3.3.6.2 to 3.3.6.4 were assembled from the literature by Dr. W. Trier).

This machine is constructed as a rotary machine with 16 blow stations which are arranged at equal distances on the continuously rotating frame. The gob falls into a fixed press mold, is pressed to a pill-shaped body by a plunger, using compressed air, and is transferred to a blow station by means of vacuum. The body is blown to the desired shape by single well controlled puffs. The shaping technique is quite similar to that of the Corning Ribbon Machine (Sec. 3.3.6.3). The output is reported as 36—48 pieces/min. The maximum diameter of the product is 105 mm, its maximum length 400 mm.

3.3.6.3 The Corning Ribbon Machine. Among all hollow glass machines the Corning Ribbon Machine takes a special position. The machine, which was developed for the mass production of lamp bulbs, works completely continuously as far as glass is concerned and by far surpasses all known glass machines in its output.

The functions of the machine are represented schematically in Fig. E 104. The glass is supplied to the machine through a feeder in a continuous stream, which flows from a 25 mm opening and is passed between two counterrotating water-cooled rollers. One roller is smooth, while the other has pocketlike recesses. The ribbon, formed by the rollers, is then redirected to a horizontal flow and placed on a plate belt which runs at equal speed. Every single plate has an opening through which the pill shaped glass sags under the influence of gravity. The sag is aided by blow heads and properly timed compressed air impulses. The blow heads are fastened to a continuous rotating chain belt. They are placed from above on the premolded glass in suitable fashion and move at the same speed as the glass ribbon. After the glass has been premolded to a considerable extent, it is enclosed by blow molds which are brought up from below and is then blown out to finished bulbs through the blow heads. The blow molds are pasted and rotate around their own axes to obtain seamless smooth surfaces. The finished bulbs are then separated from the ribbon and carried to the lehr. From the exit of the glass from the feeder to the exit of the annealed bulbs from the lehr only 8—12 min. pass. Most of this time is spent in the lehr. The production time proper is only 10—12 sec.

3.3.6.4 The Philips Bulb Blowing Machine [6]. A remarkable machine for the production of thin walled seamless hollow glassware such as lamp bulbs is represented by the Philips Bulb Blowing Machine.

This is a continuously rotating one table machine with 16 entirely independent blow units. The machine is driven mechanically; single work phases are carried out pneumatically using control valves. Fig. E 105 gives a general view of the machine. The functions of the production units are seen in Figs. E 106 a and b. The glass gobs supplied by the feeder fall into press molds *A* which can be shifted radially under table *B*.

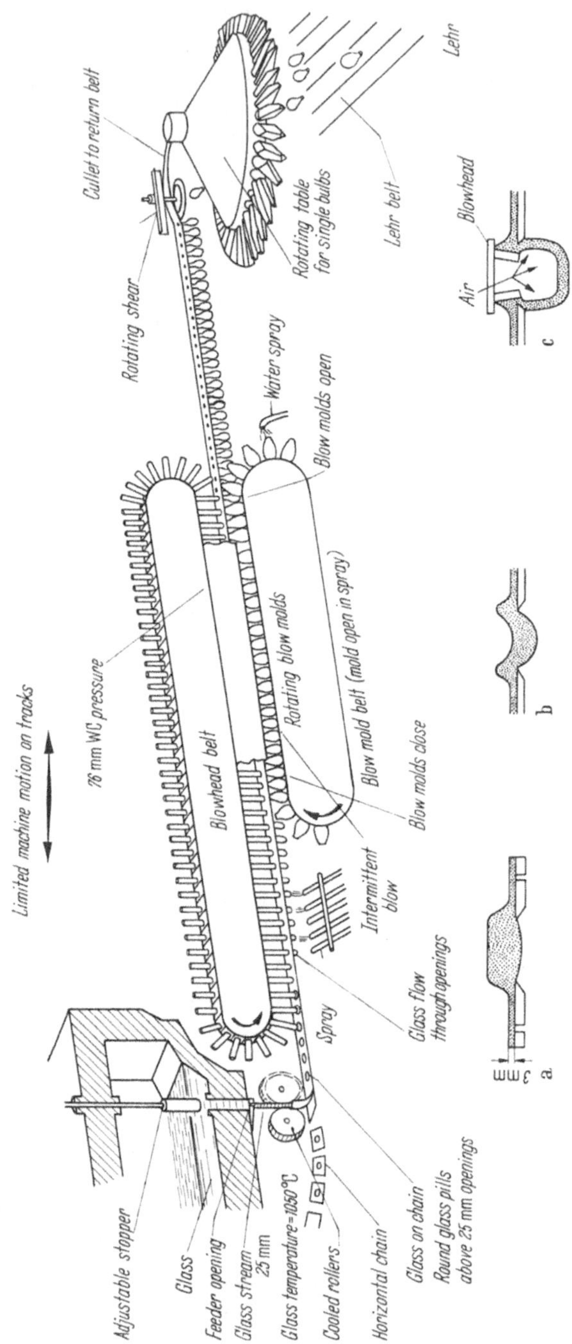

Fig. E 104a – c. Working scheme of Corning ribbon machine.

The table has suitable openings through which the glass can fall. The blow unit is pressed down on the press mold by a compressed air piston and the glass is pressed to a pellet. The glass penetrates between plunger

Fig. E 105. Total view of Philips bulb machine (Courtesy Philips).
1 Rotating table and blow units; *2* Blow unit and plunger; *3* Baffle and press mold; *4* Lever activating press mold motion with cams; *5* Control attachment; *6* Blow unit in blow position; *7* Mold halves; *8* Burners heating glass in suspension; *9* Lever activating valves; *10* Valve cams; *11* Gas and air ducts for burners; *8* Oil and compressed air for blow units; *12* Drive.

and claws, solidifies in this area, and now hangs firmly in the pipe. The blow head is lifted, the press mold is moved outward, and the blow head is pushed down through the opening the press mold has left (Fig. 106b). The blow head in this phase is set on the conical part of a rotating ring *H* and in this way is made to rotate. By means of some exactly timed blow impulses and a little flame (Fig. E 105, No. 8) the parison is length-

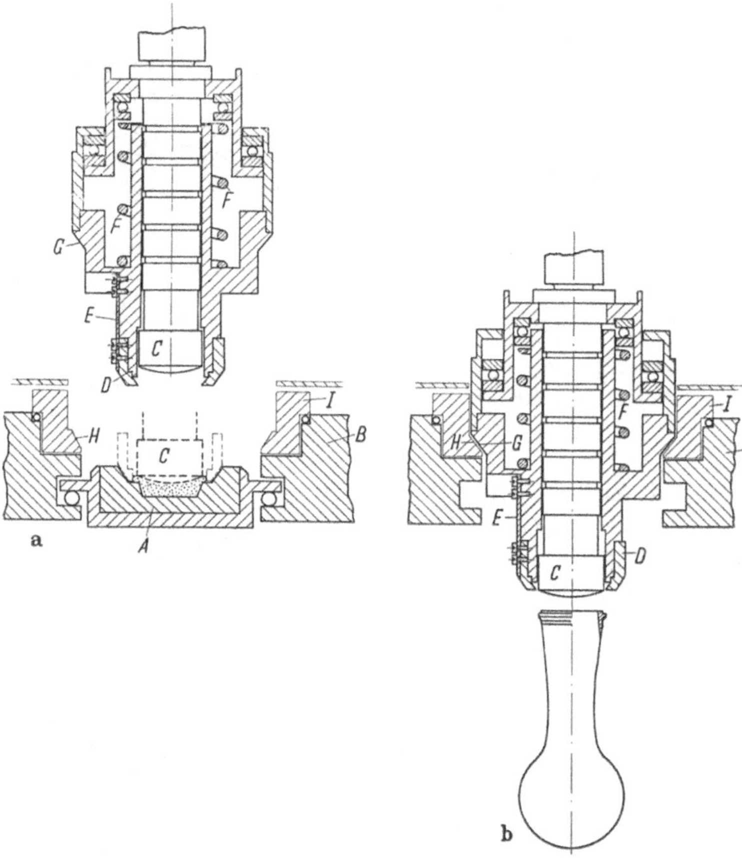

Fig. E 106a and b. Working scheme of Philips unit (Courtesy Philips).
a) Section through blowhead (with catching device and portion of table). *A* Pressing mold shifting under table; *B* in a radial slit; *C* Plunger; *D* Claw; *E* Spring; *F* Spring holding conical edge; *G* of pipe against edge *H*; of rotating ring *J*. b) Ejection of bulb by means of plunger *C*. Claws held on springs open automatically.

ened as desired and the blow mold halves close around the parison, which is blown out. Paste molds with a thin graphite layer are used. They are sprayed with water while open to produce a steam layer between mold and glass during blowing. In this manner a particularly smooth surface is obtained. To eject the finished bulb the plunger *C* is pressed down altogether and the bulb is shoved away from the spring held claws. The bulb then falls on a conveyor belt.

The pneumatic operations of the machine are controlled by valves which are arranged radially above the blow unit in the rotating part of the machine (Fig. E 105, No. *9*). The valves cooperate with cams which are attached at the under side of the fixed machine frame (Fig. E 105, No. *10*). Each production unit has 8 valves. The machine has 16 entirely independent blow units of which each makes one bulb in each revolution of the machine. The machine makes about 4 rpm and so produces about 4000 bulbs per hour.

Since each unit works independently various types of bulbs can be produced simultaneously on one machine. The only conditions for this performance are equal glass weight and equal production time.

References

[1] U.S.A.-Patent No. 1979211 (Rowe).
[2] Prospectus No. 28 (Hartford).
[3] U.S.A.-Patent No. 1790397 (Wood & Gray).
[4] Preston, F. W.: New lamps for old. Glas Ind. 12 (1931) 159—165.
[5] Tooley, F.: Handbook of Glass Manufacture. Section Feeding and Forming. New York: Ogden Publishing Comp. 1953.
[6] Zonneveld, P. van: Maschinelle Fertigung von Glaskolben für Glühlampen und Radioröhren. Philips techn. Rdsch. 22 (1960/61) 360—378.

4 Drawing of Tubes and Rods

By F. Holl, Zwiesel

Glass tubes and rods of various dimensions, made from various glasses, are required in many places. They have been made by hand from ancient times and are now extensively made by machine. The variety of tubes and rods required has preserved the manual process for special types and low quantities in spite of the development of efficient tube and rod drawing machines. For this reason the description of the most important mechanical drawing processes will be preceded by a short review of the manual methods. This also helps in better understanding of the mechanical drawing processes.

4.1 Manual Tube Drawing [1, 2, 3, 4]

The glassmaker (tube drawer) gathers glass on a blowpipe and molds it to a truncated cone shaped form by blocking it on a smooth iron plate. Air is blown into the pipe in such a way that in the gather an egg shaped bubble forms with the narrow point in front. By further overlaying of glass one sometimes obtains a nearly cylindrical surface in preparation for the "draw". In this process the body is stretched somewhat, but the air bubble must remain reduced in front. In the meantime a helper prepares glass on a gathering iron and broadens it on a plate to form a dish shaped punty. The front face is briefly dipped into water which makes the front more rigid. It is from this front that later the glass is drawn uniformly to a tube. Overlaying and blocking causes non-uniform temperature distribution. In particular, the rim zones have become too cold. By turning and swinging the glass in the furnace it is made to assume the temperature required for drawing, but its shape must not be greatly altered. The helper also briefly reheats the punty. He now drops the punty on the end of the gather and pulling begins. Both men work on a wooden tubing floor which is 80—150 m long. The glassmaker with his pipe remains at his stand and continues to blow. The helper slowly walks away, drawing the bulk of the glass from the pipe. At the same time both pipes and punty are rotated slowly and uniformly to avoid the sagging and flattening which would cause uneven wall thickness. A second helper cools the tubing with a fan or a small blower to cause it to solidify in the places where the desired diameter has been attained. The outer diameter is controlled by calipers. The helper who is walking away from the glassmaker can influence the outer diameter and wall thickness by variations in his walking speed. The drawing continues as long as there is enough glass of the desired size. The length of the drawn glass is about 15 meters for 140 mm diameter and 2 mm wall thickness, and is up to 150 meters for thermometer capillaries. A shop, consisting of a glassmaker and two helpers, can produce 200 to 250 kg of tubing per shift.

4.2 Mechanical Drawing

Today the following processes are used for the mechanical drawing of tubing:

(*a*) *Rotating pipe processes:*

Danner process: The glass is drawn horizontally from the outer surfaces of a tube. Philips process: The glass is horizontally drawn from the inner surfaces of a tube.

(*b*) *Nozzle processes:*

Vertical up: Maetz-Schuller, Corning and Korolsow-Schapiro processes.

Vertical down: Vello and Haenlein processes.

General descriptions of these processes are found in Ref. [5—13].

4.2.1 Danner Process

The process was developed by E. Danner in 1912 and was based on the manual process. [14—18, 18a]. German Patent (Libbey Glass Co., Toledo) 384,796, 1-23-1917; US 1,218,598; British Patent 110,642,

111,521). The regulable glass flow (Fig. E 107, E. 108) falls upon the outer mantle of a slanting rotating cylindrical mandrel and while air is blown through the mandrel, the forming glass tube is drawn continuously in the direction of the axis of the mandrel (Fig. E 107).

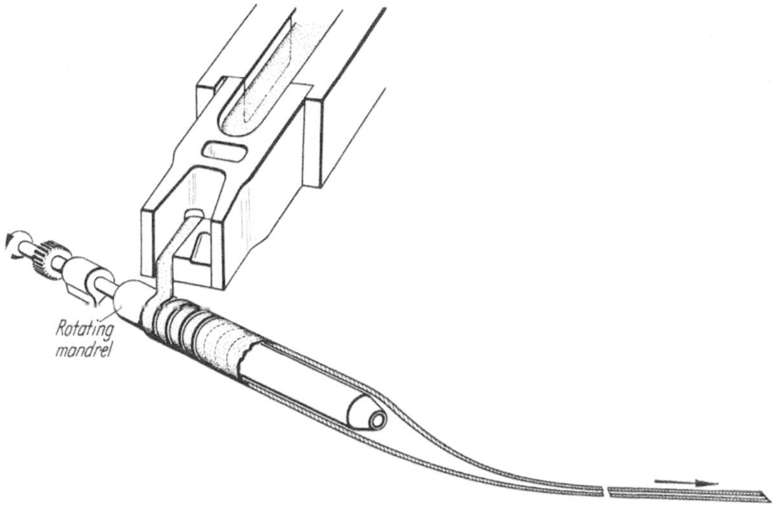

Fig. E 107. Working scheme of Danner rotating mandrel.

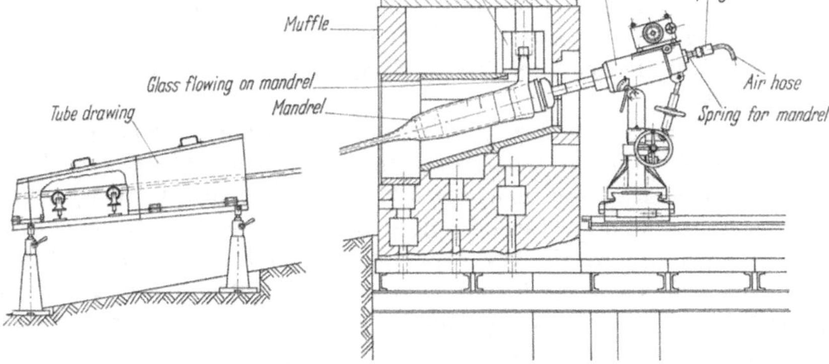

Fig. E 108. Danner machine and muffle.

A small tank shaped drawing furnace is required which in the case of tank operation is built right next to the melting tank. In the case of pot operation the glass is supplied by ladle. From this drawing furnace the glass flows along a lateral trough in a refractory channel onto the mandrel near its upper end. A movable cylindrical baffle is used for regulation. To avoid disturbing air inclusions the glass stream hits the mandrel at a slight angle. The mandrel consists of a tube shaped carrier usually double walled and water cooled, and fabricated from nonscaling (in Germany NCT 3) steel or a chromium vanadium alloy upon which an expendable tapered refractory tube is mounted. The outer face of the refractory must be reshaped from time to time. The driving end has water cooled bearings. The drive is by an electric motor through a bevel gear wheel and worm gear. The mounting of the pipe bearings and the motor permits a change in inclination by 15—20° on the average as well as exact settings for the channel stop. The mandrel is enclosed in a muffle which is open at the front and which permits temperature control by means of several adjustable burners. The glass

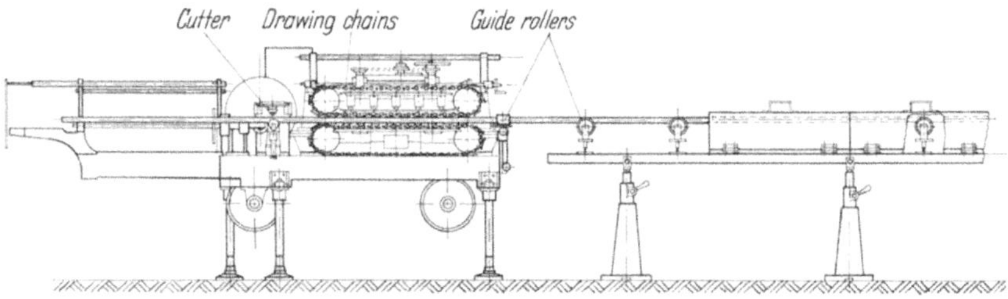

Fig. E 109. Drawing machine in Danner process.

mass is wrapped around the refractory section and flows with a fire polished smooth surface towards the lower end of the mandrel. The temperature may be 1040—1050 °C in the middle and 900—910 °C at the end of the mandrel. Today temperatures are in part regulated electrically.

The pulling mechanism of the Danner machine is equipped with endless chains. Coming from the mandrel the tube or rod runs for a short distance in air and then over a 30—50 m long path on exactly adjustable asbestos, wood or carbon rollers. Guided by two lateral rollers, the tubing is gripped by the two chains which act from above and below and drawn it away (Fig. E 109). One has to pay attention to a uniform draw. Non-uniform drawing causes variations in diameter and wall thickness. The chain distance can be adjusted at the upper chain to fit the tube diameter. In addition a series of springs maintain friction between the chains and the tubing. The single chain links are little cars on two rollers which run on guide tracks. For small tubing which is drawn rapidly, endless asbestos ribbons have been found satisfactory.

In addition to pulling the tubing, the Danner machine also cuts it to length. A rotating cutting stone (Fig. E 110) is first accelerated to the drawing speed, in the drawing direction, and is then pressed vertically on the tubing by a cam operated mechanism. The tubing, neatly and precisely scratched in this manner, can now be separated by bending it slightly by means of a spring controlled roller.

The most important dimensions of the tubing, including diameter and wall thickness, depend upon the mandrel diameter (up to 25 cm), the glass quantity, glass temperature, muffle temperature, speed of rotation, air pressure, and the drawing speed. The latter is 10—200 m/

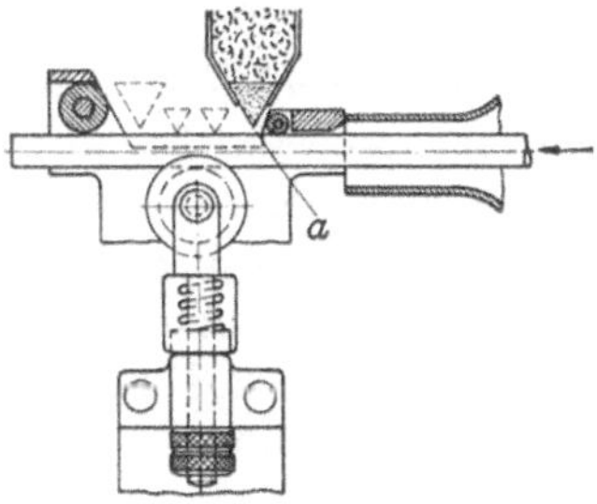

Fig. E 110. Cutting device (cutter: *a*).

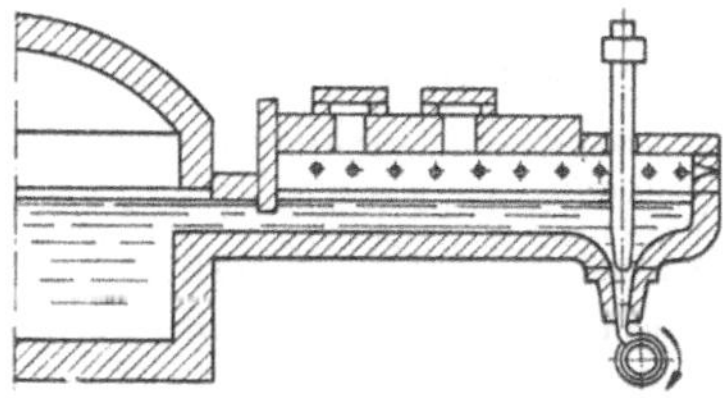

Fig. E 11. Danner process with feeder.

min. The quantity produced increases with diameter and wall thickness. These data are summarized in Table E 36.

The heat consumption of Danner installations, including that of the melting tank is between 4500 and 7500 kcal/kg of finished tubing, depending upon diameter and wall thickness. These values may have to be doubled for special glasses requiring high melting temperatures. For a tolerance of $\pm 5\%$ for diameter and wall thickness the Danner machine supplies up to 95% good tubing, while hand drawing supplies only 20% salable glass. Danner processed tubing obviously surpasses hand made tubing in regularity and precision. It is also understandable

Table E 36. *Drawing speed and output of a Danner machine*

Tubing dimensions		Drawing speed	Output		
Diam. (mm)	Thickness (mm)	(m/min)	(kg/h)	(kg/shift)	(kg/day)
1	Thinwalled	—	20	av. 1500 kg per machine (2 operators)	4000 to 6000 kg ready to ship (bundle 5 kg or 10 kg)
2—3	Thinwalled	225	—		
10	0.5	90	160—230		
16—17	0.5	62	160—230		
21—22	0.6	38	160—230		
	Thickwalled				
6—7	0.9—1	50	120—150		
11—12	1.5—1.75	30	120—150		
50	1.3	15	300—350		

that the cost of hand made tubing is about 10 times higher. The machine works most efficiently in the area of 6—20 mm diameter. Since only the mastery over glass viscosity assures high production efficiency and conformity to tolerances, the flow feeder is used increasingly (Fig. E 111). The feeder is of the type used in hollow glass production. The periodically operated plunger is replaced by a regulating needle.

4.2.2 Philips Process

The Philips process (Fig. E 112) (German Patent 442,568; Austrian Patent 95,775, 95,933) resembles the Danner process. The drawing tank is heated by a burner. The glass flow is baffle regulated. The glass enters the inside of a chrome iron alloy cylindrical pipe. The tube is heated by an electrical resistor winding and insulated against external radiation. Since the pipe rotates continuously a hollow cylinder of viscous glass is formed on the inside which can be drawn as a tube at the end of

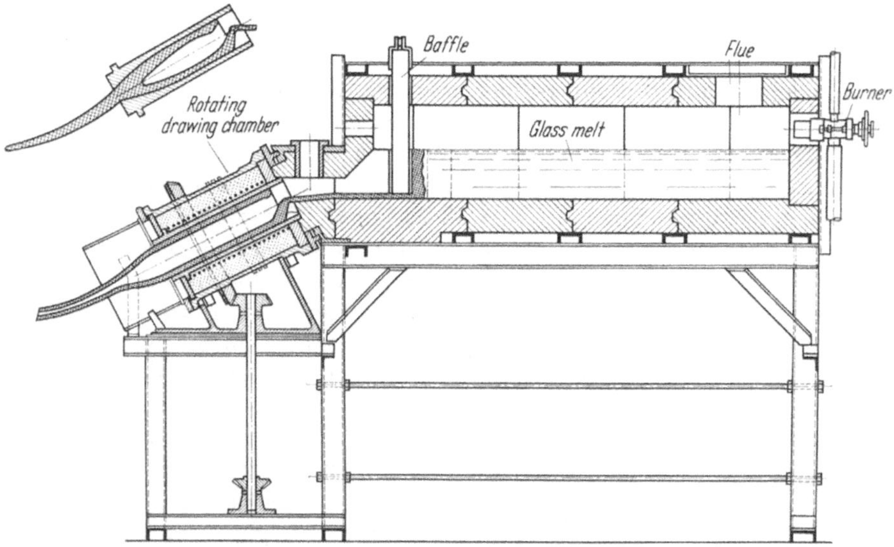

Fig. E 112. Philips tube drawing unit (upper left: pipe for rod drawing).

the pipe. In contrast to the compressed air used in the Danner process air is drawn through the hollow pipe by vacuum. If rods are to be drawn a converging chromium iron body sketched in the upper left of Fig. E 112 is used which causes the glass to flow into a rod. The drawing machine of the Philips machine differs from that of the Danner machine only by the slightly different operation of the cutting device. The basic principle is still the scratching of the tube by a rotating cutting wheel which either moves along or is fixed. The tube is then subjected to a bending force and breaks at the scratched place. The maintenance of exact temperatures requires particularly exact burner adjustments in the various sections of the furnace. Although the process has some advantages in control it is being supplanted by the more efficient Danner process.

4.2.3 Maetz-Schuller Process

The basis of this process is the German Patent 554,026-16-6-31. In the processes discussed so far drawing was horizontal. This has the advantage that the tube can be cut easily by machine. A disadvantage is the space requirement of the drawing apparatus so that only two machines can usually be connected with one tank. This disadvantage is avoided by vertical drawing. Three machines can be connected to

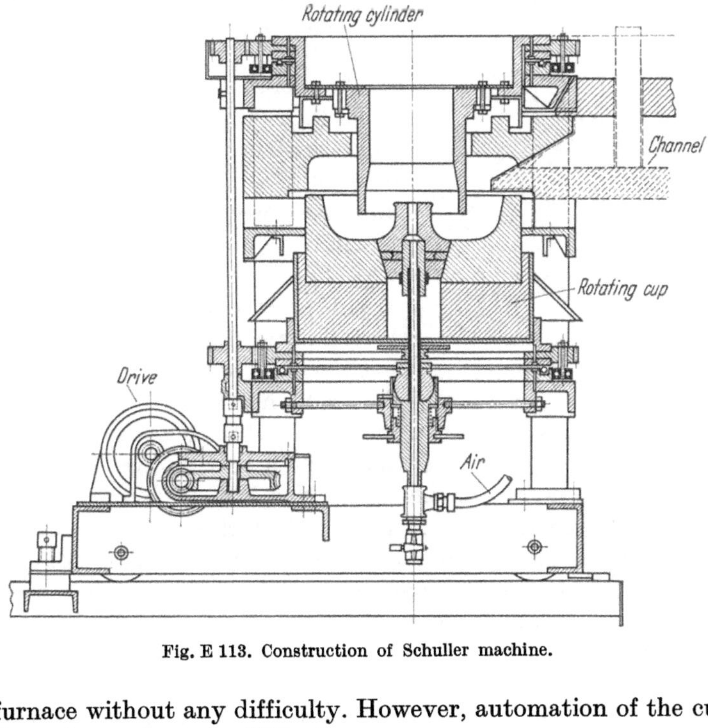

Fig. E 113. Construction of Schuller machine.

one furnace without any difficulty. However, automation of the cutting off phase is much more difficult.

The first patent claims the adjustability of the nozzles. In practice, however, the patented arrangement was changed (Fig. E 113). The original hearth was replaced by a rotating refractory cup. A tube shaped opening in the bottom of the cup serves to receive the interchangeable nozzle. Only recently has there been a trend towards changing nozzles from below. The demand for better control of cooling and homogenizing of the glass has led to screening by a synchronized refractory cylinder. The rotating cup is fed through a short channel. The cup and the cylinder are synchronized in slow (0.5—0.82 rpm) rotation. The updrawn glass mass initially has the shape of a trumpet—the "onion" — the central opening of which is maintained by the air blown from the nozzle (Fig. E 114). From a cooling ring, with staggered downward openings, cooling air is blown on the forming tube near the upper end of the onion. In this manner the wall thickness of the tubing can be controlled. Nozzle and cooling ring require exact air adjustment. The cooled and solidified glass tubing is continuously drawn up by two asbestos covered wooden rollers about 6 m above the rotating cup.

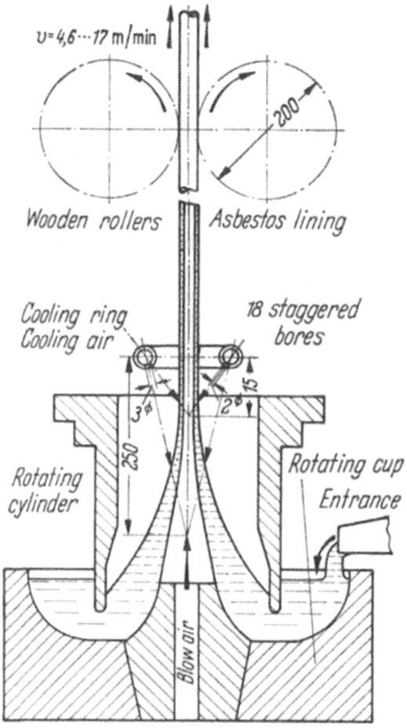

Fig. E 114. Schuller process.

In detail, the usually octagonal, separately heated working end of the tank delivers glass to the rotating cups of the machines through channels equipped whith baffles. Starting the machine requires considerable experience and skill. An "onion" must be formed around the nozzle by means of iron hooks. After drawing upwards with the hook and severing the onion the tube is drawn by hand and introduced between the rollers. The glass level in the cup must reach the nozzle. Later it must be 10 mm above the rim of the nozzle. Nozzle and cooling air are regulated to yield the desired diameter and wall thickness. The starting operation takes six hours. The movable machine (E 113) carries all necessary parts in a four column angle construction. A one HP electric motor drives a vertical shaft with two pinions through a worm gear. The lower pinion engages the gear of the rotating cup, the upper pinion the gear of the cylinder support which rests on ball bearings. Cylinder and cup rotate in the same direction at the same speed. The fixed refractory hood has a lateral opening for the glass supply and an upper opening for the tubing. The glass in the cup is typically held at 820 ± 20 °C. Burners on brackets, which permit horizontal and vertical angular changes, control the temperature. The flames must hit the cylinder tangentially above the cup and envelop it. The waste gases escape through lateral holes in the hood to avoid contact with the forming tube. Compressed air (pressure: about 60 mm water column) is fed to the nozzle through a vertical conduit. The amount influences the diameter of the tubing. Control of the cooling ring nozzles is difficult. They influence the wall thickness of the tubing because of the relation of cooling and solidification to the pull of the asbestos lined rollers.

Variations in wall thickness at any part of the cross section must be corrected, even during the drawing process. To this end 18 bores staggered regularly, are attached to each cooling ring at two different angles. The cooling ring is equipped with radial cross faces for six single

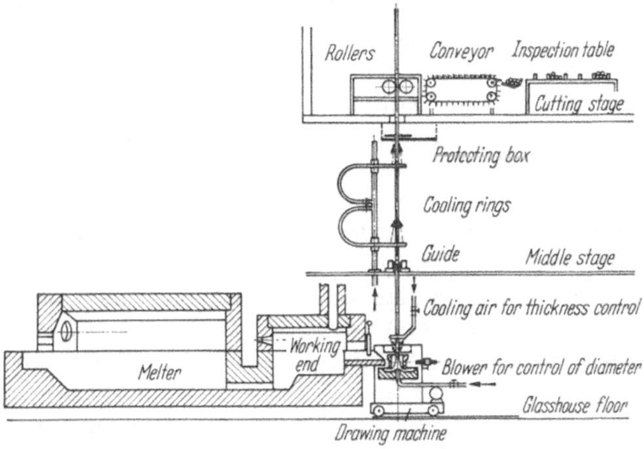

Fig. E 115. Total view of a Schuller unit.

sectors so that each sector can be supplied independently with regulated air by separate valves. The branch conduits which combine later have a common reduction valve for overall compressed air control. The tubing then goes through the ceiling and enters an intermediate floor (Fig. E 115) where it is further air cooled in a thin tube mantle. On top one finds an electrical indicator for diameter and thickness variations. The tubing next enters the cutting room, still drawn by the concave, asbestos wooden rollers. The revolution of the rollers corresponds with the drawing speed and can be adjusted to speeds varying from 4.5 (8) to 17 (25) m/min by stepless regulatable gear.

More recent studies show that non-uniform roller grip varies the drawing speed and, consequently, the external diameter of the tubing. The cutting process is simple. An adjustable wooden ledge has on one end a stop, on the other a small adjustable carborundum wheel. The cutter places the stop against the rising tube end and scratches the regulated length with the cold moist wheel. The small scratch permits the operator to break the tube with his left asbestos-gloved hand. The sorting machine groups the tubing according to external diameter; wall thicknesses are established by weighing known lengths. The diameter range of 3—50 mm can be covered by the use of three nozzle sizes and two cylinder sizes. Nozzles and cylinders can be exchanged easily. Production depends upon the dimensions. The machine may, for instance draw 360 m of tubing, per hour, with 16—17 mm diameter and 1.2—1.4mm wall thickness, corresponding to a production of 1.6 t/24 h. This means that drawing speed is quite low. The supplier of the machine (Regina Glass Co. Ltd. Stoke-on-Trent, England) equips it with three interchangeable nozzle types: for capillaries and tubes, for 6—16 mm, and for 16—50 mm diameter. Thus diameters from 3 (4) to 40 (50) mm can be obtained for wall thicknesses from 0.3 to 3 mm. Best production is from 1200—1800 kg/24 h of soda lime glass to, at best, 2500 kg/24 h of lead glass. As to fuel requirement, a single value of 7200 kcal/kg can be given for a 3 machine tank. Until 1958 about 15 machines were built of which 10 are in operation in Germany. One machine requires 8.5—11.3 m^3 coke oven gas/h and 5 kWh. Operation requires 1 foreman and 2 unskilled helpers for cutting, sorting and packing.

There is a whole series of patents aiming at the improvement of the homogeneity of the drawn glass. R. Salomon (German Patent 540,161, 7-20-1928; 542,559, 9-27-1928) uses a rotating forehearth in the bottom of which a hollow needle is supplied with blown air. The tubing is drawn from the needle to an upper muffle. The muffle protects the glass from flame gases and a water cooled ring provides cooling and the necessary viscosity. W. Schuller (German Patent 782,366, 4-19-51 British priority 11-30-1950) uses a drawing pot which rotates at 0.5—2 rpm. A drawing member placed below the glass level in the center of the pot is used for upward drawing. A cylinder dipping into the glass with its lower part is synchronized with the pot. The Regina Glass Co., Ltd. London (Austrian 192,064, 11-15-1956; British priority 11-20-1950) has devised a special shape for the wall and bottom of the drawing pot. The relative diameter of the cylinder is chosen to permit it to dip into the glass at the deepest place of the pot. E. Popp (German Patent 972,085, 3-15-1955) provides a circular collar at the lower end of a dipping rotating drawing pit. The inner edge of the collar is a circular stop for the drawn glass. I. Kitaigorodski [10] describes a process in which glass is supplied to a revolving pot, and a rotating refractory cylinder which dips 10 to 15 mm into the glass. The cylinder is to smooth out the glass at its level and to hold back surface glass. Melting tank branches permit the use of up to 9 machines. Interchangeable drawing heads are built into the bottom of the revolving pot. A short distance above the cylinder a ring provides cooling air; a second ring, about 3 m above the floor, has a cooling air baffle. Cutting is manual. Production is about 1000 kg/24 h for diameters 4—40 mm at 6—12 m/min drawing speed.

4.2.4 Corning Process

The Corning Glass Works have developed their own process for upward drawing more than 20 years ago (Fig. E 116). The process is extremely efficient and flexible. Tubing from 3 to 180 mm with wall thickness from 0.3 to 12 mm and a maximum length of 3 m can be produced. Rods can be made up to 50 mm in diameter. With special attachments normal values can be exceeded and wall thicknesses of 20 mm can be attained. Production varies between 0.7 and 6.8 t/24 h. The glass flows to the cup through a channel. The cup forms a sort of feeder cup area. The glass is drawn upward through the hollow fixed nozzle mouth piece. This mouth piece can be adjusted on a carrier which can be varied in elevation. A rotating cylinder again separates the area in which the tube is formed from the remainder of the glass. Tube formation is made possible by cooling from a water cooler. The blowing air reaches the mouth piece proper through a nozzle from below. The drawing speed is capable of stepless regulation of as much as 1:260. Cutting is mechanical or manual. Philips developed a similar uwpard drawing process.

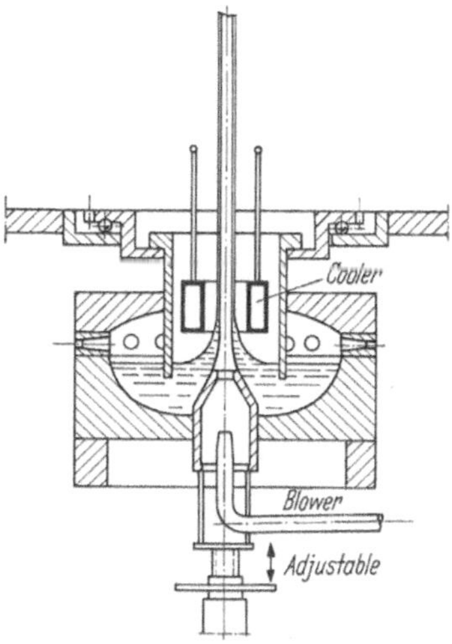

Fig. E 116. Corning tube drawing process.

4.2.5 Korolyov Process (Figs. E 117, E 118)

Korolyov developed a tube drawing process based on the Fourcault process. It uses a refractory body pressed into the glass surface (Figs. E 117, E 118). Korolyov's experiments were carried out on reconstructed Fourcault machines. Normally the nozzle has two circular openings in which the drawing cones are inserted. In this manner tube formation starts from concentric ring areas. The drawing cones are hollow and have connecting channels for air supply. If rods are drawn the cones are solid.

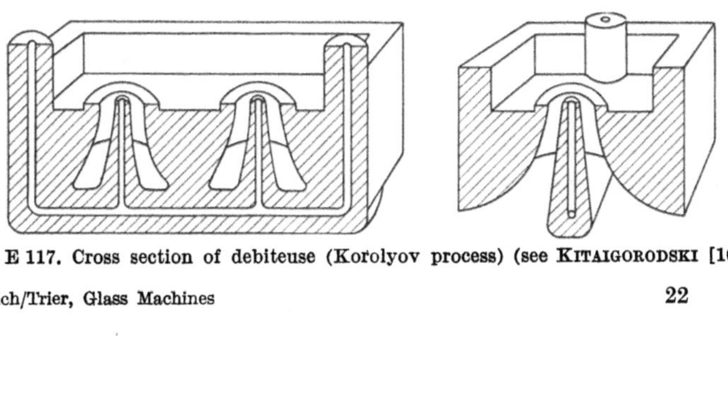

Fig. E 117. Cross section of debiteuse (Korolyov process) (see Kitaigorodski [10]).

The drawing channel of the Fourcault tank, in which nozzles are arranged in series, is separated into heating and drawing chambers by submerged refractory bridges (Fig. E 118). The drawing machines are equipped with pairs of rollers which have their own drives to obtain positive traction

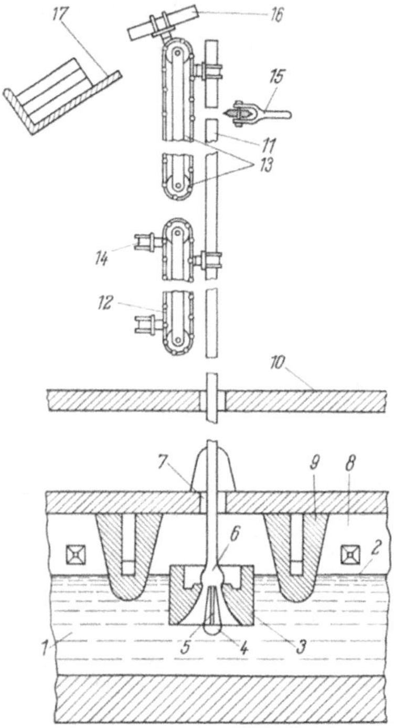

Fig. E 118. Korolyov drawing process with one nozzle.
1 Melt; *2* Glass surface; *3* Nozzle; *4* Core; *5* Air conduit in core; *6* Onion; *7* Opening in the crown of the drawing chamber; *8* Heating chamber; *9* Refractory bridge between heating and drawing chambers; *10* Crown of drawing chamber; *11* Drawn tubing; *12* Drive chain; *13* Guide rollers; *14* Jaws; *15* Cutting wheel; *16* Cut tubing; *17* Catcher.

for various tube sizes. The number of roller pairs has been increased to 16. The pit heighth is up to 9.2 m. The temperature in the hot chamber is 1150 to 1160 °C, in the onion 960 to 990 °C, and on the Rubikon 460 to 480 °C for 100 m and 560 to 570 °C for 150 mm tubing. The temperatures in the drawing pit are regulated by heating the chamber below the drawing machine and by shielding the upper pit. From this machine tubing up to 175 mm external diameter is drawn. As an example, a drawing speed of 25 m/h for 100 mm tubing has been given. The production is listed as 600 kg tubing glass per 4—6 man shift on 5 machines. In the West this process has not been used to date.

4.2.6 Shapiro Process [19]

Russian experiments (Shapiro and others) to draw glass tubing continuously in a vertical method without a drawing nozzle (debiteuse) (Fig. E 119) are worth mentioning. However they represent by no means truly free drawing from the bath, as perhaps examplified in the Lubber window glass process. A central refractory mouth piece still centers and guides the drawing. A ring chamber of 2 m extension and 0.5 m depth is supplied with glass by a broad channel. It is heated with 10 tangential gas burners. Above the chamber is an asbestos-lined cooling pit with 7 pairs of rollers and suitably distributed burners. Cutting and breaking is manual. The cooler with its truncated cone shape and cylindrical inner wall is an innovation in the control of the "onion".

The position of the cooler can be adjusted precisely. Position and temperature of the cooler greatly influence the tube and must be regulated well. In practice the approach of the cooler to the glass level effectively increases tube diameter, and, in conjunction with drawing speed, controls wall thickness. 2″ tubes are drawn at 65—70 m/h, 4″ tubes with 8—10 mm wall thickness were drawn experimentally at 50—55 m/h. Test measurements gave the following data: Glass mass at about 100 mm from the cooler 970—990 °C; onion 910—925 °C; at first rollers about 600 °C; at 5th rollers about 500 °C; at 7th rollers 480 °C; on cutting about 450 °C; at cooling water exit 40—60 °C; at air exit 400—440 °C; blow air 115—130 m³/h (2″ tubing); fuel gas 400 m³/h (1400—1450 kcal/m³) for 2″ tubing. Development of this proces has not been completed.

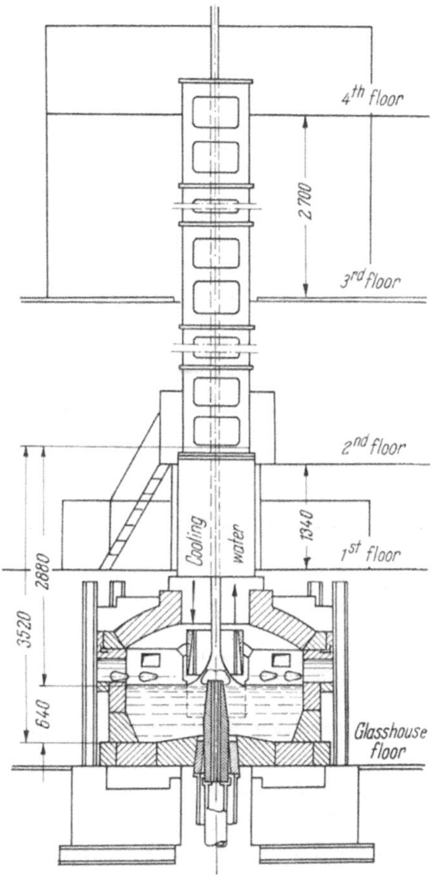

Fig. E 119. Schapiro tube drawing process.

4.2.7 Vello Process

The fundamentals of this process are recorded in German Patent 579,506 and French Patent 692,141 (6-8-1929). The hearth possesses in its bottom a cylindrical opening with a nozzle through which the molten glass can flow out over an adjustable, funnel shaped mandrel (Fig. E 120). The latter is hollow and its extension leads upward through the ceiling of the hearth. Above the ceiling it is set in bearings, and connected with the blowing air conduit. It is rotatable. The glass leaving the rotating mandrel is kept open by the blow air and is widened to the shape of the tube. A burner ring reheats the glass. Blowing air in the cone shaped widening of the mandrel comes out of slanted channels. Due to the circulation of the air and the rotation of the mandrel, the glass solidifies sufficiently to be rotated at the lower end of the mandrel. Since the bending of the tube to the horizontal drawing path causes non-uniform wall thickness the cone shaped mandrel is eccentric. In this manner the non-uniform wall thickness anticipates the change in the transition to the horizontal draw. The glass is rotated and drawn by grooved wheels and rollers and cut in the usual manner. Newer approaches are described in German 1,030,479,11-8-1955 (Owens Illinois Glass Co.). The hollow mandrel which is connected with the blowing air, and from which drawing starts, rotates in a muffle free from burner gases; to the point of solidification the drawn tubing is essentially enclosed in a second chamber containing a heated quiet atmosphere free from waste gases. In this chamber the temperature of the glass decreases uniformly with the motion of the tubing.

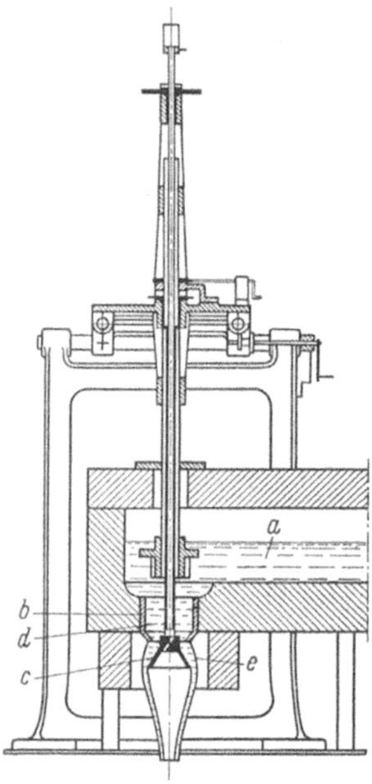

Fig. E 120. Nozzle and tube formation in the Vello process.

As a related process the overflow process should be mentioned (Fig. E 121). The glass flows over a ringshaped threshold and is drawn down into tubing. An immersion baffle lets glass flow at an adjustable rate. Compressed air for tube formation is introduced through a hollow nozzle. Eccentric adjustment controls non-uniformity of wall. The rotation of the hearth with its overflow improves homogeneity. (German Patent 604,579 (12-9-1931) to Corning Glass Works provides cooling of the inner surfaces of the overflowing glass (Fig. E 121) by water cooling the nozzle through which the blow air enters. A burner reheats the flowing glass once more to avoid striation. After this a cooling device produces the right viscosity for drawing. In this process the glass is turned from the vertical to a horizontal direction as in the Danner process.

4.2.8 Hänlein Process [20, 21]

In order to melt and draw silica tubing of precise dimensions in one operation a vertical down drawing process was developed permitting

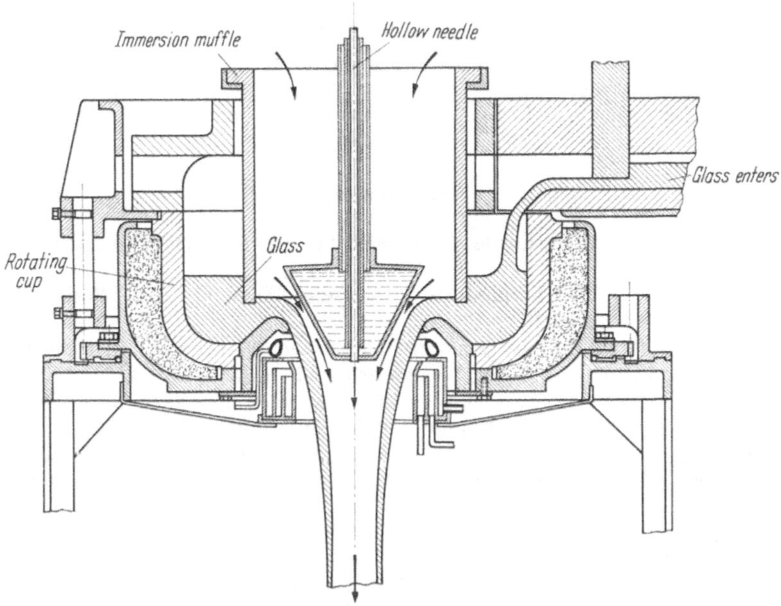

Fig. E 121. Corning overflow process.

safe operation at 2000—2200 °C (Fig. E 122). The melting furnace is an electrically heated pot furnace. The fused silica leaves the lower end of the crucible in the form of tubing. Next a tractor type drawing machine based on Danner's construction seizes the silica tubing and draws it down in continuous vertical motion. Diameter and wall thickness can be controlled by variations in drawing speed and nozzle exit temperature.

The exit temperature depends upon the position of the nozzle in the pot which can be adjusted by a device. Above the tractor the tubing coming out of the furnace is held straight by rollers. After passing the drawing machine the tubing is cut to desired length. The productivity of the installation is relatively large. With nozzles of varying dimensions tubes about 3 to 32 mm outer diameter and 0.8 to 2 mm wall thickness may be drawn. For maximum diameters and wall thicknesses the drawing speed is 12—15 m/h; for smallest 200—250 m/h. Melting capacity is 3—9 kg/h. Heat requirement is 60—80 kW. Power consumption for heating is 8.9 to 20 kWh per kg molten fused silica. The machine re-

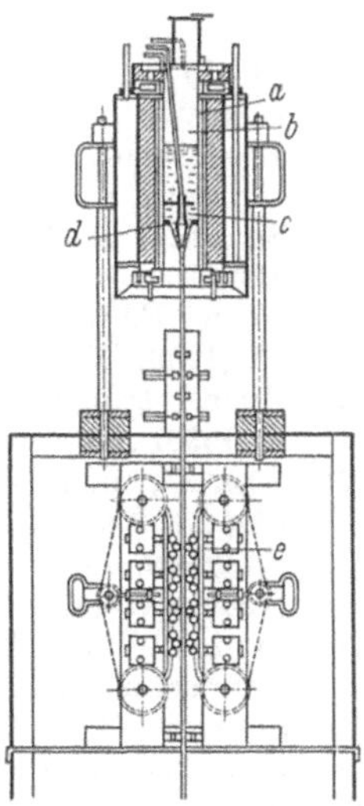

Fig. E 122. Melter and drawing machine according to W. HÄNLEIN [20, 21]. *a* Tungsten wire; *b* Molybdenum crucible; *c* Needle; *d* Nozzle; *e* Drawing machine.

quires about 0.55 kW. The installation is served by 2 men per shift. These Osram drawing machines work in West Berlin, Augsburg, Eindhoven (Holland), Wembley (England) and Fairmont (U.S.A.).

References

[1] Friedrichs, F.: Geschichte der Glastechnik. Die Glashütte 75 (1948) No. 1, pp. 7—9.

[2] Pischel, B.: Die thüringische Glasbläserei (Diss. of Univ. Berlin 1936). Weimar: F. Fink 1936.

[3] Antrich, D., and D. G. Osborne: Measuring the variation in diameter of machine-drawn glass tubing. Engineering 172 (1951) 425—427.

[4] Apel, S.: Geräte für die automatische und kontinuierliche Kontrolle beim maschinellen Ziehen von Röhren und Stäben. Silikattechn. 10 (1959) 283/84.

[5] Dralle, R.: Die Glasfabrikation (Editor G. Keppeler), 2nd Ed., München and Berlin: Oldenbourg 1926—1931, Vol. I, pp. 704—714. Vol. II, pp. 1381 to 1384.

[6] Gehlhoff, G.: Über maschinelles Röhrenziehen. Glastechn. Ber. 2 (1924/25) 205—213.

[7] Thiene, H.: Glas, 1st Ed., Vol. 2, Jena: Fischer 1939, pp. 922—926.

[8] Thiene, H.: Röhrenziehen. Glashütte 75 (1948) 7—9.

[9] Thiene, H.: Röhrenziehen. Glashütte 76 (1949) 68—70, 87/88.

[10] Kitaigorodski, I. I.: Technologie des Glases (German edition J. Winkler), 1st Ed., München: Oldenbourg 1957, pp. 229, 611—615; 2nd revised Ed., Berlin: VEB Verlag Technik 1959.

[11] Adams, F. W.: 20 years of progress in glass forming machines, II. Glass Ind. 21 (1940) 177—179.

[12] Gastew, J. A., N. A. Pentko, A. S. Spiridonow, and F. S. Entelts: Über Erfahrungen bei der Herstellung von Glasrohren (from the Russian). Steklo i Keram. 8 (1951) No. 12, pp. 3—11.

[13] Shand, E. B.: Glass Engineering Handbook, 2nd Ed., New York: McGraw-Hill 1958.

[14] Anon.: Edward Danner inventor of tube drawing machines. Nat. Glass Budg. 73 (1958) No. 48, pp. 6, 16.

[15] Anon.: Danner Röhrenziehen. Ruhrglas AG-Bericht. Z. Glas-Instr.-Techn. 2 (1958) No. 7, pp. 11—13.

[16] Res, M., and J. Michlek: Um die beste Ausnützung der „Danner"-Maschine. Sklář a Keram. 4 (1954) No. 11, pp. 285—289.

[17] Zadrazil, J.: Feststellung der Verluste bei Danner-Maschinen. Ceskosl. Sklář a Keram. 2 (1952) No. 2, pp. 24—26.

[18] Sibilia, V. E.: The automatic production of glass tubing and rod on the Danner machine. J. Soc. Glass Technol. 23 (1939) 292—307.

[18a] Solow, G.: Teilautomatisierte Anlage zur Herstellung von Glasröhren. Silikattechn. 11 (1960) 410/11.

[19] Schapiro, J. E., and coauthors: Erzeugung von Glasröhren durch maschinelles vertikales Ziehen ohne Ziehdüse (from the Russian). Steklo i Keram. 12 (1955) No. 4, pp. 4—8.

[20] Hänlein, W.: Schmelzen und Verarbeiten von Quarzglas und ähnlichen hochschmelzenden Gläsern. Glastechn. Ber. 18 (1940) 308—314.

[21] Hänlein, W.: Schmelzen von Quarzglas. VDI-Z. 85 (1941) 958/59.

5 Drawing of Flat Glass

Machines for drawing flat glass should also be discussed here. However, it should be noted that in this case the machine proper has only an auxiliary function. In contrast to hollow glass shaping, flat glass forming is not achieved through forming tools as such, but instead occurs partly in the conditioning zone of the tank and partly by itself. The drawing hearth and its particular construction must be described along with the description of the machine. Since not much space is available here, only the essentials will be discussed. Details can be found in the literature references at the end of this section, particularly in the work of J. Kitaigorodski [3] and of H. Goerk [4, 6].

5.1 Vertical Drawing Processes

(By K. Flessa, Weiden, and W. Giegerich, Heilbronn)

5.1.1 Debiteuse (Fourcault) Process

5.1.1.1. Development of the Process. E. Fourcault, about 1902, conceived the ingenious idea of compensating for the surface tension ot the glass in ribbon form by pressing the glass from the bath in the

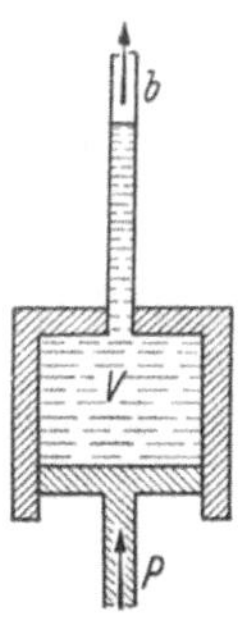

Fig. E 123. Support and lift of the sheet formed under pressure (Fourcault process).

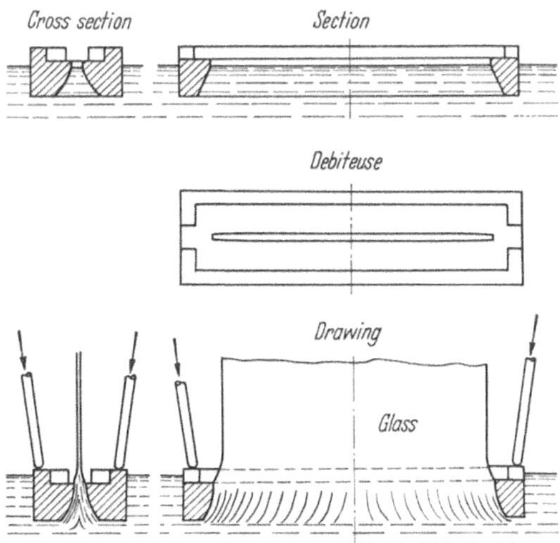

Fig. E 124. Design of debiteuse and drawing process.

shape of a very small rectangle (Fig. E 123). No less ingenious than this principle was the practical execution developed by Fourcault and illustrated in Fig. E 124. He pressed a nozzle (debiteuse) into the bath, through the slit of which debiteuse the glass swells upward capable of vertical drawing to a ribbon. This ribbon is solidified by coolers and is next transported upward by the numerous roller pairs of the drawing machine proper while continuing to cool. At the end of the machine the arriving infinite ribbon is cut to suitable length (Fig. E 125). It is absolutely necessary that equilibrium is established between the glass leaving the nozzle and that taken up by the drawing machine. The quantity of glass swelling up through the slit depends upon the depth of immersion and, therefore, upon the hydrostatic pressure, and also on the cross section of the slit and, particularly, the viscosity of the glass.

The edges of the ribbon are rounded and thicker than the center because of surface tension. This happens to be an advantage since the edges cool faster and contribute to the maintenance of the width as soon as the glass layer moves no longer by hydrostatic pressure but under the pull of the machine. Knurled edge rolls increase the stiffening effect of the edges on the ribbon with heat transfer through the knurls playing a secondary role.

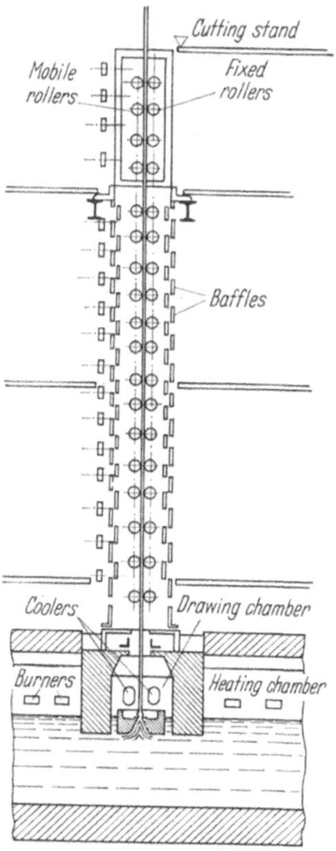

Fig. E 125. Fourcault machine.

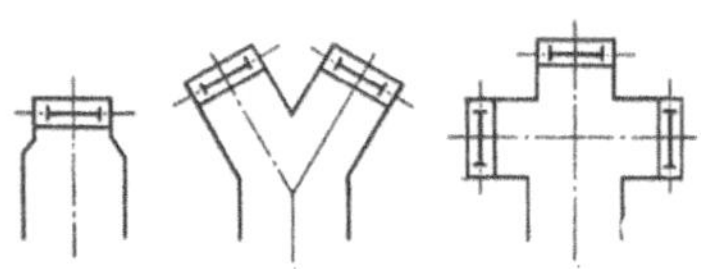

Fig. E 126. Arrangement of 1–3 drawing machines.

During past decades the Fourcault process has been greatly improved. Ribbon width was increased, drawing speeds improved and, particularly, the optical quality improved to the point of allowing the manufacture of glasses for high demands such as safety windshields for the automative industry. About 50% of all flat glass is made by the Fourcault process today. In spite of the debiteuse it remains technologically the simplest process. It also has the largest tolerance for changes in operational conditions because of the compensation for the surface tension by the debiteuse—in contrast to other drawing processes such as the Pittsburgh, Libbey-Owens Ford and others derived from them.

5.1.1.2 Arrangement of Machines. At every tank one to nine machines can be arranged depending upon the tank melting capacity. The usual groupings are shown in Figs. E 126 and E 127. This takes into consideration the fact that the mean path from the conditioning zone of the tank to each machine should be as nearly equal as possible. Under such

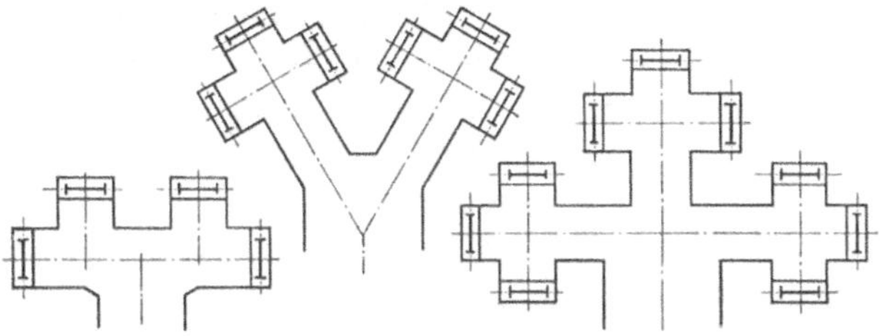

Fig. E 127. Arrangement of 4 or more drawing machines.

conditions the glass flowing to the drawing chambers of the single machines has about the same temperature. Temperature corrections corresponding to thickness variations must be made in front of the bridges or in the drawing channels.

The older serial arrangement has not been proved sound because homogeneity is disturbed by cooling and reheating. For this reason today only direct feeding of each hearth is permitted.

5.1.1.3 Construction of Drawing Chambers. Glass for each machine is taken out in a separate drawing chamber which is separated from the conditioning zone of the tank by a fixed, immersed bridge (Fig. E 128). In order to protect the delicate vertical arch against the impact of the

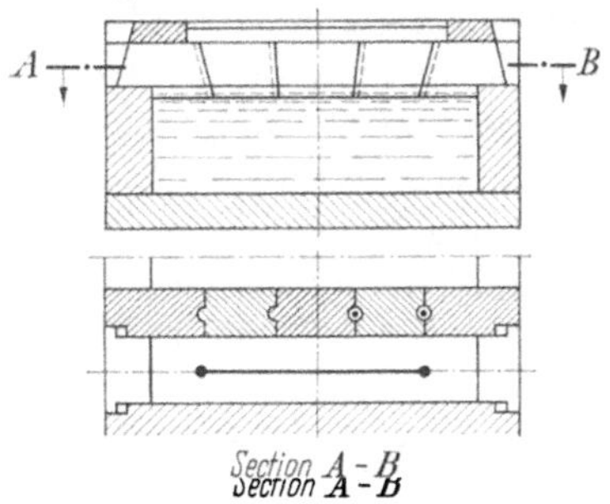

Fig. E 128. Fixed bridge. Section.

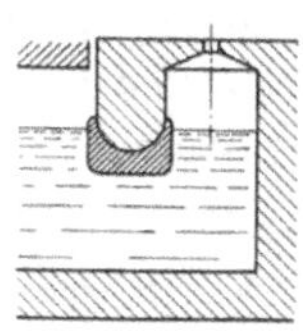

Fig. E 129. Loose bridge.

glass stream, and also for possible variation of the temperature at which one can take out the glass, a loose bridge of a certain cross section is shoved below the fixed bridge and presses up against the arch because of its buoyancy (Fig. E 129).

To limit devitrifaction tendencies in the drawing chamber various devices are used: slanting of corners, auxiliary bottom heaters, bottom and well insulation, back wall heating, etc. Since the drawing process cannot be continued beyond about 250 h because of devitrification in and at the debiteuses, one has to stop after this time and heat up the chambers. This reheating is done from the frontal parts or from one side, using gas or liquid fuel. To attain the necessary temperatures with producer gas recuperation is applied. The heat requirement for the reheating process is on the average 300,000—500,000 kcal/h depending upon fuel, chamber size, and reheating time.

Shape and dimensions of the debiteuse, particularly of the slit, are very important. They are functions of the viscosity of the glass at each place in the slit and, therefore, very dependent upon temperature. Since the temperature of the glass flowing towards the debiteuse is not the same along the entire extension of the slit, conditions are too complex to permit calculation and the most suitable shape of the slit is found empirically.

The swelling and rising glass should have minimal deviations in thickness across its width for all speeds; at least they should be within close tolerances. Obviously thick glasses need wide, and thin glasses need narrow slits. Slits 50—80 mm wide in the middle are conventional. At times attempts have been made to draw from 200 mm slits. The slit narrows towards its end 20—30 mm, depending upon glass temperature and viscosity. The transition from the usually parallel center must also be found empirically

There are a few patents aiming at improvements in the surface of the glass and in productivity through constructions of debiteuses allowing different flow patterns. The flow profile at the lower side of the debiteuse is also influenced by the radiation from the glass in the chamber to the ribbon leaving the slit. The arrangement of the debiteuses limits the possibilities for cooling the arriving glass and thus the production figures. During past years the optical quality of drawn glass has been much improved by changes in the composition of the debiteuse material and by the finishing and forming of the slit. The life of the debiteuse normally is 50—70 days, sometimes even longer.

5.1.1.4 Drawing Machines and Accessories. FOURCAULT's first drawing machines had pits with about 13 roller pairs and a height of 5 to 5.5 m. Most installations followed his example. But soon it was recognized that this height was insufficient, particularly at high speed and for thick glass. Consequently, round table after-coolers and even horizontal lehrs had to be used to minimize temporary strain.

The probable reason for not increasing the height of the pit was the desire to avoid the costly enlargement of buildings. Nor is the historical diameter of the rollers — 132 mm — technologically sound. In wide machines the rollers should be larger to give more flexibility in keeping the center of the sheet from sagging. Higher pits of 9—12 m and roller diameters of at least 170 mm appear desirable. But this could probably only be reached in new installations which do not depend upon existing cutting houses and roller inventories. At present pits are 5.25 to 6.50 m high, with cross sections (depending essentially on roller diameters) of 350 to 600 mm. The pit, of course, also follows the width of the glass ribbon, being about 200 mm wider to allow for ribbon and edge. The pit is constructed from cast iron, cast as true to form as possible. Front and side parts are screwed on. The pit is divided into single parts, the height of which is conditioned by the technology of their fabrication, by transportation limitations, and by the capability of lifting tools. Between each pair of rollers are two openings which can be closed by lids. In wide machines sometimes three openings are provided. The ledges and the rims of the lids should be machined to prevent the entering of undesirable air. There are also small openings on the front. These are closed by baffles. The pit has an inner asbestos insulation which is applied to the sheet metal for its protection. The construction of a drawing machine with the rollers and all drive aggregates is represented in Fig. E 130.

Fig. E 130. Drawing machine with Cardan drive.

The rollers consist of a steel core with high quality asbestos discs pressed on by spindles or hydraulic presses, with or without drenching in water glass. If larger roller diameters are admissible tubing can be used for the core and the asbestos discs can be swaged on.

The length of the cylindrical part of the rollers is calculated to allow the rims to safely run outside of the cylinder. The ends of the roller are either conically tapered from 132 mm to 120 mm, or end in a 120 mm diameter cylinder after a short transition.

The rollers are supported in cast iron or bronze sleeves; sometimes roller bearings are used (Fig. E 131, 132). At times the lower rollers are supported by water cooled bearings. Since all these bearings must be lubricated, and since all lubricants leave residues at such high temperatures, carbon or graphite sleeves are used which require no lubrication.

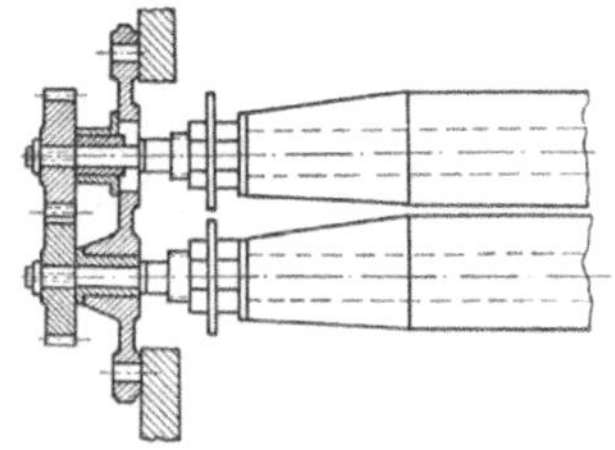

Fig. E 131. Slippage bearing of asbestos rollers.

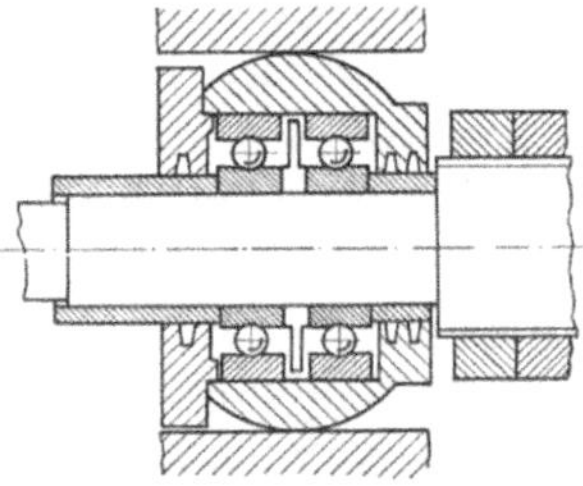

Fig. E 132. Roller bearings of asbestos rollers.

The ends of the rollers which run in these sleeves must, of course, be well polished. The bearing sleeves sit in bearing covers which are fastened on the front of the pit by screws and dowel pins. The openings in the front which close the bearing covers must, of course, be large enough to permit easy exchange of the rollers.

In earlier models the fixed rollers were driven by bevel gears and the loose rollers were moved along by a pair of cog wheels with coarse teeth. In this construction continuous engagement was not assured and a non-uniform run of the loose rollers was the consequence. For this reason a main drive shaft at some distance from the machine has been

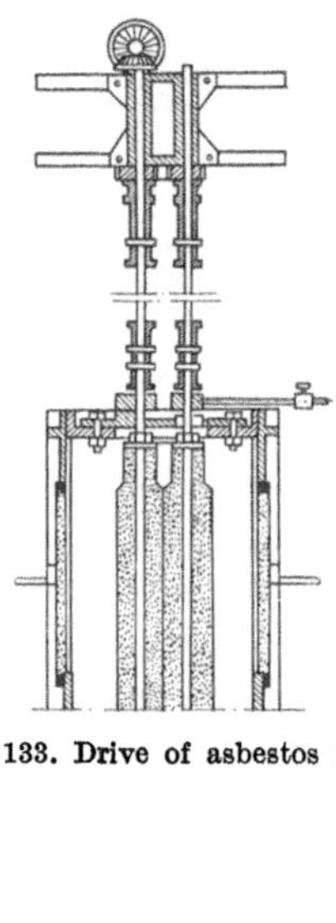

Fig. E 133. Drive of asbestos rollers.

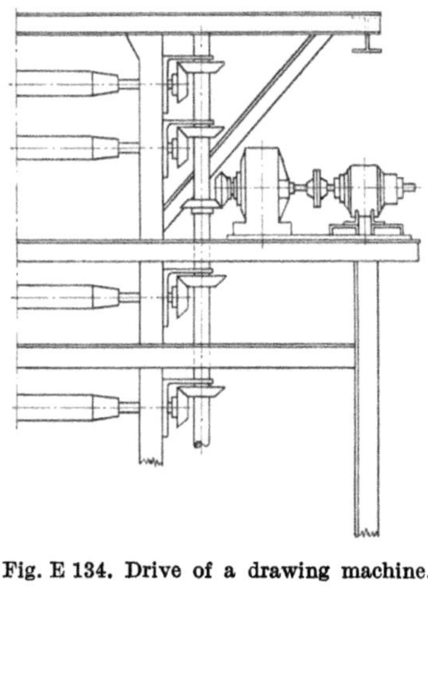

Fig. E 134. Drive of a drawing machine.

provided for some time; the rollers are connected with this drive by Cardan drive shafts (Fig. E 133). For kinematic reasons these Cardan shafts must have a length not less than a certain minimum.

To maintain the engagement of the cogwheel, slanted or spiral cogs are used. Even though the low rpm of the rollers and the low pressure on the bearings permit relatively crude fits, the increased drawing speeds used, particularly for thin sheet (<1 mm), calls for improved running properties of the rollers, which have been attained by the features just described.

The machines are driven almost exclusively by DC. The main reason is the necessary instantaneous standby power which only a battery can assure. Also the drawing process demands a large range with fine

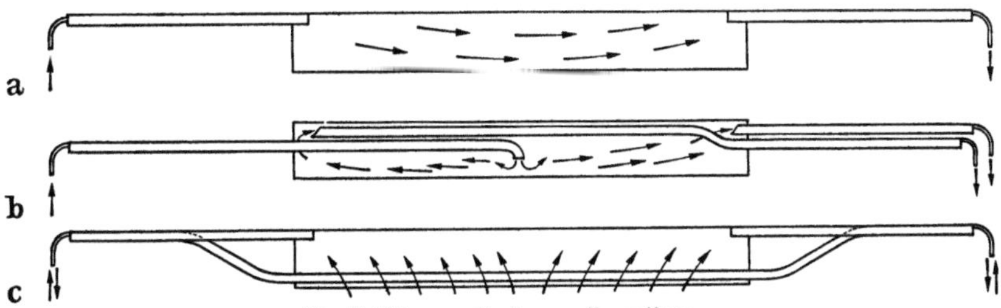

Fig. E 135a—c. Cooler configurations.

divisions from 1 : 14 or 1 : 18, particularly at the start. This is accomplished by voltage regulation or by field weakening. Fig. E 134 shows the arrangement of the DC drive with reduction gears at the drawing machine. Of course, other ways of regulation are conceivable, but these are not compatible with the rough glasshouse operation. It should be added that the current consumption of a drawing machine is relatively small (about 200—300 W). It is also nearly independent of the drawing speed, and is governed mostly by friction.

The reduction of the motor rpm to that of the main shaft can be achieved by suitable worm or gear wheel drives. The dimensions of all rotating parts including those of the motor must take account of the rough glasshouse operation and the long furnace life (at present up to six years). It is absolutely necessary to minimize repairs in the production units during the life of the furnace.

An important accessory is the cooler which is located just above the root of the glass ribbon to withdraw heat from the glass and to accelerate its solidification. Since heat transfer is almost exclusively by radiation and since it must be uniform across a large area the construction is not simple. Fig. E 135 shows some cooler shapes according to KITAIGORODSKI [3]. Water is the only coolant used. It either flows through the cooler, or enters in the middle and leaves at the side, or enters through special nozzles over all its length to flow out laterally. Water consumption is considerable (about 10—15 m³/h), with the water heating up by 10—15° per cooler pair. Depending upon drawing speed and sheet thickness 2 or 3 cooler pairs are used above each other.

At the upper end of the cooling pit the cooled glass is cut off. Various cutting devices are used which are driven electrically, hydraulically or by compressed air.

Fig. E 136. Temperature of glass sheet in the drawing channel.

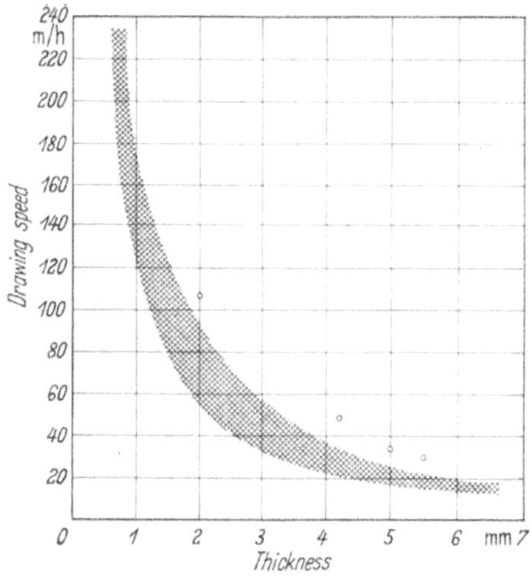

Fig. E 137. Drawing speed for Fourcault glass.

5.1.1.5 Machine Operation. Cooling in the drawing pit may be further adjusted by air currents in the lower part, and heat withdrawal in the upper part, by the opening and closing of holes. Fig. E 136 shows cooling curves which were obtained by pressing thermocouples into the sheet as it left the onion and allowing them to travel with the sheet.

5.1.1.6 Production and Quality. Sheet widths of 1400, 1800, 2300 and 2700 mm are conventional. These dimensions were not determined technologically. One simply wishes to cut standardized plane dimensions with a minimum of cutting loses. The usual drawing speeds are in the range indicated in Fig. E 137. From this the net production of a tank can be calculated depending on width of the sheet and number of machines. 25—35% should be used for losses through breakage, edges, cutting, etc. It is impossible to compare drawing efficiencies directly, because the drawing speed increases more than proportionally for thinner sheet. The reverse is the case for thicker sheets. H. Goerk proposed a factor of merit in which the product of drawing speed and glass thickness is used [6]. Goerk also investigated the effect of certain measures to increase drawing speed.

A characteristic defect of Fourcault glass are cords in the direction of drawing. Numerous devices and patents to improve the so-called optics of the glass have been developed in the fight against this defect. Examples are additional heat in the drawing chamber, blowing-in of cooling gases, etc. The quality has been improved considerably over the past ten years. The glass composition must be selected to control devitrification, speed of solidification, and durability. It is clear that a lower temperature causes an increase in viscosity and efficiency (see GOERK's paper).

5.1.2 Drawing without Debiteuse

The classical and conventional Fourcault process has its disadvantages: drawing cords and surface devitrification, both conditioned by the use of the debiteuse. These cause frequent interruptions of the machine. Processes which do not use the ordinary debiteuse may avoid those disadvantages, and the draw can still be vertical. Drawing without debiteuse slits is now essentially accomplished in two types of machines:

(a) Drawing chambers with deeply immersed bridges below the drawing location. There are some variants.

(b) Drawing chambers with shallow immersion of bridges at one side of the drawing location.

5.1.2.1 Pittsburgh Process. The Pittsburgh Process was developed in the USA and was introduced in Germany in 1931.

A refractory body immersed about 65—80 mm is pressed into the bath (Fig. E 138), the glass ribbon is freely drawn from the surface and is moved vertically upward, as in the Fourcault process. The refractory body allows more flexible temperature adjustment since it acts as a radiation shield. A slight inclination of the refractory body will influence the glass streams on both sides. In the larger chamber, usually 800—1000 mm wide, four pairs of coolers can be inserted on top of each other. This permits a working temperature higher than in the Fourcault process. The tendency to devitrification is decreased, and the Pittsburgh glass can therefore contain less alkali.

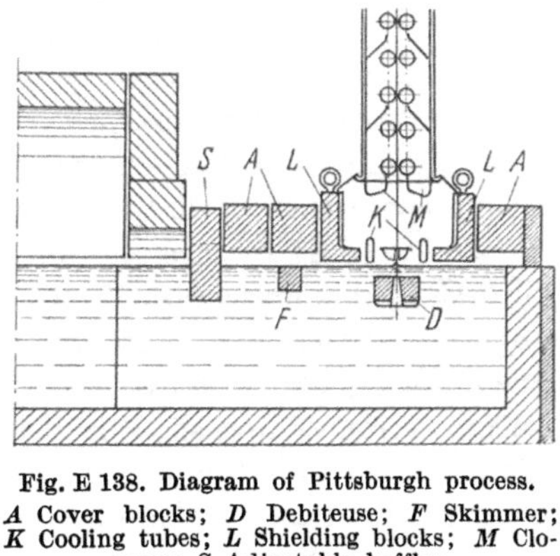

Fig. E 138. Diagram of Pittsburgh process. *A* Cover blocks; *D* Debiteuse; *F* Skimmer; *K* Cooling tubes; *L* Shielding blocks; *M* Closures; *S* Adjustable baffles.

The machine itself is similar to those described above. Sheet widths are 2 m, 2.40 m, and 3.20 m (with edge).

Drawing of glass thinner than 2 mm is difficult in this process. The quality is better than in the debiteuse process, but there are two disadvantages. Heating is more difficult, and the tank must be controlled more precisely.

5.1.2.2 Drawing Process with Fixed Bridges (Shallow Immersion). The direct drawing of glass from chambers with bridges in shallow immersion represents a further variant of the vertical processes. Cooling at the drawing location is attained by insulating bricks placed above the glass level. Variations in height and shape of such screens influence the viscosity gradient in the glass alongside the bath as well as vertically. The projected shape of the screening bricks is determined by whatever upward radiation is desired, and is the result of long operational experience. As in the debiteuse process edges are supported by tong-like clamps. Water coolers cool the glass ribbon above the drawing location.

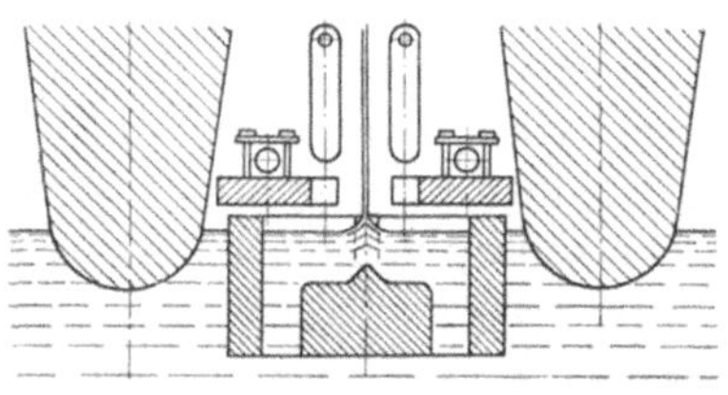

Fig. E 139. Drawing process with submerged debiteuse.

The glass is constrained from both sides by small submerged refractory plates of about 100—120 mm thickness and of the width of the chamber. To utilize homogeneous glass from deeper regions exchangeable bridges can be immersed to a total depth of 200—250 mm. In addition, refractory bodies are inserted below the drawing location to stiffen the glass where it is drawn and to regulate the streams from both sides. Because of the large influence of temperature variations this device is harder to operate. The running time between required shutdows is about 250—500 h, or about twice that of the Fourcault machine (Fig. E 139).

The process requires a quick setting ("short") glass with higher SiO_2-CaO and lower Na_2O content. The quality of the drawn glass is superior to that drawn from a debiteuse because of the avoidance of the drawing cord. Characteristic defects of this process are thin fine surface striae which show in reflected light. Production depends upon glass composition, the insulating power of the screens, the cooling conditions in the drawing chamber, the glass temperature, and the width of the slit formed by the screens. The following drawing speeds have been reported:

1.9—2.4 mm thickness: 100—110 m/h
2.7—3.3 mm thickness: 60— 70 m/h.

The reason for these high speeds is (according to Kitaigordodski [3]) the strong cooling of the glass mass which streams to the drawing process, always assuming good chemical homogeneity.

References

[1] WENDLER, A.: Maschinelle Glasverarbeitung, Leipzig: Akad. Verlagsges. 1929.
[2] KÖNIG, W., L. VON REIS, and R. SIMON: Flachglas, Leipzig: Akad. Verlagsges. 1934.
[3] KITAIGORODSKI, J.: Technologie des Glases, 2nd. Ed., Berlin: VEB Verlag Technik 1959.
[4] GOERK, H.: Neue Entwicklungen in der Tafelglaserzeugung. Silikattechn. 12 (1961) 213–217.
[5] KRÜSSMANN, G.: Stand des maschinellen Ziehens von Tafelglas. Glastechn. Ber. 34 (1961) 592.
[6] GOERK, H.: Die Ziehgeschwindigkeit bei der Erzeugung von Tafelglas nach dem Fourcault'schen Verfahren und dem düsenlosen Ziehverfahren. Informationsdienst Glastechnik Jan./Febr. 1962, pp. 1–8.
[7] LONG, B.: Einige technische Gesichtspunkte zur Herstellung von Tafelglas nach dem Ziehverfahren. Glastechn. Ber. 30 (1957) 288–291.

5.2 Libbey-Owens Process

By W. v. REIS, Gelsenkirchen

5.2.1 Development of the Process

IRVING COLBURN, a young man quite interested in engineering, was an associate of MIKE J. OWENS in the late 90's. OWENS had just developed his revolutionary bottle machine in collaboration with COLBURN's father and brother. In 1898 COLBURN began to occupy himself with the mechanization of the production of flat glass which was still mouth produced by the cylinder process. He first succeeded with a blow machine which produced glass cylinders which were cut open and flattened as in the manual process. He concluded, however, that this was not the best way to make flat glass, and worked toward a machine that would draw continuous sheet from the melt.

Before he met with final success he had drained himself financially to the point where he had to give up (1911). The bankrupcy of his company was announced. In 1912 his patents were acquired by the Toledo Glass Co. at the instigation of M. J. OWENS.

The realization of the patents proved a difficult task. Finally 1915 and 1916 brought him his first big successes. He succeeded in preventing the constriction of the drawn sheet, a surface tension phenomenon which has preoccupied FOURCAULT as well. The successes were so convincing that in 1916 the Libbey Owens Sheet Glass Co. was founded to exploit the process on a broad basis. In the following years daughter companies spread the machine throughout the entire world. COLBURN died in 1917 before he could enjoy the fruit of his labor. The process had already developed the basic form which has not been changed in essence to this day: A glass sheet is continuously drawn upwards between the divider channels of a melting tank from a forehearth, the "cooling chamber". Rotating edge holders prevent its restriction in the area of the drawing onion. At about 70 cm above the glass level the sheet is directed into the horizontal position by means of a guide roll. The glass travels through a lehr about 60 m long and leaves it at a temperature only warm to the hand. It is cut and released on a transport table (Fig. E 140).

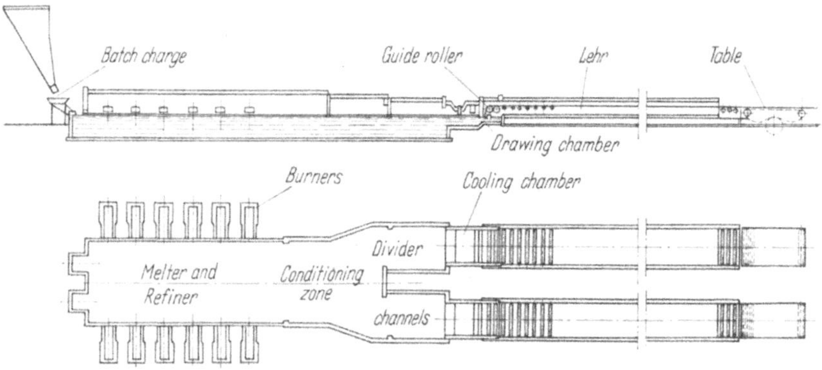

Fig. E 140. Total installation (tank, machine and lehr).

5.2.2 Construction

5.2.2.1 Furnace. The melting furnace does not differ in principle from other furnaces for flat glass production. Generally 5—7 lateral burners are used on each side, with 4 or more in operation depending on the program. Regenerators provide better heat economy.

The bath depth in the melting and conditioning zones as well as in the channels is usually 1500—1600 mm. Originally one drawing machine was used per furnace; later, two were provided for better economy. The melting tank is about 8 m wide, and becomes 11 wide in the conditioning zone. Operators sometimes speak of the refining instead of the conditioning zone. Refining, however, i. e. the removal of gas bubbles, takes place mostly in the melting zone, which might better be called the melting and refining zone. Through each of the two channels (Fig. E 140) the glass is led to a so-called "cooling chamber". The "cooling chamber" is a forehearth for the thermal homogeneization of glass. More for additional security than out of necessity, there is a three-part chain of floats at the end of the melter to deflect impurities in the surface. Frequently floats are also found at the beginning of the channel, at the end, or at both places. These, as well as the three-part float chain, usually consist of refractory material. Sometimes water cooled blocks replace these floats. All these blocks are so far away from the place at which the glass is taken up that their effect on the cleanliness of the glass is quite marginal. It is very important to keep the surface clean close to the point of use since, in contrast to the Fourcault process, the Colburn process uses surface glass. For this reason it has been proven essential to separate the furnace from all following parts by a blocking element immersed in the glass before it enters the cooling chamber, and to deflect contaminations swimming on the surface (Fig. E 141).

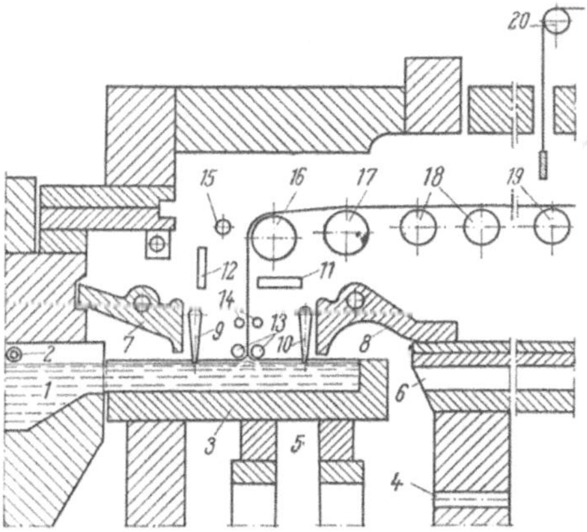

Fig. E 141. Machine construction (from lehr to drawing table).
1 Glass; *2* Burner; *3* Drawing chamber; *4* Burner; *5* Heating chamber; *6* Flue; *7* Front lip; *8* Back lip; *9* Front cooler; *10* Back cooler; *11* Roller cooler; *12* Vertical cooler; *13* Knurls; *14* Edge burner; *15* Long burner; *16* Guide roller; *17* Carrier roller; *18*, *19* Table roller; *20* Catcher.

5.2.2.2 The Machine. From the cooling chamber the glass enters the basin in which it is drawn, the so-called drawing chamber. Now the glass depth is only 150—200 mm. The drawing chamber consists of one single refractory block which is hard to fabricate because of its unusual dimensions: 2600 to 3800 mm (width) × 1000 mm (length). It would be simpler to fabricate it from several parts but the joints would be quite disadvantangeous. The refractory manufacturers do not like to produce these pieces, for understandable reasons. Flat glass producers must develop their own knowhow to make perfect drawing chambers. To avoid excessive glass cooling in the drawing chamber it must be heated from below. The burners *4* sit outside the heating chamber proper *5*. The waste gases give up their residual heat content and leave through the flue *6*. To prevent upward radiation two refractory pieces, the "lips", are provided. The English designations, front lip *7* and back lip *8*, are also used in German. The rising sheet now must be cooled strongly to avoid constriction and tearing. This is achieved by water cooled tubes or sheet boxes which in their normal position are arranged as in Fig. E 141: front cooler *9*, back cooler *10*, and vertical cooler *12*. The latter need not be present.

The roller cooler *11*, besides contributing to the cooling of the sheet, primarily affects the guide roller *16*. Of course none of the coolers touches the glass surface; otherwise tracks would originate. The cooling thus is effected by radiation. Where the sheet originates, at its onion or foot, knurls *13* are provided at each side through which the glass passes. These wheel pairs hold the edge, their speed being adjusted to guide the sheet tautly so that constriction is avoided. A separate drive adjusts their rpm to the speed of the sheet. To keep the edges malleable special "edge-burners" are provided (*14*).

Before the glass sheet is guided to the horizontal position a long burner *15*, reaching across the entire width of the sheet, reheats the glass

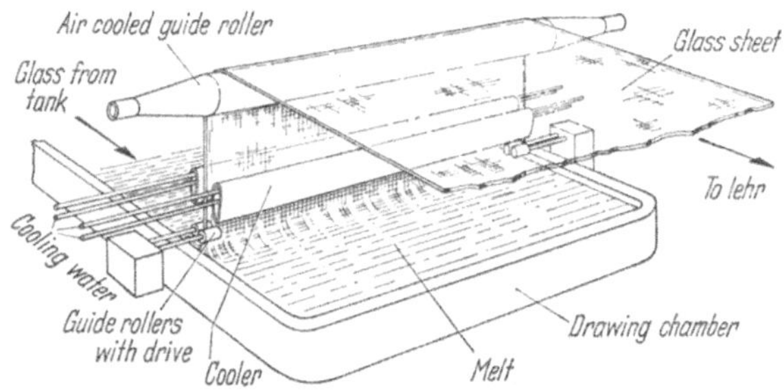

Fig. E 142. Drawing chamber.

to avoid breakage as it passes the guide roller *16*. This roller is the first solid contact on the glass and so must have a well-polished smooth surface. It consists of a special heat resistant chromium nickel alloy since it must operate at 700—800 °C. To avoid overheating and sticking it is cooled by the roller cooler and by air blown through its bore. The following carrier roller *17* must also be of high quality since the glass is still deformable at that point. The following table rollers *18* are similarly made from high quality materials.

The original construction of the drawing table did not provide transport rollers and was very complicated. Since the sheet used to break frequently, holding devices had to be provided to avoid the trouble of the remaining sheet no longer being drawn and then sliding back over the guide roller. An endless chain conveyor was used to take up the glass as soon as it left the guide and carrier rollers. A dragging chain above the glass loaded the glass to the extent that friction was exerted and the pull secured. This provision gave security in the case of a torn sheet. However, markings by the chain due to small irregularities were disadvantageous. The present system of table rollers *18* is much simpler and needs almost no maintenance. The introduction of these rollers became possible only when it was realized that the still plastic sheet could be obtained completely flat when properly transported over rollers.

The machine has a lateral enclosure to avoid contamination from the operating room. This can be achieved in two ways:

(a) The machine is covered completely and put under slight overpressure to prevent air infiltration.

(b) Another much used method is the application of blow crosses. They are attached at the two external sides of the machine in a way to blow the air way from the machine space. This creates a mist of air which prevents the penetration of dirty air and contaminating materials into the drawing zone.

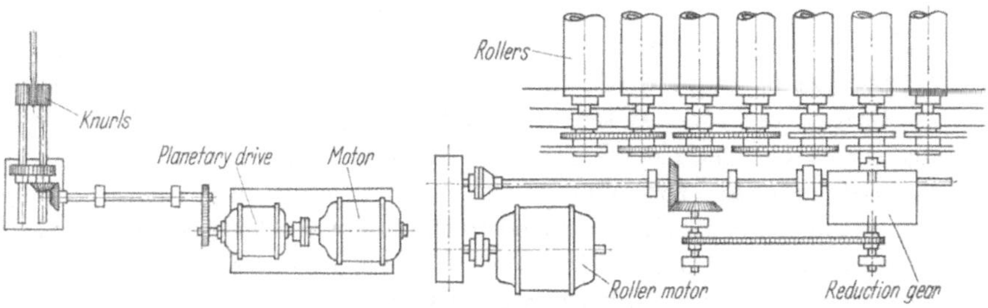

Fig. E 143. Driving arrangement.

To drive the machine (Fig. E 143) direct current motors are used. In case of current failure they are automatically switched to batteries to prevent interruption of production. Separate drives are provided for the knurls and rollers.

5.2.2.3 Lehr and Cutting Table. In a lehr about 60 m long, which should be particularly well insulated to exclude external influences, the glass sheet is transported towards the cutting table over about 200 asbestos rollers of about 15 cm diameter. A transmission shaft arranged parallel to lehr and table drives roller groups through reduction gears.

Heating elements are usually provided only in the first part of the lehr in such a way as to allow the establishment of various temperature gradients, corresponding to every glass thickness, over the length and width of the lehr. For temperature measurement the lehr is divided into 10—20 sections.

The cutting table, also driven by the transmission shaft, consists of a woodlined chain conveyor or a series of rubberized transport rollers. The wood or rubber prevent scratches. The following work phases can be performed on the cutting tables: edge cutting on both sides, and cutting to length, according to the desired dimensions. The cutting steps are fully automated. Then the glass is broken. The precut glass is transported to the cutting halls, the finished cuttings to the packing rooms.

5.2.2.4 Sheet Dimensions. To produce a sheet width of 3.00 to 3.60 m one needs a machine width of 5 to 6 m. The height of the machine including foundation is about 7 m, its length about 5 m; lehr and cutting table about 70 m.

5.2.3 Operation

5.2.3.1 Forming of the Glass Sheet. The greatest difficulty is the initial starting of the glass sheet. As soon as the glass in the drawing chamber has attained the proper viscosity, through proper heat input, one cautiously introduces into the glass mass a bait, a sort of rake made from sheet steel corresponding in its dimensions to the glass sheet, and covers it with glass. This bait is now drawn up by means of chains, copper bands and a drum (*20* in Fig. E 141). One has to be sure that no hard object touches the table rollers during this work. The smallest hit or scratch would mark the glass. Wooden boards introduced between the bait device and the rollers shortly before startup help to avoid damage. As soon as the glass sheet is lifted to the top of the hoist wooden pressure pieces are put on the sheet to assure its advance. The bait and the wooden pieces can then be removed because a sufficient length of the glass ribbon exists to maintain continuous contact with the rollers.

5.2.3.2 Glass Thickness. Drawing Speed. The desired glass thickness can be achieved by variations in the drawing speed: high speed for thin, low speed for thick glass. This dependence of thickness upon drawing

Table E 37. *Effect of glass thickness on drawing speed*

Thickness designation code	Thickness	Speed in %
1—1.2 =	1.1 mm ≙	158
ED =	1.9 mm ≙	100
MD =	2.9 mm ≙	58
DD =	3.9 mm ≙	40
4—5 =	4.6 mm ≙	35
5—6 =	5.6 mm ≙	29

speed is not linear. At larger thicknesses relatively more speed is sacrificed and the amount of glass drawn is reduced.

If the production is called 100% at ED (usual thickness = 1.9 mm on the average), the values of Table E 37 are obtained for other thicknesses. One notes the strong decrease of speed with thickness.

The absolute drawing speed is subject to local conditions and specifications so that only average values can be given. According to DAEVES

[2] the dependence of the drawing speed on glass thickness may be represented by the relation:

$$V \cdot d^n = K$$

(V = drawing speed, d = thickness, n, K = constants determined by local conditions).

For economical reasons a high drawing speed is, of course, desired. This may be achieved by the strongest possible cooling in the machine and the rising sheet. But there are limits: the glass may become so cold that it can't be brought over the guide roller and to the table without danger of breaking, or the surface is affected to the point that the optical quality suffers. A compromise must be found.

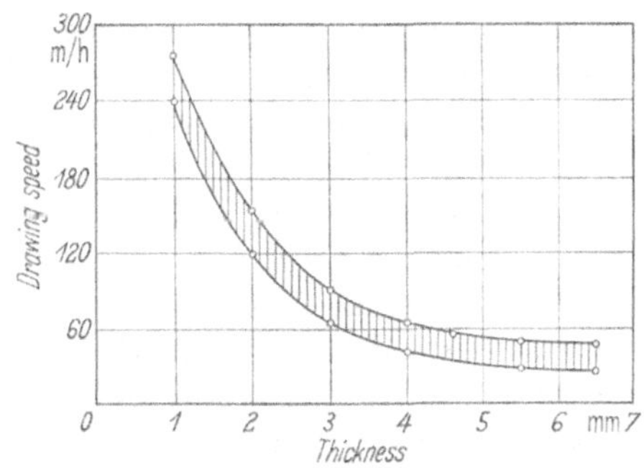

Fig. E 144. Output diagram.

5.2.3.3 Personnel and Power Requirements. One operator supervises and controls the machine. A chief operator is responsible for two machines attached to a tank and for the lehrs. The power requirements of a Colburn machine depend upon glass thickness, drawing speed, and running time. The latter comes in because during the 3—5 years life of the machine some parts cannot be refurbished and friction losses increase. 15 kW/machine may be regarded as normal.

5.2.3.4 Influences on Glass Quality. Glass defects influencing quality may be arranged in three groups: bulk defects, surface defects, thickness deviations.

5.2.3.4.1 Bulk Defects. Bulk defects primarily arise in the tank, only in exceptional cases in the machine. Today temperature is so well under control and the refractories of such high quality that such defects have been reduced considerably compared to earlier times. However, it is impossible to suppress in the machine bulk defects deriving from the melt.

5.2.3.4.2 Surface Defects. ***Markings*** arise when the polish of guide and other following rollers is no longer perfect. Also bad temperature control of rollers may be responsible. ***Foreign matter*** on the surface gets into the machine in the form of dust particles or may be caused by dropping refractory particles in the machine linings. Fuel and combustion air for the drawing onion and the machine should be absolutely free from foreign matter (see 5.2.2.2). The ***optical quality*** may be defined in many ways as it is judged subjectively. Generally the following is true: the transmission of the pane should be disturbed as little as possible. Striae, undulations, and mirroring defects are the most frequent. Among the causes are:

(a) Tank origin: compositional or temperature inhomogeneity of the glass;

(b) Machine origin: faulty cooling or faulty temperature control.

5.2.3.4.3 Thickness Deviations. These give much more trouble than in the Fourcault process because of the absence of a debiteuse which tends to keep the thickness constant. The crucial condition is homogeneity of the glass in composition and temperature. This uniformity must start in batching and melting. A uniform temperature distribution must be attained in the lehr. By skillful control of burners and coolers in the lehr, the heating chamber, and the use of a special burner below the front lip, the last nonuniformities may be smoothed out. Cooling air may also be adjusted. Many of the influences listed here are not symptomatic of the Colburn process only but will appear in equal or similar form in any glass process and for similar reasons.

5.2.3.5 Glass Composition. The composition of the glass must be adapted to the process. Solidification should progre ssjust fast enough to permit bending over the guide roller without danger of breakage. The following composition has been found best:

	SiO_2	Al_2O_2	Fe_2O_3	CaO	MgO	Na_2O	K_2O
%:	72—73	0.6—1.2	0.05—0.12	8—10	3—4	13—13.5	0.1—0.6

5.2.3.6 Special Features. There are several particularly advantageous features of the Colburn process. The machines permit very high drawing speeds and therefore, very high outputs. Per unit, 50—70 t/24 h may be drawn. The machine life may be up to $1^1/_2$ years. Because of the horizontal arrangement of lehr and cutting table very large dimensions can be made. Maximum width is 3.6 m. The length would be unlimited were it not for limitations of buildings and transportation. Lengths of 5 to 6 m are normal. Thickness, too, can be varied widely. Glass from 0.6 to 20/30 mm may be produced without any particular difficulties.

The long lehrs permit quite strainfree cooling resulting in easy cuttability. The redirection to the horizontal by the guide roller is a unique feature of the process. It causes a smoothening of the backside which thus excels in its planeness. This is a great advantage for making mirrors and multiple safety panes where the backsides are silvered or turned outside. Disadvantages are: the difficult operation, the requirement of special refractory pieces, and the high edge loss.

References

[1] KÖNIG, W., L. VON REIS, and R. SIMON: Flachglas, Leipzig: Akad. Verlagsges. 1934.

[2a] DAEVES, K.: Zur Methodik der Großzahlforschung mit Beispielen aus der Glasindustrie. In: Der Werkstoff Glas und seine Bedeutung in der Gegenwart (Vorträge des Hauses der Technik, Essen, Reprint No. 492), Essen: Vulkan-Verlag 1949, pp. 10–15.

[2b] ALLOLIO, R.: Großzahlforschung in der Glasindustrie. In: DAEVES, K.: Rationalisierung durch Großzahlforschung, Düsseldorf: Verlag Stahleisen 1952.

[3] LONG, B.: Sur quelques aspects techniques de la fabrication du verre en feuilles par étirage. Verr. et Réfract. 11 (1957) 281–286.

6 Rolled and Cast Glass

By H. SPIEKERMANN, Düsseldorf

6.1 General and History

The name rolled glass now includes the flat glass variations designated as wired glass, figured wired glass, polished wired glass, translucent wired glass, rolled sheet glass, figured glass, cathedral glass, and horticultural cast glass (Table E 38).

These may be recognized in raw sheet by the more or less distorted view of objects seen through them. The distortion is present because of the type of surface (figuring) produced in the hot rolling operation, in spite of inherent transmissivity. The surfaces are not necessary decorative in character alone, but may serve to cause optical effects as well (diffusion, reflection, absorption, refraction). Special products with special profiles may be rolled. It is characteristic of these glasses that many transitions between the extremes of strong and weak diffusion can be obtained. The designation "rolled" gains over the older word "cast" because rolling was and is the main characteristic of most products of this class.

The proper conditions for the industrial production of rolled glass arose in the 17 th century. As late as at the end of that century large quantities of flat glass were made by blowing a hollow cylinder with a glass maker's pipe which, after removal of cap and bottom, was cut open lengthwise and then stretched and smoothed. In complete departure from this blow process the glassmaker LUCAS DE NEHOU in 1688 first emptied pots containing liquid glass on a flat table of iron or bronze. He "cast" the glass and then shaped the plates directly by an iron or bronze roller moving over the cast glass. Even though LUCAS DE NEHOU just intended to improve the existing fabrication of mirror glass, particularly to obtain larger sizes, he unintentionally laid the foundation of the present-day rolled glass industry. His glass was the first which, though

Table E 38. *Rolled and cast glass*

- Plate glass, cast glass
  - Wired
    - Both surfaces smooth
      - Wired glass
    - One surface figured
      - Figured wired glass
    - Both surfaces figured
      - Translucent wired glass
  - Not wired
    - One surface figured
      - Corrugated or wavy glass, 6–8 mm, 8–10 mm thickness × 102–126 mm width and 360 cm, 420 cm, or 450 cm length
      - Rolled sheet
      - Light diffusion glass
      - Figured glass and cathedral glass
      - Horticultural glass, around 3 mm, 3.8 mm, 5 mm thickness; and in standard sizes
    - Both surfaces figured

transmitting some light, had a translucent effect because of the rough, non-parallel surfaces. From the first it was sometimes used without subsequent grinding and polishing. Originally the glass obtained its figured surface unintentionally by quenching against the cold but smooth table. About 1850 ribbed and facetted glass with similar optical effects was developed in England. The glass could be figured on either or on both sides, the latter by engraving both the roller and the table. Later attempts were made to improve quality and to remove the difficulties caused in production by some features of the casting tables. They were not only costly in construction and maintenance but also hard to manipulate and control: They warped easily, became leaky, and were uneven in heat transfer. Also the duration of contact between glass and table was too long.

It soon became apparent that on the same plate the surfaces of the "table glass" differed; the part having the shorter contact time with the roller was better than the one formed against the table. This suggested the replacement of the table by a second roller to obtain both surfaces of better quality. In due sequence roller processes of various types originated. Their common principle was that the glass was shaped by rollers only and that the rolled glass was transported to a table which had no forming function. It merely served for transportation of the ready glass to the lehr. Such rolling machines were particularly successful in the production of very flat, thin plates, and also stimulated the manufacture of all types of figured glass. The first attempts to make wired figured glass were made in Germany in 1857. Difficulty in finding wire which was both thin and strong enough to be inserted into the fluid glass delayed successful development and marketing until 1891. The differing chemical and physical properties of the two materials had no disadvantageous effects on the end product. It is not surprising that the proper insertion of the wire required more effort. Therefore various possibilities developed. It became, for instance, possible to insert wire sections between *two* separately and simultaneously rolled glass ribbons which were then rolled together with the wire inserts. In contrast a process was developed in which the wire section was rolled into *one* forming pane using additional glass for smoothening and closing furrows caused while the inserts were pressed in by rollers.

The improvement of the physical and chemical properties and the investigation of the behavior of cast glass in its practical applications led to the recognition of the broad applicability of rolled glass — like that of other flat glass varieties as — a building and working material in modern architecture and construction. Examples are facings, improvements in illumination, fire protection, etc. The effect of expert efforts and studies was that increased demands could only be fulfilled by the development of a continuous process. These efforts were successful and started an important period in the development of cast glass. The most significant results were the sandwich (Appert) process and the continuous rolling process developed under designations such as Dauerwalzen, Band, and Kontinue process. The latter process (continuous rolling) will be discussed extensively in the following section. Other process ideas have not gained footholds: the idea to let a glass ribbon stream from the bottom of a tank and introducing the wire on top of the slit where the thickness of the glass pane is formed; or the idea of a special apparatus to introduce the wire insert at the debiteuse in a debiteuse drawing process.

6.2 Description of the Continuous Rolling ("Band"- "Kontinue"-) Process

In the development of the continuous rolling process experience with earlier processes was helpful. The method of inserting the wire is particularly important. The principal method of manufacture in the

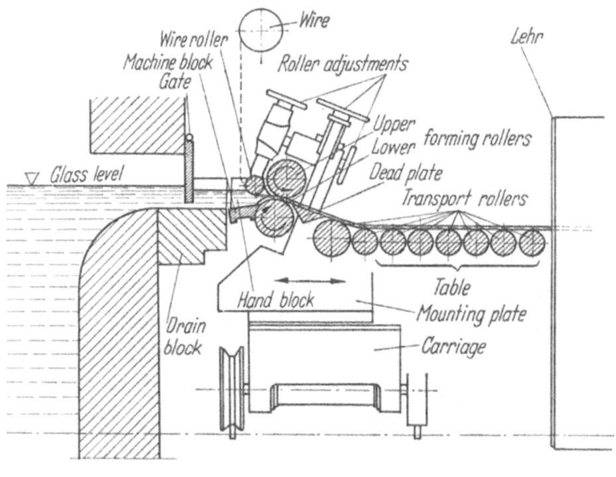

Fig. E 145. Outlet opening of the tank with the rolling machine.

United States today is to introduce the wire from below just before forming. The wire is worked in while the glass is rolled. The method of inserting the wire from above, just before the roller, is less used today. In the continuous process described here the glass melted in a continuous tank flows toward the rolling machine, at the elevation of the rollers, from a lip in the tank (Fig. E 145). The position and length of the lip is arranged so that a smooth, continuous flow of glass to the roller results. Thickness changes are made by adjusting roll spacings. Until the rolls start operating the lip is closed by a gate which is opened to allow glass flow when the rolls operate. The top roll acts as a dam holding back the fluid glass. The center lines of the rolls are usually vertically aligned for figured glass, but are often inclined as much as 45° for wired glass. The continuously flowing glass goes between the two rollers, which operate in opposite directions, thus moving the sheet toward the lehr. The wire mesh covers from above and is pressed into the glass just before it is seized by the rollers. This is accomplished by mobile or fixed cooling devices located in front of the rollers. The still plastic glass-wire ribbon is led to the lehr by water cooled rollers called apron rolls. These may be part of the machine or part of the lehr and serve the same functions as the table in earlier days. A water cooled plate is used between the forming rollers and the first transport roller. This is called a dead plate and has the function of chilling the figure on the bottom surface so that it will not be subsequently marred by the lehr rolls.

After the ribbon has gone through the lehr it reaches the cutting device where it is cut to specified dimensions. It will then be packed, loaded, stocked, or finished.

6.3 Construction (Fig. E 147)

The hollow forming rollers are made of iron and are water cooled. The bearings are supported on roller stands. Adjustments are provided by means of screws. Drives are connected to the roll shafts by chain

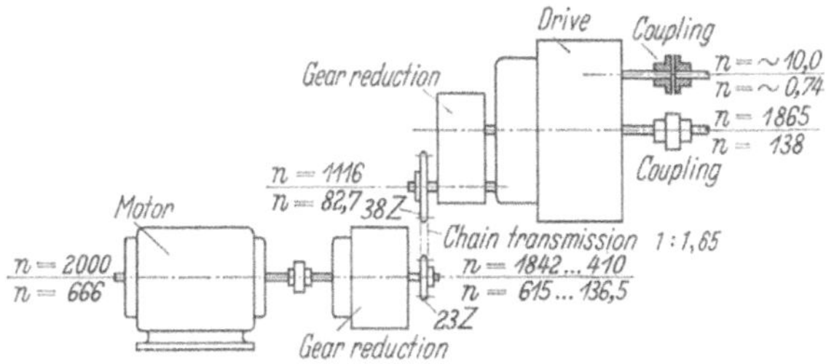

Fig. E 146. Drive arrangement for a rolling machine.

or other couplings. The relatively slow rotation of the rollers requires a high gear reduction and a means of varying speed, such as a stepless transmission (Fig. E 146). Depending upon drive and gear used, upper and lower rolls can be controlled at differing speeds.

The cooling water enters the roll through a rotating joint and into the hollow shaft. Water is fed through the rolls and out of a similar rotating joint at the other end. The temperature of the outgoing water is measured and controlled by varying the water flow. In the event of a current failure a hand crank permits at least temporary operation of the rolls. The mounting plate has space for equipment for water

[Ref. p. 369

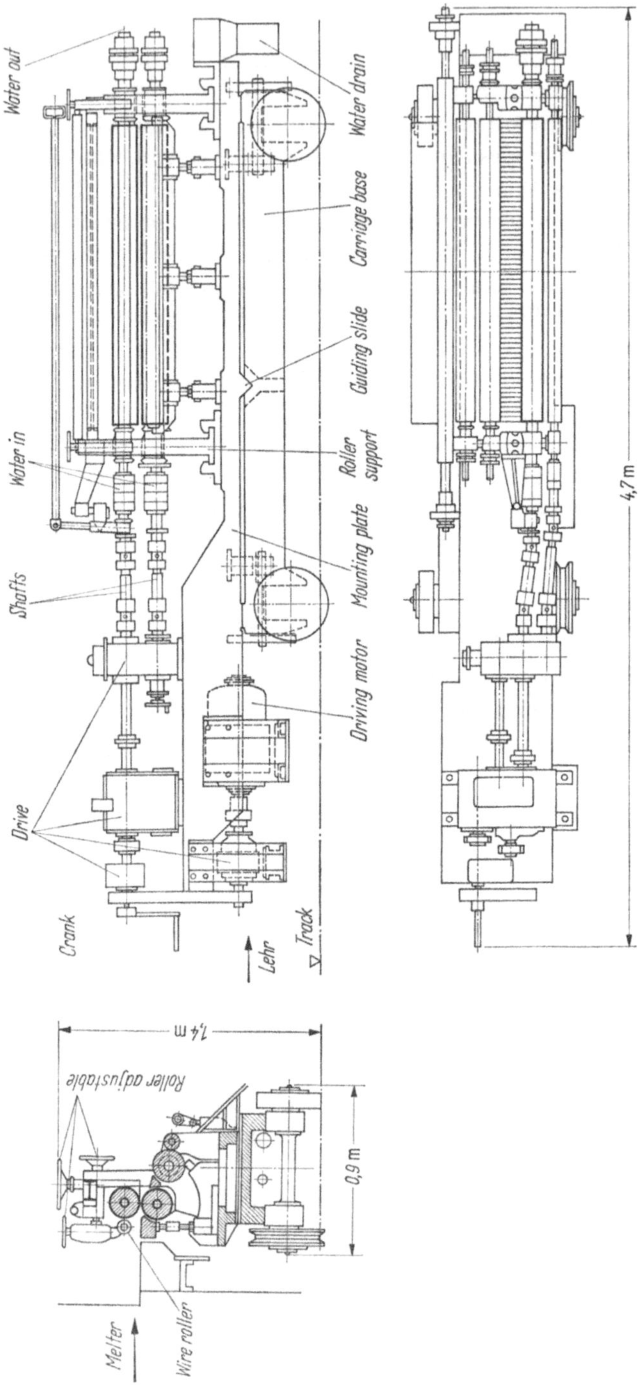

Fig. E 147. Rolling machine.

removal, cullet cleaning, lubrication, installation of instruments, safety equipment, etc. It rests on a low cart frame in guide rails. In these rails the entire glass rolling machine may be moved either toward the melting tank or toward the lehr. Other adjustments are provided to raise, lower, or align the rolls. The rails permit the complete set up of the glass rolling machine, away from the tank, even while another is in service. Quick changes of pattern are possible in this way. Water, air, electrical, and drive connections are manifolded or so arranged as to reduce changeover time to a minimum. Since the rollers must be adjusted in their relation to each other, flexible couplings are required between drive and rolls.

For the production of polished wire glass two forming rolls with smooth surfaces are used to reduce grinding and polishing to a minimum. It is also possible to have both glass surfaces figured, with or without wire. This is done by having both rolls figured with the desired patterns. The wire is supplied in the form of a welded net of crossing wires, or as a weave with four or six sided meshes. Rolls of wire mesh having about the same width as the glass are fed into the glass ribbon. Some shrinkage must be anticipated, particularly in non-welded inserts. The rolled length is variable. When a roll is exhausted the next one is hooked on without any interruption of production. Usually the wire is first guided by a guide roll above the roller mill proper and brought into the glass — just before the forming rollers — by letting it run around a little rotating roller which dips into the glass stream and allows horizontal and vertical adjustment. This (pressing) roller is usually not smooth but has a spiral-like groove which encourages the flow of the glass, influences the pile-up in front of the forming rollers, and helps to remove bubbles. The speed of this little roller can be adjusted with that of the forming roller independently. While this rotating accessory roller is found frequently, there exist also fixed devices which press the wire into the glass shortly before the glass enters the forming rollers.

6.4 Machine Sizes

The size of the machine generally depends on the desired width The speed of the glass ribbon is determined by the ability of the tank to melt or the lehr to anneal the desired glass production. One tries to select the width of the machine so as to obtain maximum economy in relation to the sizes demanded by the market. At the present time these sizes are still so diversified that compromise solutions are necessary. A preferred pane dimension is 2.00 to 2.10 m length × 1.26 m width. The determining factor is often a width suited for wired glass as well as for plain figured glass. For this reason one finds many machines whose roller length allows a 1.26 m finished width, but also some of greater width. The overall dimensions for a rolling machine, producing a ribbon of 1.26 m in width are about 5 m wide × 90 cm long × 1.6 m high. Approximate weight is between 3 and 4 t.

6.5 Personnel Requirements

The continuous process has spread to most countries producing rolled glass, particularly all over the European continent. For maintenance and operation, normally four persons are required. One serves as machine operator and is responsible for the machine in operation, including maintenance of thickness, width and speed of the glass ribbon. The assistant operator watches lehr temperature, and readies auxiliary materials for the operator and the machine, such as wire inserts. There is a foreman for supervision, and a maintenance man for taking care of small deficiencies in the operation. This does not include personnel used to operate the melting tank, make the batch, work in the shop, etc. There are also cutters, packers, crane drivers, loaders, and occasionally repair men for the rollers.

6.6 Roller Interchange

Whenever rollers must be interchanged because of change of type, production must be interrupted while the machine is taken from its location. Today it is generally not possible to interchange during operation. Either the machine taken out is allowed to cool, and the interchange takes place on this same machine, or a second machine is ready with the proper rollers already installed. The latter procedure is recommended because of the shorter interruption. In the most favorable cases, implying such a reserve as well as the absence of other disturbances, the stoppage of glass flow, moving out and in, and restarting should not take more than 60—90 min. But additional production is lost since the newly started ribbon is at first not usable.

The total run of a machine is very different from case to case, since more than technical factors determine it. Therefore unequivocal data cannot be given. Long runs promoting high efficiency are desirable, but the great variety of types now required necessitates frequent interruptions.

The life of the rollers also varies tremendously and unequivocal data again are not available. If rollers are not defective to start with, and if they are not damaged in operation, the wear is quite small and a life of 1000 h can be attained easily. This depends upon pattern, temperatures and maintenance.

One must decide whether damage or wear spots can be taken care of by simple repair work. One also has to judge whether small surface changes in particular cases, say in figured glass patterns, impair the usefulness of the end product. But the most decisive factors are condition and quality of the rolls as delivered.

6.7 Manufacture

The upper and lower rolling speed of the rolling machine is determined by the tank melting capacity and by the length of the lehrs. For proper annealing a fixed tonnage is usually drawn so that thin glass is rolled more rapidly than heavier glass. Thicknesses vary from 3 to 10 mm for wire, raw and figured cast glass. One must take into consideration thickness of the glass, deformation in the lehr, and too cold arrival at lehr for proper strain removal. The sticking limits of the roller material must not be exceeded and the patterns must be transferred to the glass free of distortion. In practice, these factors as well as the economy involved in the length of the lehr, have resulted in speeds between 0.8 and 4 m/min. the lower limit applying to the thicker types. These are only guide values. Considerable differences may be found if some factors, such as the length of the lehr, show unusual deviations. Power requirements are relatively low. In general the described rolling mills are driven by 4—5 kW motors.

Since it is a characteristic of rolled glass to appear on the market in many types and shapes, it is inevitable that the continuous run must be frequently interrupted for the production of one or another of the many patterns. Any simplification would mean progress. Increasing standardization of sizes would also be of great advantage for the construction and efficiency of rolling machines.

References

[1] SCHULZ, H.: Die Geschichte der Glaserzeugung, Vol. I, Leipzig: Akad. Verlagsges. 1928, p. 73.
[2] SPEIKERMANN, H.: Erweitertes Gußglastabellarium, Schorndorf bei Stuttgart: Hofmann 1958.
[3] KÖHNE, K.-H.: Die Herstellung von Drahtglas. Glas-Email-Keramo-Technik 10 (1959) 1—6, 46—49, 89—93.
[4] SIMON, R.: Maschinelle Herstellung von Drahtglas. Z. VDI 68 (1924) 509 to 528.
[5] SPIEKERMANN, H.: Neuzeitliche Draht-, Roh- und Ornamentglasherstellung. St. Lucas: Allgemeine Glaserzeitung (1952) 305—309, 319—322.

7 Blowing, Centrifuging and Drawing of Glass Fibers

By H. ALBRECHT, Düsseldorf

Glass filaments and fibers today are produced primarily by blowing, centrifuging, or drawing. The classification is not always simple since combinations of two processes are also used. In this area, the product of the first process may become the input materials to the second process.

All these different methods coexist producing considerably different products which can be adapted to the technical requirements of a specific end use. The differences are mainly concerned with the type, length, diameter, and arrangement of the fibers; and the temperature resistance, purity, and composition of the glass used. According to P. A. Koch, German technology defines as glass fibers the relatively short, thin, strong glass structures. The thin, practically infinitely long glass structures, however, are referred to as single filaments. (This distinction is not as definite in the U.S.A.)

Fibers and filaments may or may not be twisted into a yarn. This depends upon their diameter and their physical arrangement. Single filaments can be twisted if their diameter is below 12 microns and if several filaments are combined with a suitable sizing into a strand that can be unwound without difficulties from a spool or tube. The end product will be twisted textile yarn. Glass fibers of equal diameters can be twisted into a yarn, if they exceed a certain minimum length, have small fiber-to-fiber diameter variations, and are parallel to each other. These are combined with a suitable sizing into a tow of sufficient strength that can be unwound without difficulty from a tube or spool. These fibers are called staple fibers and are twisted into staple yarn.

Glass fibers that do not meet the above requirements are not suitable for textile applications. Due to their random arrangement these fibers enclose large volumes of air and therefore are effective thermal and sound insulators. They are designated as non-continuous fibers and their products are known under different names, such as glass wool, glass cotton, etc.

For reinforcement of plastics, only textile fibers (continuous filaments) are in use. Insulating type products have not yielded a sufficiently uniform improvement in the strength of plastic materials.

7.1 Blowing

7.1.1 Owens Steam Blowing Process for the Production of Non-Continuous Type Fibers

The efficiency of this process is based on the production of large quantities of long fine fibers with a minimum of technical effort. The largest part of all insulating type glass fibers, glass wool, glass fiber felts, and sheets for thermal and sound insulation, is produced by this process.

The glass batch is mechanically fed to a melting tank, where it is melted at temperatures of 1400—1500 °C (Fig. E 148). From the frontend of the melting tank extends the so-called forehearth. The refined glass is introduced through a flow channel into the forehearth, in which a constant glass level is maintained. The bottom of the forehearth is constructed to hold two rows of bushing blocks, which are provided with slot-type openings. Bushings fabricated from a platinum-rhodium alloy, and set into a thermal insulating refractory, are fastened to the bottom of the bushing block. The bushings are heated electrically to keep the glass in the bushing, and especially in the tip orifice, at the temperature required for fiberizing.

Mounted below each bushing is a steam blower, which fiberizes the glass stream flowing from the tips of the bushing. The fibers fall through a spout into a hood. The action of the reduced pressure produced by the high volume blower evacuates the steam and air from the hood and deposits the fibers uniformly onto a conveyor chain belt which transports the fibers in the form of a felt or mat from the hood. Inside the hood, steam and air cool the fibers rapidly, while a steam injector sprays binder onto the moving fibers. The binder envelops the fibers and causes them to adhere to each other. A special oil is used for lubrication of white wool. In the manufacture of the more rigid glass fiber boards and mats the oil is replaced by a plastic binder, e.g. phenolic resin. After leaving the hood, the resin sprayed or saturated fiber deposit is conveyed into a drying oven, is compressed and the binder is cured. The material is then trimmed, cut to shape, and packaged. The mat is first cut into parallel strips which then are divided into single boards.

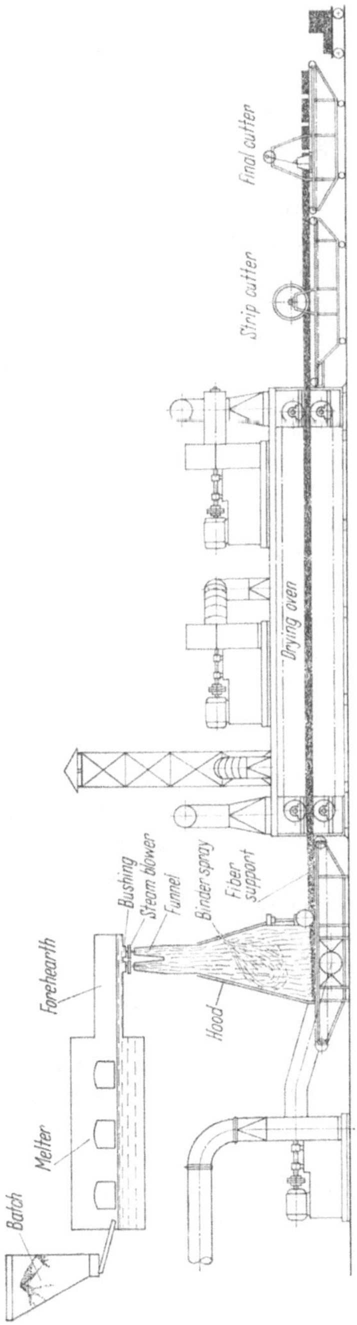

The best material for the bushing has proven to be a platinum -10% rhodium alloy. The rhodium content improves the mechanical strength, temperature stability, and chemical resistance. The weight of a bushing used in the production of insulating wool is approximately 1100—1300 g. The trough shaped bushing body has a V-shaped cross section with a flange attached to the upper end. The flange serves as a seal to prevent glass leakage between the bushing and the bushing block. The glass seal is improved through water cooling, built into the refractory block immediately below the bushing flange. The ends of the bushing are equipped with terminals, serving as contacts for the electrical connectors. Uniform temperature distribution over the bushing body and increased heating at the tips is obtained by varying the thickness of the platinum which affects current density. The temperature conditions necessary to obtain uniformly thin fibers are rigorous. To assure good fiberization up to 55 tips are pressed out of the reinforced bushing bottom. They are provided with 1 to 2.2 mm bores. The thin glass streams flowing from these tips are fiberized by the steam blower. The tips keep the glass from spreading along the bottom of the bushing and prevent the combining of several glass streams. The diameter of the bores is a determining factor for fiber diameter, length of single fibers, and bushing productivity. The entrance to the bushing is protected by a perforated metal sheet serving as a strainer to hold back impurities that would clog the tip bores. The bushings are heated electrically by forming a comparatively high resistance in a circuit of about 4000 A. Electrical connections are established by flexible busbars at the transformers, water cooled clamps, and connectors at the bushing. The bushings are either connected in series using a single AC transformer, or each bushing is heated separately by stepless control from one transformer, or, finally by ring core regulator-transformers with 10—15 coarse taps. A variable transformer connected to the line side of the main transformer can be used as a fine control of the current between each coarse step. This permits practically stepless control. The transformer is rated at 15—17 kVA (transmission ratio 380/4 V and 46/370 A). This rating applies to the operation of one bushing.

The shape of the steam blower (Fig. E 149) is important in the production of glass fibers [3]. Under each bushing is mounted a steam blower with special devices controlling its exact alignment relative to the bushing. Essentially the steam blower consists of two steam chambers connected through spacers and end pieces. One of the end pieces is equipped with steam fittings. The high pressure steam is distributed uniformly over the chambers and it exits approaching the velocity of sound through small slots at the inner sides of the chamber. In these slots the steam is aimed at an acute angle, approaching a flow parallel to the glass streams discharging from the tips. The steam mixes with the induced air, divides the glass streams into many filaments, and attenuates the viscous glass into fine finite filaments through the acceleration it imparts. The fiber diameter can be varied between 3 and 30 microns, depending upon the end use, but there is always a certain spread in diameter.

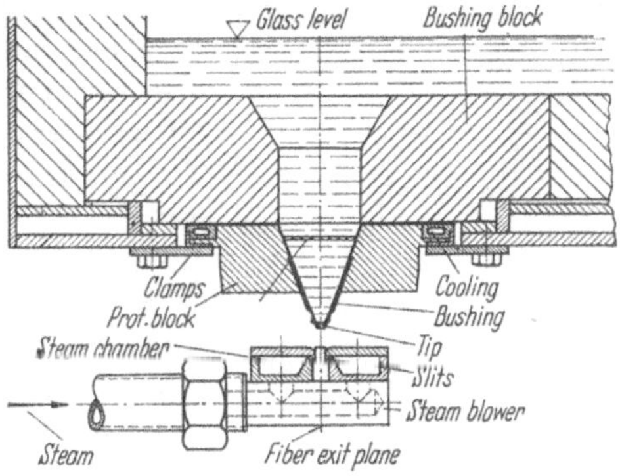

Fig. E 149. Bushing steam blower.

Actual installations show the following requirements and production figures:

Size of installation	12—90 t/24 h
Number of bushings	4—18
Tips per bushing	36—55
Tip bore	1.2—2.2 mm
Steam pressure	9—14 atm
Steam quantity	200—300 kg/h
Mean fiber diameter	5—14 microns
Mean fiber length	80—100 mm

7.1.2 Steam Blowing Process for the Production of Textile Staple Fibers

This process for the production of staple fibers corresponds to the one just described, except that glass of higher viscosity is processed and a lower steam pressure is used. This permits the attenuation of long fibers with smaller diameter variation. An electrically heated platinum rhodium alloy bushing (Fig. E 150) is uniformly supplied with glass marbles or cullet. This raw material is then melted in the bushing. The thickwalled bottom of the bushing has 35 tips with holes varying from 0.8 to 1.2 mm in diameter from which thin glass streams issue, which then are drawn into fibers by the steam blower. If suitable blower skirts are used the calculated speed is 250 m/sec for fibers 50—350 mm long. As the fibers leave the vicinity of the steam blowers or blower skirts, they are sprayed with a binder which improves the bonding of the fibers to each other and avoids abrasion when the fibers contact each other. Subsequently the fibers are dried and collected on a perforated drum. This is accomplished by means of a reduced pressure generated by a blower acting upon one segment of the receiving drum. After leaving the collector the fibers are drawn off the drum in the form of a web; at the same time they are oriented approximately parallel and drawn into a sliver. To improve subsequent processing, projecting filaments are moved parallel to the sliver surface as the sliver is pulled through the hollow shaft of a fast rotating motor (Nitschel-motor). The sliver is wound onto a spool in crosswise layers. Textile technology is applied for further processing (twisting, plying, weaving, etc.). The productivity of a bushing producing fibers in the diameter range from 8 to 10 microns is between 1.25 kg/h to 2.5 kg/h.

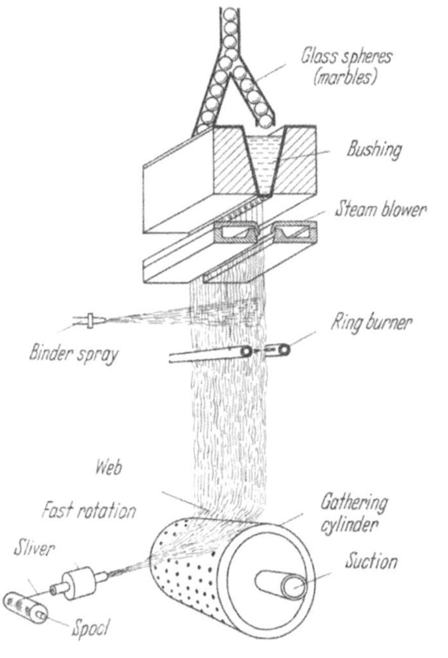

Fig. E 150. Nozzle blowing process (bushing).

The same process can be adapted to produce a glass fiber mat. In this application several bushings feed onto one conveyor which replaces the drum. The fibers drawn by the steam blower are collected on the conveyor by means of a reduced pressure. The resulting glass fiber mat with its long randomly arranged fibers is saturated in a sizing bath and then cured or hardened in a drying oven. The edges are trimmed and the mat is cut into the desired width. Standard mat weights are 30—60 gm/m² for fibers in the diameter range from 6—16 microns and length range from 30—50 mm.

7.1.3 Flame Blowing Process

Some technical specialties require glass fibers with a maximum diameter below 5 microns. These fine fibers are prodeced by a combination of the attenuating and blowing processes. Continuous infinitely long glass fibers (primary fibers) are mechanically drawn from the tips of several electrically heated platinum bushings. Glass is supplied to the bushings from a small tank. The primary fibers are carefully advanced into the flame of a high temperature, high velocity, gas burner. In this flame the primary fibers are softened and blown into superfine bead free fibers of finite length. They are then sprayed with a binder, collected on a receiving conveyor, compressed and cured in an oven, resulting in a dimensionally stable felt. The productivity is 5—7 kg/h, per bushing and burner, for fiber diameters ranging from 0.75 to maximum 5 microns. The products yielded are relatively expensive since the process consumes much energy and requires much care for relatively small production quantities.

7.2 Centrifugal Processes

7.2.1 Hager-Rosengarth Process

Another process for the mass production of glass fibers uses the centrifugal force of a fast rotating disc. A molten glass stream is fed in the center of this from where it is hurled in the form of fibers [5]. In an

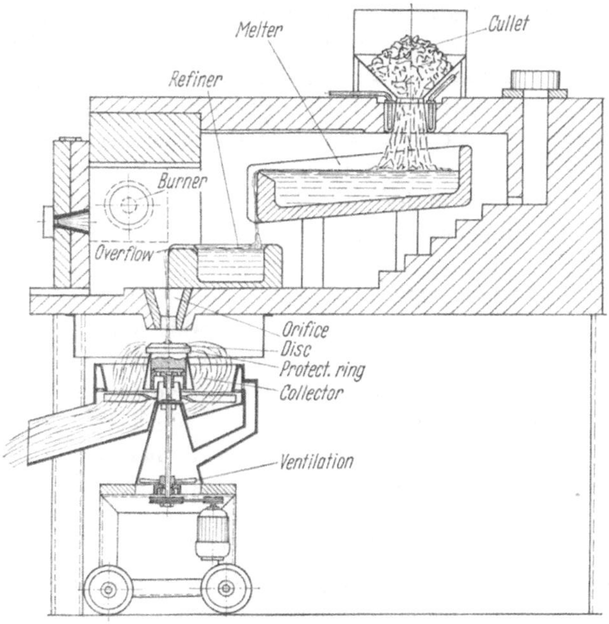

Fig. E 151. Centrifuge process.

installation of a horizontal centrifuge disc (Fig. E 151], the raw material — mostly cullet — is melted in a small oil or gas fired furnace at a temperature between 1200—1300 °C. The molten glass discharges over a lip and into a refiner, from which it leaves in a thin stream passing through a bottom orifice, 100 mm in diameter, onto the center of a disc which is rotating at 3000 to 4000 rpm and is heated by waste gases from the tank. The glass distributes in a thin and uniform layer over the face of the disc and is hurled off the edge by centrifugal force in the form of little droplets. Each droplet pulls a fiber [6], one end of which keeps adhering to the periphery of the disc. The fiber so formed is attenuated by the motion of the droplet as well as by the rotation of the disc. The droplets or beads are separated mechanically from the fibers. The fibers drop around the shaft of the disc in the shape of a hollow cylinder. The cylinder is either cut open by a rotating knife or is chopped to staple length by the air stream. Often several machines work on one conveyor. The further treatment and processing of the mat resembles that used in the steam blowing process (Sec. 7.1).

The construction of the centrifuge is simple. On a movable frame an interchangeable centrifuge disc is mounted on a vertical shaft. It is driven by a motor through a V-belt drive. A collector is connected around the disc in such a manner that beads fly over it and thereby are separated from the fibers. By means of directional air streams the fibers are guided to the collector, gathered and transported. The centrifuge disc is 190 mm in diameter and is made of a ceramic material. On its upper face 1 mm deep grooves are formed in a radial arrangement becoming more shallow as they approach the center. This increases the surface area in contact with glass and improves the centrifugal action. For safety the discs are enclosed in a steel protection ring. The disc is tempered by heating with gas burners before installation. Production rate and requirements are as follows:

Production rate	55—60 kg/h
Disc Speed	4000 rpm
Drive	2.2 kW
Furnace Gas Consumption	35 cubic meters/h
Furnace Temperature	1200 °C
Average fiber diameter	20 microns
Average fiber length	50 mm

7.2.2 TEL Process

A new fiber process was developed during the last years in France and in the U.S.A. This process combines the centrifugal force approach of the centrifugal process (Hager Process) with the attenuating approach of the flame blowing process. A glass with special temperature characteristics and a low crystallization rate [7] is melted in a tank. From there it flows into a forehearth equipped with electrically heated platinum bushings. Each bushing has one orifice, which is 10 mm in diameter. The exit velocity of the glass is controlled by the bushing temperature. Under each bushing is a centrifuge-like fiberizing device. The continuously flowing fine stream of glass is fed into the distributor, or slinger-cup, in the inside of a hollow shaft (Fig. E 152). The distributor and the centrifuge device, also called a "spinner", are mounted at one end of the hollow shaft which rotates at about 3000 rpm. The glass is distributed by centrifugal force from the slinger-cup as a thin film onto the inner surface of the spinner. The spinner hurls the glass in fiber from through the holes of the spinner sidewall. Immediately on leaving the spinner, the relatively coarse fibers are subjected to the pulling action of a high velocity, high temperature gas stream [9] which emerges from a ring shaped burner chamber. The opening of this chamber surrounds the spinner and is porpendicular in relation to the centrifuged fibers. Under the influence of the gas stream the fibers change direction by 90° and are drawn into very long thin fibers. The fibers then fall down in the direction of the centrifuge axis in the form of a hollow cylinder, are sprayed with a binder, collected on a conveyor and taken away for further treatment. Usually several fiberizers work with one conveyor. The fibers fabricated in this manner are used in the building and transportation industries. The density of the product ranges from 12—20 kg/m³.

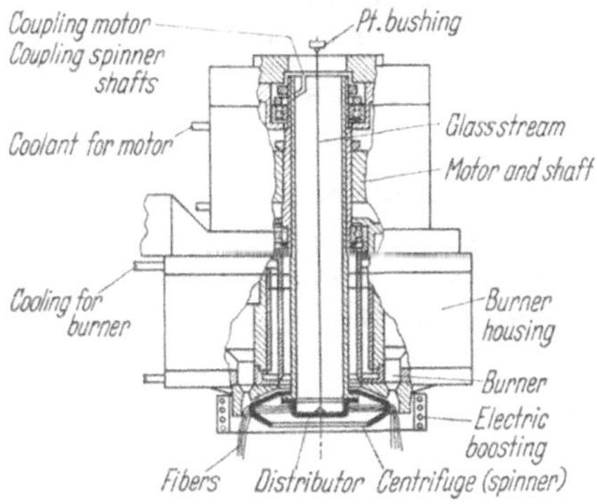

Fig. E 152. The TEL machine (Pat. Appl. 1070792).

The rotor of the liquid-cooled drive motor for the TEL machine is formed as a hollow shaft, into which a second shaft is inserted. They are connected by a heat resistant quick disconnecting coupling. The inserted shaft is closed at its lower end by the distributor, which may be designed like a basket, permitting glass to flow over its sidewall or through perforations in the sidewall. The basket material is a platinum-rhodium alloy. The distributor is enclosed by the spinner, which is also fastened to the end of the hollow shaft. This arrangement has the advantage that the entire rotor, consisting of hollow shaft, distributor and spinner, may be removed easily for maintenance or repair. The spinner is fabricated from a heat resistant chromium-nickel alloy. On the surface of the spinner are up to 6000 holes, arranged in several rows, with diameters ranging from 0.8 to 1.2 mm. A burner with a ring-shaped combustion chamber is located between the motor and the spinner. The ring opening of the burner surrounds the rim of the spinner and is constructed in such a way that the gas stream is uniform in velocity and temperature throughout the fiberizing zone. The correct construction of the burner, in conjunction with its proper adjustment, is very important for this process. The perforated rim of the spinner may be kept at the desired temperature through gas firing or by an auxiliary high-frequency coil [11].

Production data are:

Output of the TEL machine	250 kg/h
Spinner Speed	3000 rpm
Drive	2.2 kW
Burner Output	300,000 kcal/h
Fiber Diameter	6 microns
Average Fiber length	500 mm

The special feature of the process is the production of practically shot-free and very long fibers of uniform diameter. The machines can be adjusted for the production of fibers ranging in diameter from below 1 micron, 2—4 microns, 5—6 microns, or higher. At small diameters the productivity is considerably reduced.

7.3 Drawing

7.3.1 Rod Drawing Process

This process is now much used for the production of glass fiber mat or tissue — and also of textile yarn. The principle is simple: the end of a glass rod is heated by a flame until softened, and finally a drop is released. This falling drop pulls a fiber which adheres to a rotating drum. Melting of the rod and winding on the drum are synchronized, so that a continuous fiber is wound on the drum. This principle, drawing fibers from a rod, represents the oldest known fiber making process. At present, for industrial production, 120 glass rods of as uniform diameter as possible are arranged at equal distances in one row. The rod diameters range from 3—4 mm.

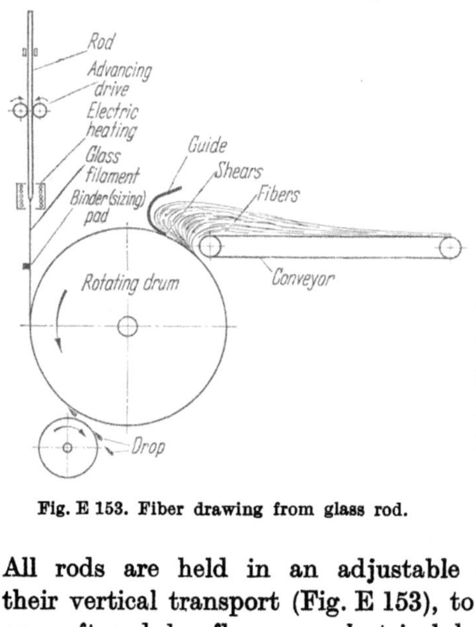

Fig. E 153. Fiber drawing from glass rod.

All rods are held in an adjustable advancing device which guides their vertical transport (Fig. E 153), to the heating zone where they are softened by flame or electrical heating. After the formation of a drop (bead), a fiber originates at each rod. All fibers pass a binder pad. Next they are caught by a drum 1 m in diameter, rotating at 800 rpm. In order to perfectly separate beads and fibers during breakouts, and to facilitate starting, another fast rotating drum is located beneath the drawing drum.

In newer installations this second drum is constructed as a separate winding machine [12], which can be operated to take fibers from the drawing drum in case of trouble. In this manner a uniform mat deposit can always be obtained from the installation proper. The fibers are in contact with the drum for about 240° of the circumference. Then they are removed with the knife edge of a guiding device, using the rotation of the drum as a transporting force. This allows automatic deposition of the fibers on a conveyor. The deposit of mat is approximately 1100 mm wide, with a thickness ranging from 0.3 to 2 mm. The thickness depends on the adjustable speed of the conveyor. The maximum permissible fiber diameter is 12 microns for an average length of 1500 mm. The fiber diameter depends upon the diameter of the glass rods, the melting temperature and the winding speed. The productivity of a unit is about 6 kg/h. Not only can the fibers be converted into a finished mat, which is accomplished by saturating the web with binders and subsequent curing, but the fibers can also be pulled from the conveyor into the form of a sliver. From this sliver textile yarns can be produced through twisting, and, if necessary, drawing (stretching) of the sliver [13].

Before this continuous process came into use, fibers were wound onto the drawing drum and after reaching a thickness of 6—10 mm were cut off at the drum circumference, since they could not be unwound. Depending upon the end use, they were then cut into specified fiber lengths and further processed or drawn into mat.

7.3.2 Gossler — or Drum Process

In a small furnace constructed as a drum (length 0.5 m, diameter 0.25 m) cullet is melted by means of oil, gas or electricity. The bottom of the furnace consists of a tip section made of heat resistant steel, with up to 200 holes 3 to 4 mm in diameter. Through these holes a sufficiently fluid melt flows out first as a drop which then pulls a fiber. The drops are manually attached to a pulling drum which is located 1 m below the furnace, has a diameter of 1 to 1.2 m, and rotates at about 270 rpm. To facilitate mat forming, the furnace is traversed, so that the fibers pulled are uniformly distributed onto the drum at a small cross angle. The length of the furnace and the traversing distance determine the width of the fiber deposit on the drum. After the fibers have been deposited to a thickness of 10 mm the drum is stopped, the layer is cut and removed carefully. By subsequent stretching of the entire length, which is determined by the circumference of the drum, a mat with a fibers crossing at an angle between 60—90° is obtained. The fibers normally range from 20—25 microns in diameter and the productivity is 5—6 kg/h/furnace.

7.3.3 *Bushing Drawing Process*

In an enlarged electrically heated platinum-rhodium bushing automatically fed marbles are melted. The bottom of the bushing (Fig. E 154) usually consists of a tip section with 102—204 tips arranged in a double row, with hole sizes ranging from 1—1.2 mm.

From these holes a corresponding number of fibers are pulled and gathered into a single continuous glass strand. The single fibers pass over a pad which is saturated with a suitable size. The sizing glues the fibers, so that they adhere to each other; later the individual strands can be pulled away from the surface of the tube to be spun (twisted, plied). The mechanical pull originates from a rotating tube, which is driven by the winder. The winding speed is 60 m/sec. A fiber guide mounted above the winder traverses the strand to achieve the desired winding angle. As soon as a tube is filled it is exchanged for an empty tube. The tubes are held in place by centrifugal force applied through movable cups which are attached to the drum of the winder. Recently, automatic winders have eliminated the nuisance of interruptions introduced by tube changes, which are referred to as "doffing". With the simple winders, it used to be necessary to start the winding process with a pick in order to pull the fibers from the bushing tips. The speed necessary to obtain the desired fiber diameter was reached only after a predetermined speed-up time (acceleration), otherwise the fibers would break. Only after the desired speed was reached, was the correct traversing of the strand initiated. The operation has to be stopped again when the tube was changed. The bushing drawing process produces continuous fibers with diameters ranging from 5—6 microns. The productivity of a bushing ranges from 0.5—0.9 kg/h, with a power consumption between 6—13 kWh, depending upon the glass composition used. Each strand of 102 single filaments results in an untwisted yarn of the metric number Nm 180, which means 180 m of yarn weigh 1 g. Then the yarn is further processed by the usual techniques of the textile industry, depending upon the end use.

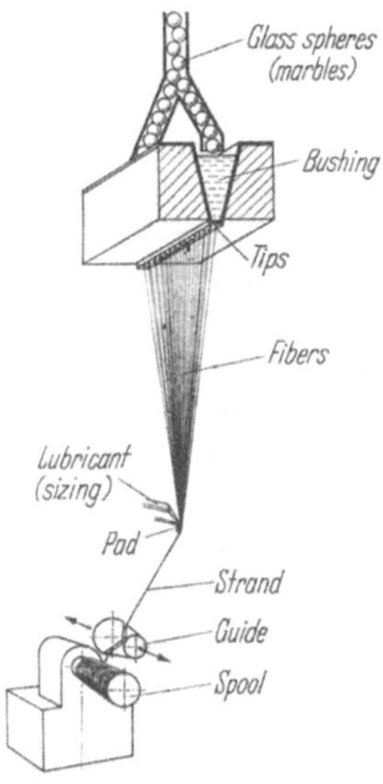

Fig. E 154. Bushing drawing process.

References

[1] DRP No. 723330 of 18. 6. 42. Metallenes trogartiges Auslaßmundstück von etwa V-förmigem Querschnitt an Glasschmelzöfen.
[2] KASWANT: Neuzeitliche Anlagen zur Herstellung von Mineralfasern nach dem Düsenblasverfahren. Sprechsaal 91 (1958) 577–586.
[3] DRP No. 626436 of 26. 2. 36. Verfahren und Vorrichtung zum Herstellen von Glaswolle.
[4] RIEDEL, L.: Verspinnbare Glasfasern und Glasfäden. Z. VDI 91 (1949) 227 to 233.
[5] DRP No. 539738 of 26. 11. 31. Verfahren zum Herstellen von Fasern oder Gespinst aus Glas, Schlacke und ähnlichen in der Hitze plastischen Stoffen.
[6a] VON REIS, L.: Herstellung und Verwendung von Glasgespinst. Glastechn. Ber. 15 (1937) 219–228.
[6b] VON REIS, L.: Glasfaser. Umschau 42 (1938) 463–467.
[7] DBP No. 1061977 of 31. 12. 59 (Inventor: STÉPHANE DE LAJARTE). Glas, insbesondere zur Verwendung für die Herstellung von Glasfasern in Schleudermaschinen.
[8] AHLBERG, R.: Tillverkning av glassfiber. Glastekn. T. 13 (1958) 127–132.
[9] DBP 1037085 of 2.4.59 (Inventors: MARCEL MABRU and PAUL PIOT). Vorrichtung zur Herstellung von Fasern aus thermoplastischen Stoffen, wie Glas.
[10] DBP 1070792 of 30. 6. 60 (Inventors: DALE KLEIST and HENRY J. SNOW). Vorrichtung zur Befestigung von mit hoher Geschwindigkeit umlaufenden Rotoren zur Herstellung von Fasern aus Glas od. dgl. durch Schleudern.
[11] Auslegeschrift (DAS) 1030529 of 22. 5. 58. Verfahren und Vorrichtung zur Herstellung von Fasern aus thermoplastischen Massen, insbesondere von Glasfasern.
[12] Brit. Pat. No. 842410, reg. 25. 8. 58.
[13] BOBETH, W., W. BÖHME, and R. TECHEL: Anorganische Textilfaserstoffe (Editors W. BOBETH, W. BÖHME, and J. TECHEL), Berlin: VEB Verlag Technik 1955.
[14] SHAND, E. B.: Glass Engineering Handbook, 2nd Ed., New York: McGraw-Hill 1958.

F. Quality Control

By W. GIEGERICH, Heilbronn

Special and routine control of glass machines is generally similar to the control of other types of machines [1]. Within the scope of this book it is considered sufficient to give a brief summary with emphasis only on special instruments.

1 Pressure

Air pressures are usually from 0–5 atmospheres, gas pressures from 0–5000 mm water column, and oil pressures from 0–3 atm. Control equipment includes simple U-tubes filled with water or mercury, spring type tube and plate instruments, and dial gauges.

2 Temperature

In addition to the obvious surveillance of the temperatures of gears and bearings, the glass temperature must be determined as near to the machine as possible [2]. In special studies glass and mold wall temperatures [3, 4, 5] at various process stages [6, 7] are determined.

3 Quantity

It is very difficult to measure the required working quantities because of pulsation. The mold cooling air can be measured with relative ease. The control of product quantities is achieved by counters or tachometers at the machine or photocells at the conveyor.

4 Time

Time is controlled by the simple measurement of machine speed with the tachometer. The usual instruments, stop watch, light stroboscope, recording clock, are used for the exact examination and analysis of production rates and of single production phases.

5 Power and Output

Occasionally measurements are made of vibrations in machine parts and in building areas, or of force and acceleration in drives, by using dial indicators, spring dynamometers, etc. Motor loads are checked by the usual means.

6 Heat Consumption

The heat consumed in heating is calculated from the amount and BTU of the fuel used. Heat losses — such as through glass cooling when worked — are measured by calorimeter (e.g. for bottles see Ref. 6.7).

7 Technical Analyses

Technical analyses are made by usual methods. Such analyses comprise fuel and waste gas studies, lubricant testing, testing of contamination of press and vacuum air metallography of mold materials, etc.

8 Working Properties of Glass

Unfortunately there are few practical methods for testing the working properties of glass, although such methods would be very helpful for the control of the machines. As described in earlier sections, viscosity, total heat conductivity (radiation and conduction proper), surface tension, and thermal expansion are of fundamental importance for the forming of glass. Simple tests are needed. In special studies heat transfer, heat capacity, density and gas content (SO_3, F_2 etc.), volatilization, and surface tension in the working range are all significant.

References

[1] GRAMBERG, A.: Technische Messungen, 7th Ed., Berlin 1953, and Maschinenuntersuchungen, 3rd Ed., Berlin 1924.

[2] MEISTER, R., WEBER-KLEIN, F. WEFER, and J. HARTWIG: Temperaturmessung in Glashütten. Fachausschußbericht No. 55 der Dtsch. Glastechn. Ges., Frankfurt (Main) 1955.

[3] TRIER, W.: Temperaturmessungen in Glasformen. Glastechn. Ber. 28 (1955) 336–351.

[4] BABCOCK, C. L., and D. A. MCGRAW: Application of glass properties data to forming operations. Glass Ind. 38 (1957) 137–142, 144–146, 148–151, 161.

[5] HOLSCHER, H. H., J. C. COLEMAN and C. C. COOKE: Study of glass and mold temperatures ... Glastechn. Ber. 32K (1959) II/1–II/17.

[6] UNGER, L.: Bestimmung des Wärmeinhaltes und der mittleren Temperatur von Flaschen während des Formgebungsprozesses. Glastechn. Ber. 32 (1959) 153–157.

[7] GIEGERICH, W.: Über technische Grundlagen der vollautomatischen Flaschenherstellung. Glastechn. Ber. 33 (1960) 441–449.

[8] Hubzähler für halbautomatischen Betrieb. Fragekasten Gr. Glas. Sprechsaal 89 (1956) 109.

[9] HARTSCH, R. F.: The use of slow motion photography in improving the glass making process. Glass Ind. 22 (1941) 11–16.

[10] BAUMER, H.: Die Photozelle bei der Preßglasherstellung. Sprechsaal 84 (1951) 458.

[11] ADAMS, O.: Photo-electric measurement in the glass industry. Engineer 189 (1950) 290/91.

Subject Index

(The page numbers in italics indicate sections of the book)